Monitoring and Evaluation of Soil Conservation and Watershed Development Projects

Monitoring and Evaluation of Soil Conservation and Watershed Development Projects

Editors

Jan de Graaff
Wageningen University and Research Centre, The Netherlands

John Cameron
University of East Anglia, United Kingdom

Samran Sombatpanit
WASWC, Thailand

Christian Pieri
Formerly of the World Bank, France

Jim Woodhill
IAC, Wageningen University and Research Centre, The Netherlands

with assistance from
Annemarieke de Bruin
Wageningen University and Research Centre, The Netherlands

CRC Press
Taylor & Francis Group
Boca Raton London New York

CRC Press is an imprint of the
Taylor & Francis Group, an **informa** business
A SCIENCE PUBLISHERS BOOK

First published 2007 by Science Publishers

Published 2018 by CRC Press
Taylor & Francis Group
6000 Broken Sound Parkway NW, Suite 300
Boca Raton, FL 33487-2742

First issued in hardback 2018

ISBN 13: 978-1-138-46863-4 (hbk)
ISBN 13: 978-1-57808-349-7 (pbk)

Library of Congress Cataloging-in-Publication Data

Monitoring and evaluation of soil conservation and watershed development projects/editors, Jan De Graaff ... [et al.].
p. cm.
Includes bibliographical references and index.
ISBN 978-1-57808-349-7
1. Soil conservation projects--Evaluation. 2. Water resources development--Evaluation. 3. Watershed management--Evaluation. I. Graaff, J. de.

S627.P76M66 2007
631.4'5--dc22

2007038165

Foreword

Much concern is being expressed about the state of the world's soil and availability of water. Interventions to conserve soil and develop watersheds are being undertaken globally by a wide variety of agencies, but we need to have instruments to assess the access of these interventions. This means identifying the variables that the interventions are aiming to affect, indicators of those variables, and the people who are the intended beneficiaries.

It is therefore important that effective monitoring and evaluation (M&E) techniques are identified to ascertain that these interventions contribute to halting further environmental degradation, but also to determine their wider socio-economic impact and to learn from these experiences. This book offers evidence that better practices in monitoring and evaluation systems, with differing procedures and tools, have already been developed and applied in many countries.

In soil conservation and watershed development activities it is difficult to disentangle the roles of physical factors and human factors with respect to effects on soil and water availability and quality and the impact of interventions from underlying processes. For instance, sedimentation rates may be affected by both lower rainfall intensities and soil and water conservation activities. And tree survival rates may relate to climatic and biological conditions, and also to management and better care by the local population.

Monitoring and evaluation activities should encompass the entire range of relevant factors that bear on the performance and results of project activities, this book therefore focuses on both physical and socio-economic factors and indicators. In short, this book gives an overview of how monitoring and evaluation systems have been applied and can be applied in various types of soil conservation and watershed development projects in many different countries in the world, and can serve as a reference book for all those involved in designing M&E systems for future projects.

MIODRAG ZLATIC
President
World Association of Soil and Water Conservation

Preface

Large numbers of development projects in the last thirty years have had soil conservation and watershed development as major components. Arguably there has been some ascent on a steep learning curve through experience of failure, given Norman Hudson in his acclaimed study on 'the reasons for success or failure of soil conservation projects' (FAO Soils Bulletin 64, 1991) could claim that only 25 percent of the projects started in the 1970s could be rated as successful, while those started during the 1980s the figure rose to 56 percent.

The editors of this book are convinced that continuing improvement in project performance requires even more effective Monitoring and Evaluation (M&E) techniques to inform decision-making. The writing of this book was therefore initiated by the World Association of Soil and Water Conservation (WASWC) in 2002 after publications on two essential soil conservation topics, Extension and Incentives, had been executed in 1997 and 1999, respectively. The exact purpose of this book is to help all stakeholders, either directly or indirectly, to plan and execute more effective soil conservation and watershed development projects/ programmes.

'Effectiveness' is a rather vague term, especially when assessing development projects/ programmes. It may cover a very wide scope indeed, ranging from 'getting the budget spent', to 'seeing interventions implemented/infrastructure constructed', to 'seeing soil material kept in its place and watershed looks nice', to 'many or most people living in the area attain a better livelihood.' In this book we are primarily concerned with getting a healthy mix of environmental and livelihood impacts. The M&E techniques described in this book vary in the mix offered. But in all cases they are intended to offer practical advice to active practitioners.

Production of this book has been done through inviting academics and professionals who have long years of experience to contribute. Altogether we have received 25 papers from many countries in all habitable continents, which we had grouped into four categories to facilitate analysis and easy reading, i.e. Principles, M&E in Practice,

Physical Tools, and Socioeconomic Aspects. Readers are welcome to contact and interact with any of the authors for further discussion or cooperation. E-mails of all senior authors are provided to make such contact an easy task.

In editing this book, the editors have done their best to get examples from various locations and environments. To help the reader through the chapters, the Introduction is a straightforward description and we recommend this should be used to scope the range of material on offer here. The Epilogue, on the other hand, offers a more analytical synthesis of all chapters and indicate various principles that are crucial in getting to appreciate the methods or systems being used in monitoring and evaluating any soil conservation and watershed development intervention.

During the course of production of this book we have received guidance and comments from our advisers, David Sanders and Tage Michaelsen, formerly of FAO and Thomas Hofer of the Department of Forestry, FAO, Rome, our deep thanks to them. We also thank Annemarieke de Bruin of Wageningen University and Research Centre, The Netherlands, who helped substantially in compiling appendices concerning training courses and recent M&E literature. Last, and not the least, we thank Science Publishers of Enfield, New Hampshire, U.S.A. for facilitating us in all steps of this book's production.

THE EDITORS

Contents

List of Contributors

Ago Hideo

Sub-Director JGRC Project, Sucre, Bolivia.
E-mail: hideo_ago@chushi.maff.go.jp

Bergsma Eelko

Agricultural Engineer, formerly of the Soil Science Division, International Institute for Geo-information Science and Earth Observation (ITC), Enschede, The Netherlands.
E-mail: bodcm@wanadoo.nl

Bodnar Ferko

Erosion and Soil & Water Conservation Group, Wageningen University and Research Centre, The Netherlands.
E-mail: ferko.bodnar@wur.nl

Bonn Ferdinand

Professor and Chairholder, Canada Research Chair in Earth Observation, University of Sherbrooke, freelance consultant in environmental assessment, Sherbrooke, Quebec, Canada (deceased).

Cameron John

Reader in the School of Development Studies/Overseas Development Group, University of East Anglia, Norwich, UK.
E-mail: john.cameron@uea.ac.uk

Critchley William

Centre for International Cooperation, Vrije Universiteit Amsterdam, de Boelelaan 1105-26, Amsterdam, The Netherlands.
E-mail: WRS.Critchley@dienst.vu.nl

Curtis Allan

Professor, Integrated Environmental Management and Director of the Institute for Land, Water and Society, Charles Sturt University, P.O. Box 789, Albury, NSW 2640, Australia.
E-mail: acurtis@csu.edu.au

Das D.C.

Member, Soil Conservation Society of India. B-19, Parijat Apartments, Pitampura (West), New Delhi 110034, India.
E-mail: dinesh_ranu2003@yahoo.com

Evans Robert

Research Fellow, Mellish Clark Building, Room MEL 105, Anglia Ruskin University, East Road, Cambridge CB1 1 PT, UK.
E-mail: r.evans@anglia.ac.uk

Farshad Abbas

Soil Scientist, ESA, International Institute for Geo-information Science and Earth Observation (ITC), Enschede, The Netherlands.
E-mail: farshad@itc.nl

Graaff Jan de

Erosion and Soil & Water Conservation Group, Wageningen University and Research Centre, Soil Science Centre, Atlas Building, Droevendaalsesteeg 4, Room B. 413, Wageningen, The Netherlands.
E-mail: jan.degraaff@wur.nl

Gunawardene E.R.N.

Department of Agricultural Engineering, University of Peradeniya, Peradeniya, Sri Lanka.
E-mail: nimalgun@pdn.ac.lk

Gupta K.K.

Member of Soil Conservation Society of India, Delhi, India.
E-mail: kkgupta2@yahoo.com

Healy Sean

French Ministry of Agriculture/CERDI, Clermont-Ferrand, France.
E-mail: s.healy@u-clermont1.fr

Hernandez Edgar

Universidad de Los Andes, Mérida, Venezuela.
E-mail: edgarah@icnet.com.ve

Herweg Karl

Centre for Development and Environment, Berne University, Berne, Switzerland.
E-mail: karl.herweg@cde.unibe.ch

Hessel Rudi

Soil Science Centre, ALTERRA Green World Research, Wageningen University and Research Centre, P.O. Box 47, 6700 AA Wageningen, The Netherlands.
E-mail: rudi.hessel@wur.nl

Hoang-Fagerström Minh Ha

Department of Soil Sciences, Swedish University of Agricultural Sciences, P.O. Box 7014, SE-750 07 Uppsala, Sweden.

Howeler Reinhardt H.

CIAT, Cassava Office for Asia, Department of Agriculture, Chatuchak, Bangkok 10900, Thailand.
E-mail: r.howeler@cgiar.org

Jayakody A.N.

Department of Soil Science, Faculty of Agriculture, University of Peradeniya, Peradeniya, Sri Lanka.
E-mail: anaja@pdn.ac.lk

Kessler Aad

Technical Coordinator JGRC Project, Sucre, Bolivia. Present address: Eykmanstraat 7, 6706 JT Wageningen, The Netherlands.
E-mail: Aad.Kessler@wur.nl

Kotru Rajan

Indo-German Changar Eco-Development Project. Palampur, H.P., India.
E-mail: rkotru@gmx.net, rkotru@gtzindia.com

Li Rui

Institute of Soil and Water Conservation (ISWC) CAS and MWR, Northwest Sci-Tech University of Agriculture and Forestry, Yangling 712100, Shaanxi, China.
E-mail: lirui@ms.iswc.ac.cn

Liniger Hanspeter

Centre for Development and Environment, University of Berne, Berne, Switzerland.
E-mail: hanspeter.liniger@cde.unibe.ch

Merzouk Abdelaziz

Professor of Soil and Water Conservation at Hassan II Institute of Agronomy and Veterinary Medicine, Rabat, Morocco.
E-mail: merzouk@mtds.com

Messing Ingmar

Department of Soil Sciences, Swedish University of Agricultural Sciences, P.O. Box 7014, SE-750 07 Uppsala, Sweden.

Mihara Machito

Tokyo University of Agriculture, Tokyo, Japan.
E-mail: m-mihara@nodai.ac.jp

Napier Ted L.

Professor of Environmental Policy, Department of Human and Community Resource Development; The School of Natural Resources; and the Graduate Programme in Environmental Science of the Ohio State University; 2120 Fyffe Road, Columbus, Ohio 43210, USA.
E-mail: napier.2@osu.edu

Ngoan Tran Ngoc

Thai Nguyen University of Agriculture and Forestry, Thai Nguyen, Vietnam.

Nibbering Jan Willem

Natural resource management specialist who worked as consultant advising the NWFP Forest Department; presently with Ministry of Foreign Affairs, P.O. Box 20061, 2500 EB, The Hague, The Netherlands.
E-mail: jw.nibbering@minbuza.nl

Nichols Mary H.

Southwest Watershed Research Center, 2000 E Allen Rd., Tucson, AZ 85719, USA.
E-mail: mnichols@tucson.ars.ag.gov

Nourallah Mounif

Professor of Agronomy and presently Country Portfolio Manager for the Near East and North Africa Division, International Fund for Agricultural Development (IFAD), Rome, Italy.

Ritsema Coen J.

Soil Science Centre, ALTERRA Green World Research, Wageningen University and Research Centre, P.O. Box 47, 6700 AA Wageningen, The Netherlands.
E-mail: coen.ritsema@wur.nl

Sang-Arun Janya

The University of Tokyo, Tokyo, Japan.
E-mail: janyasangarun@yahoo.com and janyasangarun@hotmail.com

Sheng Ted C.

Department of Forest, Rangeland, and Watershed Stewardship. Colorado State University, Fort Collins, CO 80523, USA.
E-mail: teds@lamar.colostate.edu

Sombatpanit Samran

World Association of Soil and Water Conservation, 67/141 Amonphant 9, Sena 1, Bangkok 10230, Thailand.
E-mail: sombatpanit@yahoo.com

Stolte Jannes

Soil Science Centre, ALTERRA Green World Research, Wageningen University and Research Centre, P.O. Box 47, 6700 AA Wageningen, The Netherlands.
E-mail: jannes.stolte@wur.nl

Tejwani K.G.

Member of Soil Conservation Society of India, Delhi, India.
E-mail: faxaid@ndb.vsnl.net.in

Walle Robert

Senior Water Resource Specialist, CPESC 2134, P.O. Box 93, Tegucigalpa Honduras, C.A. (Zamorano-DSEA Social and Environmental Development).
E-mail: robertow@ibw.com.ni

Wang Fei

Institute of Soil and Water Conservation (ISWC) CAS and MWR, Northwest Sci-Tech University of Agriculture and Forestry, Yangling 712100, Shaanxi, China.
E-mail: lirui@ms.iswc.ac.cn

Wanigasundara W.A.D.P.

Department of Agricultural Extension, University of Peradeniya, Peradeniya, Sri Lanka.
E-mail: dpwanige@pdn.ac.lk

Watananonta Watana

Field Crops Research Institute, Dept. of Agriculture, Chatuchak, Bangkok 10900, Thailand.

Wen Zhongmin

Institute of Soil and Water Conservation (ISWC) CAS and MWR, Northwest Sci-Tech University of Agriculture and Forestry, Yangling 712100, Shaanxi, China.
E-mail: lirui@ms.iswc.ac.cn

Wennink Bertus

Royal Tropical Institute, Amsterdam, The Netherlands.
E-mail: b.wennink@kit.nl

Woodhill Jim

Head, Social and Economic Department, International Agriculture Centre, Wageningen University and Research Centre, The Netherlands.
E-mail: jim.woodhill@wur.nl

Workman Sarah W.

312 Hoke Smith Bldg., University of Georgia, Athens, Georgia 30602-4356, USA.
E-mail: sworkman@uga.edu

Yamaji Eiji

The University of Tokyo, Tokyo, Japan.
E-mail: yamaji@k.u-tokyo.ac.jp

Yang Qinke

Institute of Soil and Water Conservation (ISWC) CAS and MWR, Northwest Sci-tech University of Agriculture and Forestry, Yangling 712100, Shaanxi, China.
E-mail: lirui@ms.iswc.ac.cn

Zapata Felipe

Soil scientist, Joint FAO/IAEA Programme, International Atomic Energy Agency, P.O. Box 100, Wagramerst. 5, A-1400 Vienna, Austria.
E-mail: f.zapata@iaea.org

Zhang Xiaoping

Institute of Soil and Water Conservation (ISWC) CAS and MWR, Northwest Sci-Tech University of Agriculture and Forestry, Yangling 712100, Shaanxi, China.
E-mail: lirui@ms.iswc.ac.cn

Introduction

Since the 1960s many soil conservation and watershed development SCWD projects have been undertaken in several forms in different countries, everywhere in the world. These projects were usually aimed at reducing soil erosion and preventing watershed degradation and at the same time at increasing watershed productivity. To reach these objectives a multitude of activities were undertaken, ranging from terracing, gully control, reforestation and fruit-tree planting to zero-grazing, irrigation, off-farm employment and various socioeconomic support activities. However, when evaluating these projects, at their completion or some years afterwards, it was often found that no firm conclusions could be drawn, among others, since insufficient attention was paid to the monitoring of project outcomes and impacts.

SCWD projects are multi-sectoral in essence, which makes the always difficult and time consuming process of Monitoring and Evaluation (M&E) even more complex to implement cost-efficiently. The challenge is not only to evaluate the performance of any project in terms of inputs and outputs delivery, but to assess the added-value of an integrated project approach which hinges on the interactions and synergies among the institutional, social, economic and technical driving forces to reverse or prevent a trend in soil and environmental degradation of a specific watershed.

Several handbooks and guidelines on M&E for development projects were written by various development organisations, and the UN Administrative Committee on Coordination (ACC) Taskforce on Rural Development issued in 1984 its guiding principles (ACC, 1984). Recently, new guidelines have been issued to improve M&E systems.

This book seeks to illustrate how M&E could be undertaken theoretically and how it has been undertaken in practice by various projects and development organisations in different agro-ecological and social conditions in the world. The lessons drawn in this book about the various ways and means of conducting monitoring and evaluation will enable future projects and programmes to establish proper M&E systems, which eventually will improve project performance.

Monitoring can be considered as an internal project activity and the responsibility of the project management. Information should be continuously or at regular intervals collected about the various project activities and about the conditions surrounding these activities, in order to ascertain whether these contribute to project objectives and to make timely adjustments when necessary. Without such monitoring the evaluation of project activities, usually undertaken by external organisations on behalf of the financing agencies, cannot be properly done.

In setting up a classical monitoring system for subsequent evaluation, one looks at the project cycle. The project objectives can be derived from the project identification and preparation reports. To be able to verify to what extent the activities in the implementation phase contribute towards the objectives, clearly defined indicators need to be established for these objectives. Since objectives often relate to both physical factors, such as erosion and hydrological status, and socioeconomic and sustainability factors, including local institution development, capacity building, participation rates, financial performance and resource leveraging, a wide range of direct or proxy indicators has to be established.

The project has then to create an organisational unit and a monitoring system to collect information about these indicators, starting with initial or baseline values and followed by progress values at successive intervals. The system will make use of various tools. For the determination of scores on physical indicators, such as erosion rates and hydrological features, measurement devices or research plots should be installed, while for socioeconomic indicators various types of interview rounds and participatory methods should be arranged.

While the above seems logical, monitoring and evaluation can be hampered by various pitfalls. When project objectives are not clearly defined, it is hard to establish the right indicators and monitoring systems may then either become insignificant or yield too much data for a proper evaluation. And the information collection process is not that easy and may be subject to various errors, as is shown in Box 1.

STRUCTURE AND OUTLINE OF THE BOOK

The book is divided into four parts. After the introduction, attention is first paid in **Part I** to the general principles of M&E, going from more classical to recently developed approaches. This is followed by several papers in **Part II** that focus on the application of M&E under different circumstances in various areas in the world: M&E in practice. Within monitoring and evaluation there is a need for tools to measure the physical effects brought about by SCWD project interventions. A selection of these tools is presented in **Part III**. Monitoring and evaluation also involves various

Box 1. The Information Collection Process

The challenge of any information gathering activity in practice is to combine a concern for all the individual sources of error with an overview of the whole information gathering process. Good practice in information gathering also has principles, which apply to all methods from the more quantitative to the more qualitative.

The individual sources of error can be usefully sub-divided into a number of types which are associated with stages in the information process that appear to come at the information gatherer in a chronological order:

a. formulating the deep issues - the risk of mis-specifying concepts and causalities
b. identifying objectively measurable variables - the risk of perversely moving indicators/ proxies in relation to underlying concepts
c. designing the measuring instrument - the risk of measuring inaccuracy
d. identifying the relevant population of cases - the risk of mis-specifying who/ what is being represented
e. deciding which cases actually to investigate - the risk of bias and the handling of formal sampling error
f. negotiating contracts for detailed specific funding/ resources - the risk of under-resourcing and influence of vested interests
g. employing, training and supervising data collection people - the risk of poor quality interviewing or cheating
h. processing data - the risk of transcribing and data-inputting error
i. analysing data to produce information - the risks of inadequate and inappropriate tests and interpretation error
j. communicating information - the risk of misunderstanding by the reader of the implications of the results.

Thus each of these stages has its own potential as a source of error. The design of each stage also has a capacity for increasing or reducing error at later stages in the information gathering process. Therefore, good information gathering demands attention to the process as a whole as well as the individual parts.

The most telling example of the interconnection between later errors and earlier decisions is the division of labour in which an external consultant designs a survey up to the point of field-work and then leaves only to return for analysis and reporting. Lack of concern for the intervening interviewing and processing errors at the earlier stages of measurement instrument and sampling design can produce the classic RIRO (rubbish-in/rubbish-out) inaccurate results which remain undetected until someone else attempts to use the results practically for policy or research.

In conclusion, even the most perfectly designed information collection exercise is subject to Murphy's Law - *if it can go wrong it will*. Each problem anticipated, contractually agreed, and budgeted is a problem virtually solved. Each problem left to be 'sorted out later' is a hostage to ill-fortune with a potential high ransom payment in cash and good-will!

social, economic and institutional aspects, such as the institutional set-up, participatory approaches, economic evaluation and local development. Such aspects are dealt with in **Part IV**. Hereunder a more detailed overview is given of the contents of the papers within these four parts.

The book also includes appendices that provide information about existing manuals on and training courses in M&E of SCWD projects.

PART I: PRINCIPLES OF M&E IN SCWD PROJECTS

The initial papers relate to the standard M&E systems, that follow more or less similar recommended guidelines, established in the 1980s (ACC, 1984). The last two papers present the most recent approaches.

The first paper in this part (Sheng) shows the classic principles of monitoring and evaluation as it has been used for many SCWD projects. It discusses the design of monitoring programmes, the selection of key indicators, data requirements and assessment methodologies and institutional aspects of M&E. The second paper (Das et al.) focuses on the biophysical and socioeconomic aspects involved in the monitoring and evaluation of watershed projects and activities at different scales, with an example from India. It distinguishes four broad components of watershed development projects: biophysical status, primary production, socioeconomics and institutional aspects. The next paper (Sang-arun et al.) starts with defining SCWD projects and M&E purposes and dimensions, and subsequently focuses on the selection and detailed listing of indicators of M&E for such projects. While the attention within M&E was in the past given to attaining levels of project inputs and outputs and to some extent to project purposes, attention has recently shifted in the direction of impacts. A step-wise approach hereto is provided in the fourth paper in this part (Herweg). For this impact monitoring and assessment (IMA) instrument, six steps are distinguished, starting with the involvement of stakeholders and ending with the impact assessment. The last paper in this first part (Woodhill) argues that for M&E to make a useful contribution to improving the impact of SCWD there must be a greater focus on learning. And it asks those involved in development initiatives to place the indicator and information management aspects of M&E in a broader context of team and organisational learning.

PART II: M&E IN PRACTICE

This second part shows how practitioners in different parts of the world have adapted the principles to their local situation, and provides several interesting applications of M&E. The book and this introduction generally refer to SCWD (Soil Conservation and Watershed Development) projects and programmes, which embrace a wide variety of natural resource management and development activities. These are undertaken at different scales, in short- or long-time periods and under a great variety of agro-ecological, socioeconomical and institutional conditions.

The first paper (Hernandez) focuses on sustainability and on the design of an impact oriented M&E system. It distinguishes nine steps, from problem analysis upto information dissemination, and it shows how impact monitoring has been applied in a case study area in Venezuela. In the second paper (Walle and Workman) attention is given to the use of M&E in a soil and water conservation project aimed at agricultural restoration after the disaster caused by hurricane Mitch in Nicaragua. The ProCuencas project demonstrated that with active M&E components the community participation in soil and water conservation (SWC) can be increased and quality of conservation impacts improved. The third paper (Gupta and Das) discusses a people-centred M&E data collection and analysis approach with village and household level data forms, and its application in the Punjab in India. The fourth paper (Jayakodi et al.) gives an overview of different M&E approaches and methods applied for different types of SCWD projects in Sri Lanka. This is followed by a paper (Li et al.), presenting a more physical oriented and regional M&E application, including erosion modelling, for the Loess Plateau in China. The next paper (Kotru) focuses on impact monitoring (IM) as it was applied to the Changar project in India. IM is considered as the key for testing the long-term watershed development project hypothesis. It was shown that simultaneous intersectoral activities can result in a culmination of impact at micro-watershed level. The last paper (Bodnar et al.) provides - with use of the logical framework - a review of M&E of the national soil and water conservation activities in southern Mali. It argues that M&E systems should contribute to a more timely evaluation of projects impact.

PART III: PHYSICAL PARAMETERS IN M&E

Attention is paid in this part to various methodologies or tools to monitor soil erosion over shorter or longer periods of time and to assess spatial differences of changes of erosion. In the objectives of SCWD projects mention is often made of both increasing production and decreasing soil erosion. In monitoring due attention should then be paid to the effects of project activities on production and erosion. By means of farm surveys, combined with yield measurements, a good impression can usually be obtained of changes in production, but effects of project activities on hydrology and erosion are much more difficult to assess.

The first paper (Evans) presents an alternative way to assess water erosion of cultivated land, with results from a field-based survey of erosion in the United Kingdom. The second paper (Bergsma and Farshad) also deals with on-site monitoring of erosion, while focusing on micro-topographic features. Data from case studies in several countries have demonstrated the use of seven of such features for monitoring erosion intensity. The third paper (Nichols) focuses on the downstream effects, and presents results of 50 years of watershed monitoring and research within the Walnut Gulch Experimental Watershed in the USA. Data from such long-term research can be very useful to develop rainfall-runoff-erosion prediction technologies, and serve as 'benchmark' data. The next paper (Merzouk et al.) discusses Geomatics: Geographical Information Systems, Remote Sensing, Global Positioning Systems and other new technologies, and the use of these technologies in M&E of SCWD projects, with examples mainly from Morocco. This is followed by a paper (Zapata) presenting the use of environmental radionuclides - of which Cesium-137 is the most well-known - to monitor soil erosion and sedimentation on field, landscape and watershed levels. The final paper in this part (Hessel et al.) deals with integrated conservation planning, with use of erosion measurements and modelling, as applied in a small catchment on the Loess Plateau in China. A participatory approach was followed to develop potential land use scenarios.

PART IV: SOCIAL, ECONOMIC AND INSTITUTIONAL ASPECTS

In this part papers have been brought together with such different aspects as participatory monitoring, adoption behaviour, SWC movements, economic evaluation and the institutional set-up of monitoring and evaluation systems.

The first paper (Kessler and Ago) discusses the socioeconomic preconditions that have to be attained before SCWD projects could be successfully initiated. This is referred to as 'laying the foundation' and it is based on project experiences in Bolivia. The second paper (Napier) deals with the environmental assessment of conservation initiatives within the Darby Creek watershed in Ohio, USA. Evidence from the monitoring studies in the watershed revealed that voluntary conservation programmes have been unable to encourage adoption of conservation practices. The third paper (Curtis) discusses the national Landcare programme in Australia, which provides a powerful example of how to establish watershed organizations. It provides recommendations on ways and means to monitor and evaluate watershed groups, whereby the focus should be on learning and on sustaining watershed groups. The next paper (Cameron) describes the principles and practice of social cost-benefit analysis as a tool for the economic evaluation of projects. It gives two interesting case studies of small-scale irrigation and micro-hydroelectric projects in north-west Pakistan. In the fifth paper (Nibbering) attention is paid to the institutionalisation of M&E in natural resource management. It describes the participatory approach in M&E followed by the forestry authorities in North West Frontier Province in Pakistan, through the establishment of Village Development Committees. The sixth paper (Howeler et al.) gives a detailed account of a farmer participatory research approach to enhance the adoption of soil conserving practices and to improve the sustainability of cassava production. After its initial development on a limited number of sites, the methodology has now been extended to about 99 villages in Thailand, Vietnam and China. The final paper (Critchley and Liniger) discusses the methodology, tools and worldwide results of the WOCAT (World Overview of Conservation Approaches and Technologies) network. Its mission is to provide tools that allow soil and water conservation specialists to share their valuable knowledge in soil and water management that assist them in their search for appropriate SWC technologies and approaches and that support them in making decisions at the planning level and in the field.

In conclusion, this book provides diverse information and critical know-how to implement appropriate methodology and cost-efficient monitoring and evaluation systems better suited to assess the impacts of soil conservation and watershed multi-sectoral development activities. It draws on a worldwide experience of specialists and a large array of ground-truthing projects and programmes. This book will meet its objective if it contributes to convince financing institutions and project managers that integrated watershed management activities have the potential to generate highly desirable impacts for the society at large, which have to be accurately measured by adequate M&E systems.

AN OVERVIEW OF THE MAJOR THEMES

Taken together all the chapters in this book clearly show how complex are the challenges of monitoring and evaluating watershed and soil conservation activities today. This is because real problems are widely seen as having become more pressing as environmental degradation processes have increased in both breadth and intensity. It is also because we, as a species, have become more demanding in what we expect from monitoring and evaluation in all developmental activities.

Part I brings out the complexity of constructing M&E systems that serve multiple stakeholders, going beyond conventional concerns with informing the administrative and technical officials of donors and governments to ensure that project blueprints have been delivered. To satisfy growing concerns about both environmental degradation processes and associated deepening poverty, M&E has had to become more concerned with reporting to multiple human scales of decision-making on multiple time scales of sustainability. On both types of scale, watershed and soil conservation activities stretch information systems more than most. At its most ambitious, the chapters in Part I suggest M&E systems should aspire to high transparency, offering accountability to everyone with a potential interest in understanding indefinite change processes marked by numerous milestone outcomes.

The papers also bring out inevitable trade-offs that are required in the real world. The most obvious trade-off between accuracy and costs of data, but more subtle trade-offs also have to be made between using external consultants, with claims to independent expertise, rather than in-house staff, with claims to local sensitivity. But a trade-off that seems to be missing is that between changes in the physical environment and changes in poverty and gender relationships. There is a common implicit assumption throughout the book that conserving the physical environment benefits the poor and women at least proportionately. More attention in M&E systems to this assumption and explicitly connecting with the Millennium Goals, so important to major donor agencies, may be needed to developmentally mainstream watershed and soil conservation activities. But, on the positive side, the chapters do describe a range of institutional arrangements that widen communication while showing concern with broadly defined feasibility and not letting the unachievable perfect stand in the way of achievable good practice.

The chapters in Part II are mainly concerned with specifying both physical and socioeconomic variables and their indicators in specified local contexts and the challenges of combining physical measurements and socioeconomic observations in practice. It is interesting to note how often external donors are mentioned as driving forces in these complex M&E systems. Presumably this is because the high cost of such systems requires the level of resources available to donors. Multi-disciplinary complexity and substantial costs are explicit or implicit common challenges for all the M&E systems described in Part II. The way forward for M&E in watershed and soil conservation activities must face up to both challenges.

Part III provides a range of impressive physical measurement techniques and analytical tools. Together they demonstrate cutting edge utilisation of the physical sciences applied to watershed and soil conservation activities. The various methods and analysis require differing highly specialised expertise and measurement and data processing equipment. In general they make little mention of socioeconomic processes, which makes them seem somewhat dated given the recent move towards inclusion of local, indigenous people's knowledge and interests in conservation activities. But this neglect is more than off-set by the impressive rigorous hard science contained in the chapters. A truly comprehensive M&E system must keep room for the inclusion of techniques drawn from 'state of the art' physical science research.

Part IV shifts the focus to socioeconomic observation. None of the chapters has the confidence in knowledge closure shown in the physical science chapters. Even the most economistic chapter is very aware that many of the parameters in social cost-benefit analysis require judgements to be made and that such judgements need to be made explicit and explored in sensitivity tests. But it is not just the problems of estimation and inaccuracy that make all the chapters less closed and deterministic.

Collecting information from people closely affected by watershed and soil conservation means engaging with them as agents exercising choice over the use of resources in addition to being data banks. It is possible to try and filter out information on people's values, motivations and aspirations as non-scientific. But in a world where grassroots participation in decision-making is seen as both desirable and necessary, such filtering appears counterproductive. Several of the chapters in Part IV raise problems that evoke variants on 'tragedies of the commons' in which people's judgements on their best interests are crucial to understanding consequent physical processes. Sadly, most of the resulting actions work in the direction of, at best, neglect or, at worst, accelerated degradation of shared physical environments. The chapters do not describe innovatory techniques, but do raise current issues of how human agency can be incorporated in M&E systems and negotiating the frontiers between observation and deliberation.

USING THIS BOOK

The editors are very aware that readers will come from a wide range of policy and disciplinary backgrounds. There will be readers concerned with global environmental issues looking for techniques to produce evidence to present to international fora producing protocols aimed at setting targets for national policies. At the other end of the scale will be local non-governmental activists seeking ways to gauge the impact of their interventions with a view to securing national and/or international resources for further activities.

In disciplinary terms, readers' foundational disciplines will vary from backgrounds in agronomy, hydrology, soil science, physics, chemistry and biology to economics, political science, sociology and anthropology/ethnography. And each discipline will bring its own views on the most useful methodology and associated observation techniques.

We have attempted to give every reader an entry point chapter that will more immediately be accessible in both policy and disciplinary dimensions. We look forward to the discussion this should induce among the experts in every field. But we are also hoping that readers will then read chapters that are more remote from their particular entry point and gain insights into other principles and practices on Monitoring and Evaluation in the field of soil conservation and watershed development activities.

We therefore suggest in approaching the book that you look at the abstracts provided for each chapter and initially select those chapters that are closest to your own knowledge and experience. Having read and critically assessed those chapters, we then suggest you look at some chapters that seem most distant from your own knowledge and experience and see what they offer in terms of fresh insights. We must say at this point, that one of the chief tasks we set ourselves as editors was to make all the chapters widely accessible while maintaining integrity in terms of rigorous argument and appropriate technical language. We intentionally offer a wide range of positions on Monitoring and Evaluation and privilege no particular position in order both to encourage discussion on particular positions and to widen the general discourse.

We also intend this book to be a reference book. Though the book has a focus on Monitoring and Evaluation methods, almost all of the chapters apply these methods in particular contexts and offer both data and interpretations on the processes operating in those contexts. We are delighted to have included authors writing on so many different countries - but we not only have chapters on many different countries, but differing circumstances in those countries. The book includes effects on soils and watersheds of long-term agronomic changes, relatively new settlement, and natural disasters in a variety of geo-ecological contexts under contrasting policy regimes. We have edited the abstracts to ensure they all include locations and contexts as well as Monitoring and Evaluation methods as an aid to using this book for reference purposes.

Reference

ACC. 1984. *Monitoring and Evaluation Guiding Principles.* United Nations ACC Task Force on Rural Development, IFAD, Rome.

Part 1

Principles of M&E in SCWD Projects

1

Principles and Practices of Monitoring and Evaluation for Watershed and Conservation Projects

Ted C. Sheng

Department of Forest, Rangeland, and Watershed Stewardship. Colorado State University, Fort Collins, CO 80523, USA. E-mail: teds@lamar.colostate.edu

ABSTRACT

The necessity for monitoring and evaluation (M&E) of watershed development and soil conservation projects is well recognized today. However, M&E is not yet a definite science and its application often encounters difficulties and constraints. Consequently, between doing nothing for M&E and doing it perfectly some practical methods must be developed.

This chapter, based mostly on the author's experience, discusses briefly the definitions, principles, designs, and methodologies of M&E work. Institution and training needs are also explained.

It is impossible to monitor every aspect of a project, and emphasis has been given to the selection of suitable indicators. Examples have been given and the ways and means of monitoring are introduced.

For evaluation, this chapter has highlighted some important principles and techniques for general reference. Stress has been put on the publishing of M&E reports so that lessons can be learned by others.

INTRODUCTION

The main objectives of monitoring and evaluation work are to record, review, analyse, and assess the work progress, accomplishments, and impacts of a project or a programme.

Monitoring and evaluation are often mentioned together as 'M&E'. Each, in fact, has a distinct function. By definition, monitoring is an arrangement or an action to continuously observe or record the operation of a system or a project. Evaluation is an assessment of a project's performance or a determination of its worth.

For a watershed or a soil conservation project, as with any other type of project, monitoring and evaluation is applied to answer two questions:

- Are we doing the project right? (Monitoring)
- Are we doing the right project? (Evaluation)

Monitoring and evaluation are closely related. Without monitoring, evaluation will become difficult, if not impossible. On the other hand, without evaluation, the monitoring work will be fruitless.

Lack of monitoring work and especially evaluation results is said to be one of the hindrances to further investment in watershed or soil conservation projects in developing countries. There are difficulties in many countries in undertaking M&E work, due mainly to a lack of data and methodology. However, between doing nothing and doing a perfect job there should be some practical methodology that can be used to implement such an essential task.

PRINCIPLES

Projects vs Programmes

Many countries in the developing world do not have the resources to establish institutionalized watershed or soil conservation programmes. Often they receive international aid to carry out a specific watershed development or soil conservation project in a chosen area. Such projects are usually short-term, over a period of several years. Since many of the indirect benefits will occur beyond the project life, the M&E work has to be concentrated on the project's physical accomplishments, its short-term as well as direct benefits.

Countries with established programmes in watershed development and soil conservation should undertake long-term and continuous M&E work with properly established units and trained personnel. A central unit with several regional sub-units is ideal for carrying out the work.

Project-specific M&E

Both watershed and soil conservation projects can be multidisciplinary and multi-sectoral. Some projects emphasize environmental protection and others agricultural production. Some aim at integrated development and others for a single purpose, all depending on the original project design. Therefore, there are no universal criteria for M&E work.

One important principle, however, is to monitor and evaluate the project against the project plan (original or revised). The monitoring items should be agreed upon with the project plan and evaluation work must be tied to the project objectives.

Simple and Useful Design

In principle, design of monitoring work should be simple and useful for the project. There is no need to monitor all aspects and activities of a project. Only the major ones should be included. With this principle in mind, the project manager and technical staff should discuss and determine the items, processes, and techniques of monitoring at the very beginning of a project. Also, there is a tendency to collect more data than can be digested and useful. This should be avoided because it will waste time and money.

Cost-effective M&E

Information is expensive to collect, analyse, and evaluate. After the monitoring items have been decided, only necessary data is to be collected and analysed for each item. In many cases, the objectives do not require overly detailed and precise data. The benefits from the information must outweigh the costs of obtaining it (Brooks et al., 1990). An approximation or a general estimation with sound professional judgment is often deemed sufficient.

Use of Existing Data

Much of the information needed for monitoring and evaluation may already exist. Managers and technical staff should not neglect available information and should try their best to locate and collect it. It is usually cheaper to use existing information than to collect fresh information. For instance, hydrologic data and erosion rates from the same climatic region or nearby experiment stations can safely be used.

An Iterative Dynamic Process

Finally, managers of watershed or soil conservation projects should view M&E as an iterative dynamic process. From time to time, the results of

M&E will provide feedback that can be used by managers to change an ongoing project or to modify current policy. Through continuous efforts of M&E, lessons can be learned for improving future project design and performance. A country with an established programme will gain the most from this kind of M&E results.

MONITORING

Design of Monitoring Programmes

Monitoring Categories

For watershed and soil conservation projects, there are normally four types of monitoring categories to be considered, as follows:

1. Monitoring project implementation: This category is most obvious and every project needs to include it in its overall monitoring programme. The main objective is to see whether project implementation has proceeded as planned with respect to time, schedule, targets, and cost. Detailed design would depend on project requirement.
2. Monitoring environmental impacts: This category is specific to watershed and soil conservation projects, most of which are intended to minimize erosion, sediments, water pollution, or floods. Collection, analysis, and comparison of biophysical data with and without project will give clues to the project impact.
3. Monitoring socioeconomic improvements: Watershed and soil conservation projects are intended, directly or indirectly, to improve people's living standards and the welfare of the rural area. Therefore, data on income, crop pattern, yields, housing, marketing facilities, and other indicators need to be collected and monitored. In addition to formal socioeconomic surveys, beneficiaries are to be sampled, frequently contacted, and monitored to comprehend their problems, attitudes, progress, and results.
4. Monitoring project sustainability: This category of monitoring is usually neglected by project management. A project, by definition, is a certain type of work to be carried out in a designated area and period. Often, it does not look beyond project life. However, from a governmental point of view, sustainability of the project work after the project is terminated is of utmost importance. Therefore, this category needs to be included in an overall monitoring programme.

Selection of Key Indicators

It is not possible to monitor every aspect of a project. Careful selection of key indicators for monitoring is cost-effective. Managers should also realize that collecting, sorting, storing, analyzing, and presenting data can be tedious, time-consuming, and expensive. Therefore, only useful and necessary data are to be collected for monitoring uses. If appropriate indicators are chosen, they will greatly help the monitoring and evaluation work (Casley and Kumar, 1987).

Some general criteria for selecting key indicators are suggested as follows:

- They reflect the main objectives of the project.
- They can be clearly defined or measured no matter who does the work.
- They are sensitive to change or responsive to the project work.
- The data needed for the indicators are relatively easy to collect.

According to the plan of a project, the key indicators should be determined by the project manager, technical staff, and the monitoring unit at the very beginning of the project. They may vary depending on the types and objectives of the project. The following indicators are suggested (Sheng, 1989) for watershed and soil conservation projects:

1. Physical outputs: This includes, for instance, hectares of trees planted, terraces completed, metres of stream bank protected, and kilometres of road built or improved, and their progress, costs, inputs, and accomplishments against the original plan and time schedule.
2. Land use changes: Land use changes over time are usually a major concern of watershed or soil conservation projects. By comparing the data of different periods the project can get a picture of land use trends as well as the impact of the project work.
3. Erosion, sedimentation, water quality, and runoff: These are also major concerns of watershed or soil conservation projects. One or two may be more important than the others depending on the main objectives of the project. For instance, if on-site erosion is of greater concern than flooding and sedimentation, then the monitoring work should be concentrated on the former. Likewise, if heavy sedimentation of a reservoir is the major concern, the task should include monitoring of both erosion and sedimentation rates.
4. Farm income, production, and land productivity: Increase of farm income through an increase of crop or animal production or marketing may be one of the major objectives of watershed conservation projects. In terms of natural resources protection,

however, land productivity in the long run may be more important than short-term income. Whatever the major concern is, both need to be monitored.

5. Sustainability and development: Whether the watershed or soil conservation work will be sustained after the project is terminated is an important concern of both the government and, in the case of a foreign aid project, the donor. The success or failure of a project is normally judged by its sustainability. Many projects have rural development components. Monitoring of development work, such as kilometres of road built and maintained, and numbers of houses improved, is also necessary.

Baseline Information and Data Needs

In the process of formulating watershed or soil conservation projects, there must be some inventory data already collected that can serve as baseline information for monitoring uses. These may include soils, rainfall, land use, land capability, physical problems of the watershed, infrastructure, and socioeconomic conditions. Such data are usually essential to serve the needs of the subsequent monitoring work.

As a project proceeds, additional data should be sought as required by the monitoring programme. Contact with farmers, interviews, farm planning, and supplemental surveys are usually required. For collecting new data, close attention should be paid to the availability and usefulness of the data as well as the time and resource limit. Data that is too expensive to collect or takes too long to obtain should be avoided.

Monitoring Methodology

Using the above key indicators as examples, their respective monitoring methodologies are briefly described in the following sections.

For Physical Outputs

Of all the monitoring work, monitoring of physical outputs is the most essential and straightforward. Each project should design and establish a sound database that includes, for instance, planned work, yearly schedule, overall targets, sub-targets, and accomplishments. The database is better linked to various maps and diagrams. Geographic Information Systems (GIS) is essential to establish such monitoring data. In addition, cost of project work needs to be closely monitored.

Experience shows that the database should be designed to allow for possible future revisions of the original plan. Also, space should be provided for simple explanation or reasoning if certain work cannot be completed in time or is abandoned.

For Land Use Changes

Proper land use in accordance with the capability of each piece of land is probably one of the most important goals for both watershed and soil conservation projects. Therefore, monitoring land use changes in a watershed or an area is necessary. In addition to surveying of land use using remote sensing techniques, a land capability map should also be produced. To overlay the land capability map with the land use map, a new map of current land use conditions (proper or misuse) can be produced. The ideal scale is 1:10,000 for watershed monitoring. This can be done either manually or by GIS (FAO, 1990; Sheng and Barrett, 2000).

Repeated survey of land use is required. The period can be around five years depending on the life of the project. The result of a new survey can then be compared with the original surveys to obtain the trends of land use and the impacts of the project work.

In designing such a survey for monitoring purpose, similar criteria for land use classes, mapping unit and interpretation techniques should be employed. Otherwise, comparison will become difficult, if not impossible.

For Erosion, Sedimentation, Water Quality, and Runoff

Monitoring of all or part of these indicators is normally required for a watershed and soil conservation project. However, these kinds of environmental data are not easy to collect. Often, the project life is too short to obtain a meaningful result, or the area is too small to produce a significant impact downstream. Generally speaking, to make hydrological data usable for monitoring purposes, a decade of records is required.

There are, however, some practical methods that can be used to cope with time or budget problems. The following are some suggested strategies:

- Collect existing data on erosion and runoff rates and sedimentation conditions from nearby experiment stations or reservoirs. If these data are sound and useful, fresh research will not be needed.
- Use erosion prediction models to estimate erosion and runoff rates. Select a model that is targeted for or developed under a similar environment (climate, slope, soil, etc.) to monitor these data. For developing countries in the humid region with steep slopes, for instance, the Simplified Process (SP) Model can be useful (Hartley, 1987).
- Establish simple erosion and runoff plots to obtain erosion and runoff data, if necessary and affordable.
- Make simple sediment surveys of ponds or check dams of a treated sub-watershed or area against a control.

- Coordinate with water supply companies or irrigation districts collect water samples for water quality analysis.
- Only countries with established watershed or soil conservation programmes are apt to set up a hydrologic network (weather stations and gauging stations) in and around a targeted watershed or area. The network should be included in the national system and should seek close coordination with it.

For Income, Production, and Land Productivity

A baseline socioeconomic survey is usually done at the inception of a project. Periodic and repeated surveys will provide information on the project's impact and benefit. Beneficiary contact monitoring should be carried out routinely. For more actual data on income and crop production, individual farming records should be kept at key farms for monitoring uses. In selecting key farms, sampling methods such as the random method or stratified method should be applied. Project staff needs to help the key farms with book-keeping to get reliable costs, labour, and results.

Land productivity is related to resource conservation, which is both a government and a public concern. Farmers may deplete soil resources in order to make a living. There are many ways to increase farm production and income; yet, for the benefit of the nation as well as future generations, maintaining or improving land productivity is a fundamental task. Few projects in the past undertook this kind of monitoring work due to resource or time limit. It is suggested that countries with established watershed or soil conservation programmes should help to check soil fertility from the project area.

For Sustainability and Development

Project sustainability can be monitored by the number of farmers who participate in the project, their maintenance work, and the rates of adoption of new techniques. A higher percentage of farmers adopting the recommended work would indicate higher sustainability. Farmers outside the project area or farmers within the area who voluntarily join the project without government help can also be a good indicator.

For watershed or conservation projects aimed at rural development, a close monitoring of development work is necessary, which may include housing, road building and maintenance, domestic water, small irrigation, fuel-wood supplies, and marketing facilities. If required by the project, a further monitoring of employment, public health, and/or education is needed. These activities should be carried out in close coordination with other specialized agencies.

Table 1 shows a summarized view of key indicators and suggested monitoring methodologies.

Table 1 Key indicators of watershed and soil conservation projects and suggested monitoring methodologies.

Key indicators	*Suggested methodologies*
Physical outputs	Establish database mainly on:
	- Plan of work (including yearly schedules)
	- Targets and sub-targets
	- Accomplishments and costs
Land use changes	Conduct initial survey of land use and land capability
	Conduct periodic surveys to compare:
	- Current land use conditions (proper, misuse)
	- Land use trends
Erosion, sedimentation, water quality, and runoff	Collect and use existing data as first step
	Apply erosion prediction model Establish simple runoff plots if necessary
	Make sediment surveys of pond and check dams
	Monitor water quality with other agencies
	Establish hydrologic network if necessary
Income, production, and land productivity	Establish socioeconomic database and conduct periodic surveys
	Conduct beneficiary contact monitoring
	Help farmers to keep records on key farms
	Compare crop yields with and without project
	Measure and compare soil fertility
Sustainability and development	Record farmers' participation rates and voluntary work inside and outside project area
	Monitor accomplishments on various types of development work
	Conduct special surveys when needed

Establishing Monitoring Units

Unit Responsibilities

Depending on project size and requirement, a monitoring unit with several personnel should be established. An officer with a biophysical or socioeconomic background could be the leader of the unit. The responsibilities of the unit are normally as follows:

- To carefully design a monitoring system with project manager and technicians concerned.

- To be responsible for data collection, store, analysis, retrieval, and presentation.
- To coordinate monitoring work among related sectors of the project.
- To liaise monitoring activities with other agencies.
- To assist the project in preparing M&E reports.

Training Needs

In addition to watershed management and basic knowledge of soil conservation, the training of unit staff and related personnel as a team requires the following basic subjects:

- Computer techniques including database management, GIS, and use of watershed and erosion models.
- Remote sensing techniques and use of Global Position Systems (GPS).
- Statistics including sampling methods, analysis of variance, correlation and regression, and tests of significance.
- Farmer interview techniques and farm planning.
- Economics and cost-benefit analysis.

The training can be conducted as a formal course in four to five weeks or as an in-service training for a longer period.

EVALUATION

Time of Evaluation

Evaluation is usually applied at a project's mid-term and at its end. The mid-term evaluation should be carried out when the project already has some results, or when it encounters some implementation problems. Often, a project changes course or plan after a mid-term evaluation.

The terminal or final evaluation is a necessity. From it, the people concerned can learn lessons of success or failure. It may result in an extension of project time or area. Post-terminal evaluations may sometimes be conducted to find out a project's long-term impact.

Independent Mechanism

An independent mechanism is desirable for evaluation work. It can be either an established mechanism in the central or regional government or one organized temporarily when an evaluation is needed. The advantage of an established mechanism is that it can help a project to design a monitoring programme at its initial stage.

An independent mechanism will be less biased in the course of evaluation. It could continuously evaluate the project's impacts after its termination. In a country with a well-established programme in watershed or soil conservation, setting up a central unit or regional units for evaluation is most desirable.

Use of Monitoring Data

If the monitoring work is properly and sufficiently done, evaluation can be carried out smoothly and swiftly without the need for much additional data. However, in many cases, additional data is required for evaluation. While it is not a problem to collect small sets of additional data, it is difficult to collect larger sets of data in a short time. To avoid this, it is desirable to involve the evaluation personnel in the design of monitoring from the beginning.

Only reliable and valid data are to be used for evaluation. The sources of such data and the methods of obtaining them should be well examined. Casely and Kumar (1988) spelled out in great detail the use of monitoring data for evaluations.

Collection of New Data

If more data are needed for evaluation, the project, especially the monitoring unit, needs to collect and present them in a timely manner. Therefore, the unit should be well aware of data availability in other related agencies. This may require keeping a file or a list of reports (annual or special), photos, maps, and digital data from other agencies, and renewing or updating them frequently to obtain the latest information.

Evaluation Principles and Techniques

Generally speaking, evaluation needs to be conducted in accordance with the following principles:

- Evaluate a project closely against its plan of work.
- Use as much of the project's existing monitoring data as possible.
- Remain objective rather than subjective.
- Make goodwill recommendations as well as constructive criticism.

Techniques for evaluation may vary; however, the most common ones are outlined as follows:

- Study the project documents and reports thoroughly before visiting the site.
- Listen attentively to the project's report during briefing.
- Interview individual staff with understanding and encouragement.

- Evaluate the big picture of the project (i.e. performance, achievements, efficiency, and impact) rather than smaller items.
- Closely evaluate each of the key indicators.
- If shortcomings are found, determine whether it was due to project design flaws or implementation.
- Use cost-benefit analysis to examine the economic efficiencies of the project, if necessary.
- Compare areas with and without the project to evaluate the impact and sustainability of the project.
- Draw up reasonable and operable conclusions and recommendations.

Reports

Evaluation reports can be of two types. The usual type is for the project management and, in theory, should be made available to anyone who wants to read it. Sometimes a different type of special report is needed for presenting to high-level administration when policy or confidential matters are concerned.

A regular evaluation report is usually composed of the following sections:

- Executive summary
- Introduction and terms of references
- Description of the project (including scope, objectives, goals, and plan of work)
- Monitoring work sufficiency
- Performance and accomplishments
- Evaluations (including methods, findings, problems, and impacts)
- Conclusions and recommendations

Except for confidential reports, regular evaluation reports should be published or made available to the public so that past experiences and lessons can be learned by others.

CONCLUSIONS

Few watershed development or soil conservation projects in the past have adopted proper monitoring and evaluation techniques and approaches. Most of the M&E work was done in a hasty or superficial manner without the support of sound data, and the results were kept for the project alone.

As resources become more and more limited and competition for investment becomes more active nationally and internationally, proper

M&E of such projects will be necessary to show their real benefit and impact .

Monitoring and evaluation is not yet a definite science and its implementation may present many difficulties. However, between doing nothing and doing it perfectly there must be some practical methodology which can be applied and adopted. To find the appropriate indicators and determine the proper methodology is the first step towards a successful monitoring and evaluation programme.

References

Brooks, K.N., H.M. Gregersen et al. 1990. Manual on watershed management project planning, monitoring and evaluation. A publication of ASEAN-US Project. University of Minnesota, St. Paul, Minnesota.

Casley, D.J. and K. Kumar. 1987. Project Monitoring and Evaluation in Agriculture. Published for the World Bank. Johns Hopkins University Press, Baltimore and London.

Casley, D.J. and K. Kumar. 1988. The Collection, Analysis, and Use of Monitoring and Evaluation Data. Published for the World Bank. Johns Hopkins University Press. Baltimore and London.

FAO (Food and Agriculture Organization). 1990. Watershed Survey and Planning. FAO Conservation Guide 13/6, Rome.

Hartley, D.M. 1987. Simplified Process Model for water sediment yield from single storms. Part I and Part II. Trans. ASAE 30(3): 710–23. American Society of Agricultural Engineers, St. Joseph, Michigan.

Sheng, T.C. 1989. Watershed project monitoring and evaluation. *In:* Sheng, 1990. Watershed Conservation II: A collection of papers for developing countries. Colorado State University, Fort Collins, Colorado.

Sheng, T.C. and R.E. Barrett. 2000. Using GIS for land capability classification and conservation farming management. *In:* Laflen et al. (eds.) Soil Erosion and Dryland Farming. CRC Press LLC, Boca Raton, Florida.

2

Monitoring and Evaluation of Watershed Development Projects in India

D.C. Das, K.K. Gupta and K.G. Tejwani

Members, Soil Conservation Society of India. B-19, Parijat Apartments, Pitampura (West), New Delhi 110034, India. E-mail: dinesh_ranu2003@yahoo.com

ABSTRACT

Monitoring and evaluation (M&E) of a watershed development (WSD) Project is seen here as an exercise that has two interlinked components. Monitoring relates to watching the progress of the implementation against set objectives, which includes collection of some observations to build data series and offers opportunities to carry out mid-course correction. It is generally an in-house exercise. Evaluation, on the other hand, is an external study. It covers performance assessment in terms of outputs and consequent outcome or effect. It also provides a measure of durable effects or overall impact of the project and the prospect of its becoming sustainable. The information that the exercise provides should stand scrutiny by a third party and, therefore, must have some quantified results and outcome. This helps a WSD project to compete with other projects in attracting investment. An M&E study has to assess four broad or component areas, i.e. biophysical status, primary production systems, socioeconomics and institutional aspects, and nine important sub-areas. Indicators are needed for all these sub-areas. These indicators should be unique, quantifiable, and directive and should have point source. They can be individual or composite.

The methodology for an M&E study has four steps: (1) sampling of micro watersheds, villages, households or respondents, sites, etc.; (2) data collection and tabulation; (3) analysis; and (4) drawing of inferences. Primary data is collected through interactive surveys of sampled households with a questionnaire, village profile by a participatory rural appraisal and participatory village transect survey, direct measurements with conventional methods, and application of remote sensing data and images as well as GIS. Secondary data are gathered from secondary sources. There are three sub-studies of an evaluation study: a baseline survey that provides the pre-project scenario; a mid-term evaluation that gives an intermediary scenario and helps identify areas for making mid-term corrections; and a final evaluation study that reveals durable effects and is thus referred to as impact analysis. Data analysis and interpretation are carried out for sub-areas as well as whole watersheds. Some illustrations are given in this chapter, which relate to climate, land use, soils, soil and water conservation, improved ground water and surface water, growth and development of community-based organizations and institutions, social and economic upliftment, environmental upgradation, and sustainability of the project implementation. The questionnaires and schedules are diverse, voluminous, and project-specific. Assessment of sustainability is a separate concluding exercise. The principles and methodology are illustrated with a case study.

INTRODUCTION

A watershed development (WSD) project contains a package of interventions. These are designed to use a set of inputs (goods and services) to yield a set of outputs, which are transformed into different goods and services. The outputs offer opportunities to change the production, use and consumption patterns of the watershed resources. They also put into operation processes of re-enrichment or regeneration, which are called effects or outcome. When some of these become recurrent and lasting, especially in terms of quality of life and environment, they are called project impact. An understanding of these sequential transformations is gained through monitoring and evaluation or M&E. Monitoring is watching implementation vis-à-vis set project objectives, a package of interventions, an operational time schedule and a sanctioned budget. It helps make mid-course corrections and generate data that provide scenarios which get lost through project interventions. Evaluation, on the other hand, covers performance assessment determining the form, quantity and quality of outputs and the consequent outcome or effect. Evaluation also gives a measure of the impact on the natural resource base and the target group. It also includes cause and effect analysis or process analysis that helps modify planning and design norms and procedure to achieve greater effectiveness. Though M&E historically followed a deterministic approach, it now combines tools to assess some non-quantifiable outcomes by a process of grading and using unit-neutral scales. The steps showing data acquisition, processing, analysis and interpretation of a typical M&E exercise are illustrated in Fig. 1.

APPROACH AND STRATEGY

It is necessary to set out basic concepts in clear terms. Monitoring and evaluation should provide useful information and data that can lead to

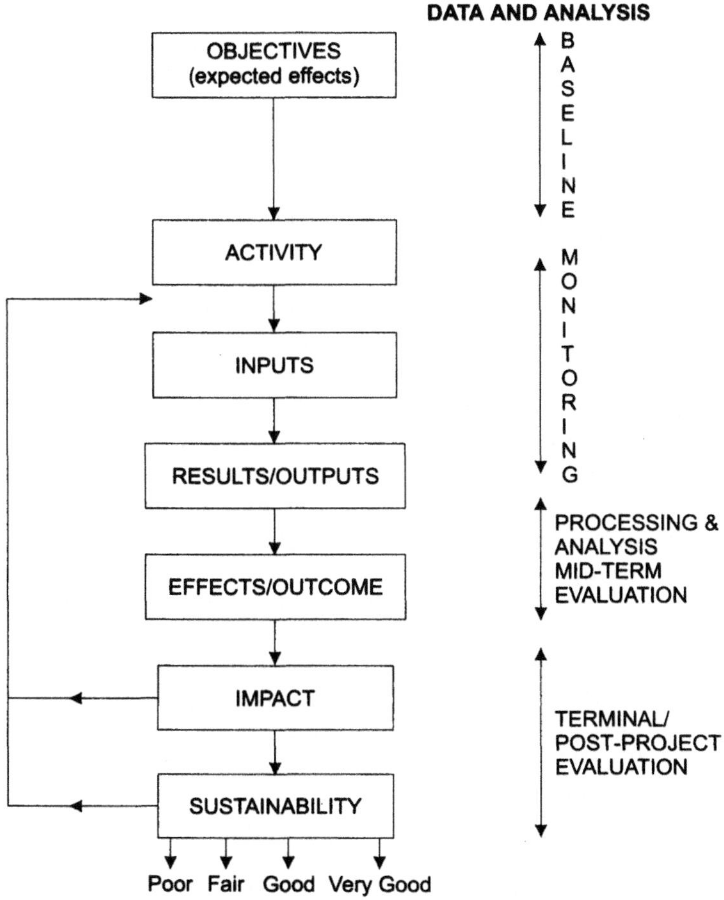

Fig. 1 Linkages over time between project input and output vis-à-vis effects/impacts and sustainability.

improvement in project processes from planning to impact evaluation for future projects. This exercise could help achieve greater cost effectiveness and thus facilitate rational investment decisions as information is derived from data through appropriate processing, collating and analysis. Collection of data and building data series over time and space form the foundation of any M&E exercise. The benchmark or reference scenario is developed from data obtained through a pre-project baseline survey, which are normally included in the project document. These are supplemented by data collected during the project implementation as part of the monitoring process. Data on changed scenarios are collected during the evaluation process from the beneficiaries and the area in which project activities have been carried out. These are primary data. To these are added secondary data and information from available documents such as reports and published literature. The difference between the two scenarios provides insight into outputs, outcomes as well as impact. If the changes are to be watched during the project implementation especially with a view to providing mid-term corrections, then one mid-term or concurrent evaluation can also be carried out.

The set objectives of the project lead to adoption of a package of interventions. These determine the areas of concern and outputs and outcomes to be monitored. In India, watershed management programmes were introduced as early as the late 1950s in the Damodar Valley Corporation area. Since then, set objectives have steadily enlarged and packages of interventions have been augmented. In the 1950s and 1960s, increasing agricultural production was the main goal. At this point of time, ways to transform losses due to various degradation processes into financial gains were not considered. Impact evaluation mainly rested on increased production and used benefit-cost ratio or internal rate of return as the main tools. Quantification of other benefits, ecological as well as social, was attempted in the 1970s. These were further refined and presented with detailed methodology and computations in the ISCO conference held in 1980 in the UK (Das and Singh, 1981). More details of these developments were published in *Soil Conservation in India* (Gupta and Khybri, 1986). Thereafter, independent agencies such as the Agricultural Finance Corporation in Mumbai, the Administrative Staff College of India in Hyderabad, and the Indian Institute of Management in Ahmedabad carried out evaluation studies in various parts of the country (Das, 1998). Many new approaches and tools, especially for the ecological, environmental and social components, were subsequently provided by Lal (1992), Maudgal and Kakkar (1992), Chopra and Subba Rao (1996), Ram and Dhyani (1998), Dhyani et al. (1997), Agnihotri (1999) and the World Bank (1999).

During the 1990s a large number of external aid agencies introduced new tools especially for assessing the impact of institutional innovations. Most of these, due to preoccupation with social and institutional aspects, paid little attention to biophysical aspects. Further, many avoided quantified data series and followed more qualitative and participatory assessment techniques. Land Use Consultant International (LUCI, New Delhi, 1999), in their assignment for RODECO, Germany, showed how participatory tools can be combined with scientific ones as well as how social and institutional data can be brought into similar benefit streams to carry out analysis to assess sustainability using a unit-neutral performance scale. The opportunities offered in the M&E studies of the Integrated Watershed Development project supported by the World Bank in the Indian states of Gujarat (AFC, 1998), Himachal Pradesh, Punjab, and Uttaranchal have been helping the team of experts from Consulting Engineering Services (India) Private Limited (CES, 2002a,b), New Delhi, in testing some of these synthesizing and integrating efforts. In future it should be possible to take into account biophysical, social, economic and institutional outputs on equal priority to evaluate WSD projects, while M&E as a strategy should combine scientific norms and tools with participatory ones.

COMPONENT AREAS REQUIRING EVALUATION

A WSD project has several distinct sets of interventions:

1. Biophysical status: These concern protection and development of natural resources of land, water, forests and other vegetation.
2. Primary production systems: These include agriculture (crop cultivation), horticulture, animal husbandry and management of forests and grazing grounds.
3. Socioeconomics: These cover demography (human and animal), occupational pattern, income and expenditure, health, literacy, assets and empowerment.
4. Institutional aspects: These relate to type, composition and level of community-based organizations, linkages, representations and empowerment.

SUB-AREAS

In a typical WSD project, M&E would thus have to get data and information on the following sub-areas as experienced by the teams of experts from Agricultural Finance Corporation (AFC, 1998) and Consulting Engineering Services (CES, 2001, 2002):

1. Basic: Climate, land use details, soil attributes, ecological or environmental hazards and risks.
2. Arable land production systems: Agriculture (annual crop cultivation), horticulture (fruits, vegetables and flowers), livestock and fodder development.
3. Non-arable lands: Forests (government, community and private), pasture or grazing grounds, plantation crops,
4. Others: Piggery, poultry, fishery, on farm vs off farm, tertiary livelihood, etc.
5. Soil and water conservation: Conservation practices on arable and non-arable land, drainage line treatments, improving water resources (surface and groundwater), water harvesting, irrigation (with surface or groundwater), land slide area treatment, water table and channel flow pattern.
6. Social aspects: Human population and its distribution among ethnic groups, economic categories, between land-owning and landless, households and holding-size classes, literacy, occupational groups, full and under employment.
7. Economic: Annual income of households and individuals and their sources, households in different income classes, expenditure, indebtedness, distress migration and sale.
8. Institutional: Type and form of community-based organizations such as watershed village development committees, user groups, self-help groups, forest protection committee, village environment committee; statutory institutions at village, development block and district levels; linkages, composition, administrative and economic empowerment as well as technical, administrative and financial management skill.
9. Environmental: Soil erosion and land degradation, surface water flow pattern (seasonal/ perennial), runoff and flash floods, water stress and drought, soil loss, green area adequacy in extent and quality, land use alienation, weed infestation, soil fertility loss, pollution and contamination of land, water and air, biodiversity (*in situ* and *ex situ*).

INDICATORS

The changes or effects will be on many fronts. These have been segregated into nine broad groups or parametric areas. Each can be classified into several sub-groups or sub-parametric areas. Assessing the changes brought about in these sub-groups would demand a large number of indicators. Gathering data from secondary sources or generating primary ones by direct measurements at representative sites and collecting from respondents is costly, time-consuming, and unmanageable. Thus, as a first step, sub-parametric areas are short-listed. Then appropriate indicators are chosen for each sub-parametric area . To be appropriate an indicator should have the following properties (World Bank, 1999; Das et al., 2000):

Unique: An indicator should be distinct and represent clearly the parametric or sub-parametric area.

Quantifiable: An indicator should be quantifiable directly or through indirect means.

Directive: An indicator should lead or draw attention by its own attributes to the resulting output and/or outcome.

Interpretative: An indicator should help clarify the process involved or the cause and effect relationship and thus help explain the transformation of input into output and the outcome.

Point source: An indicator should be identifiable to a point source on space or time or both that can help develop a consistent data series.

An indicator should be simple and easy to collect data on. It should be quickly processed and computed in the desired form for documentation and subsequent use. Collection of observations on it should not normally call for large investments, sophisticated equipment and specialization or skill as needed for research. In other words, an indicator should be cost-effective. At the same time, the relevance and sensitivity of a sub-area under a given situation should be preserved. Thus, there may be a compulsion to select more than one indicator for an identified sub-area.

Similarly, collection and documentation of observations on any indicator should receive serious consideration for adopting proper methods and tools and techniques and for ensuring minimum adherence to scientific standards or any other acceptable norm set through precise research or in-depth study. Because the entire exercise has strong elements of applied research and judicial objectivity, it must be uniformly fair and free from bias and prejudices (Chakravarty, 1992; Chopra and Kadekodi, 1992; Das, 1999; World Bank, 1999). Besides, absence of back-up data does not help a third party to critically examine the context and causes for variable outputs. Similarly, an indicator that may help carry out the exercise through beneficiaries alone and adopt qualitative descriptions or consensus, using the participatory mode, cannot meet the serious demands for an investment decision amidst competitive demands from other proposals. Thus, one cannot rate an indicator as good enough just because it is cheap or does not need equipment or skill or demands less time. The reverse is also not applicable. A judicious combination is required.

An indicator can be individual when it represents a single sub-parametric area. It can be integrated when it attempts to represent the whole parametric area substantially. It can also be a composite one covering many sub-areas of one or more parametric areas. All three sets complement each other and help in summing up the total effect of a package of project interventions. Hydrologic Performance Index, which represents the total effect in terms of reduced runoff, increased perenniality, increased recharge of ground water, and reduced flooding, is a composite index. It is computed from the values of the individual indicators by transforming them into a unit-neutral matrix scale. Similarly, Watershed Eco Index provides a combined assessment for improved green or perennial cover that takes into account the effect of enlarged area under green cover, number of plant populations per hectare, and quality of standing stock in terms of height and girth.

METHODOLOGY

The methodology involves three broad processes:

1. Participatory mode. This is to give adequate weight to the concerns of the target group.
2. Some scientific principle, technique and tools, if necessary with substantial innovation. This is to reduce high demand for time, skill and cost.
3. Interactive dialogue with different stakeholders and acquaintance with the area and activities. This is to get a perception of the context and ground realities.

SAMPLING

Watershed development works are planned and implemented on the basis of geo-hydrologic units, i.e. watersheds (WS), sub-watersheds (SWS) and micro-watersheds (MWS). Sampled areas must represent all WS, major groups of SWS and a minimum number of MWS within the representative SWS. In choosing appropriate MWS, the following criteria are to be applied:

1. Concentration of project activities in each MWS;
2. Location of each MWS within the WS such as whether on left or right bank or in upper reach or lower reach;
3. Altitude, slope, soil, vegetation, social groups and type, extent and severity of prevailing degradation.

AN EXAMPLE OF SAMPLING

In five watersheds located in the Siwaliks of Himachal Pradesh, there are 32 WS and 109 MWS. In 59 MWS, works were carried out till the time of sampling. All these MWS were considered for sampling. For farming systems such as agriculture, horticulture, and livestock development as well as for socioeconomic and institutional aspects, primary data was collected from the respondents. Thus, 10% of the villages located in these MWS or a minimum of one was randomly chosen. Within each sampled village, 20% of the households or families were sampled as respondents through the principles of stratification on social and/or economic categories and randomization. The abstract of the process and results are shown in Table 1.

For soil and water conservation measures, drainage line treatments, water harvesting and other engineering works, the primary data are collected on geo-hydrologic unit of MWS. In order to keep sample size of each activity around 10% of the total magnitude, a smaller number of MWS were chosen according to the three criteria discussed earlier (Table 2).

The 10 MWS finally selected were Amriyon, Bankala, Tikri, Koti, Dhang, Adual, Puloh, Bhadsali, Thora and Bari. For afforestation, agroforestry and other tree planting works, 10% of the sites were sampled in the selected MWS or villages for vegetation and biometric surveys. However, a minimum of 0.01% of the areas under this treatment was sampled.

LIVE OR ILLUSTRATIVE CASES

In order to focus on a specific effect of a particular intervention or a combination of interventions, a real or ground case becomes more useful. Often such cases spontaneously put together a number of durable effects

Table 1 Abstract of sampling of MWS, SWS, villages and households sampled for an evaluation study.

Watershed	*SWS*	*Sampled MWS*	*No. of sampled villages*	*No. of sampled households*
Markanda	5	9	9	95
Ghaggar	5	12	12	65
Sirsa	9	17	17	136
Swan	8	8	8	163
Chakki	5	13	13	186
Total	32	59	59	645

Table 2 Selection of sample micro-watersheds in a project where total MWS are 109 in 32 SWS of 5 WS.

Sl. no.	*Watershed/ Sub-watershed*	*Micro-watershed*	*Project activities taken up (% of total)*	*Location of MWS within a WS*	*Ranking*
1	*Markanda*/ SWS-5		11 (50.0)	Upper	1
2	Do	Amyrion	8 (36.4)	Middle	**3**
3	SWS-11	Bankala	9 (40.9)	Upper	**2**
	Ghaggar				
1	SWS-5	Tikri	14 (63.6)	Middle	**1**
2	SWS-9	Koti	12 (40.9)	Lower	**2**
3	SWS-8	Bansar	12 (54.5)	Upper	3
4	SWS-7	Batol	11 (50.0)	Upper	4
	Sirsa				
1	SWS-6	Dhang	10 (45.5)	RHB lower	**1**
2	SWS-1	Adual	9 (40.9)	RHB upper	**2**
3	SWS-2	Retwari	9 (40.9)	RHB upper	3
4	SWS-23	Majheli	9 (40.9)	LHB lower	4
	Swan				
1	SWS-15	Ghanari	15 (68.2)	RHB upper	1
2	SWS-11	Jadla-1	14 (63.6)	RHB middle	2
3	SWS-27	Puloh	14 (63.6)	LHB upper	**3**
4	SWS-8	Bhadsali	13 (59.1)	RHB lower	**4**
Total	**8**	**20**			
	Chakki				
1	SWS-19	Bhadorya	12 (54.5)	LHB lower	1
2	SWS-2	Thora	12 (54.5)	RHB middle	**2**
3	SWS-8	Lahru	11 (50.0)	LHB upper	3
4	SWS 17	Bari	10 (45.5)	LHB lower	**4**
5	SWS-9	Chakki	9 (40.9)	RHB upper	5
Total	**6**	**28**			
G. Total	**32**	**109**			

as a consequence of undertaking a number of activities concurrently. This is difficult to capture in one or even a number of indicators. For any WSD project there may be a wide variety of such live or illustrative cases, such as: transformation of a farming system; ponds and trees for improved farm economy; institutional facilities ushering in improved village resource management; networking effect of a combination of soil erosion control and water storage structures; protection of natural village forest for reviving perennial flows in a seasonal channel; inter-village group for soil and water conservation works and for achieving all-round prosperity.

In short, these live stories help focus on much sought-after conjunctive effect of biophysical interventions with those of socioeconomics and institutions. These should form an integral part of an evaluation study.

TOOLS AND TECHNIQUES FOR DATA ACQUISITION

Primary Data

For household or respondent survey, the technique used is questionnaire/ schedule and interactive methodology is followed.

For developing a village profile at any chosen time, the participatory rural appraisal technique is used.

For a set of structures, infrastructural facilities created institutions, carefully designed questions are posed to initiate free discussion and to arrive at a consensus.

For validating rationale of an intervention and gathering onsite output-outcome, detailed Participatory Transect Survey technique is followed where a combination of questionnaire/schedule and free issue-based discussion is used as a tool.

Direct measurement is used to assess: groundwater recharge and water availability from water table depth and well dimensions; surface water availability from discharge and flow duration (giving degree of seasonality vs perenniality); water storage from dimensions of ponds, tanks and water depth; erosion and degradation damages from soil loss, sediment deposit, bank cutting, gradient changes.

Data on soil characteristics (physical, chemical, hydrologic and biological) are obtained by analysing soil samples from the representative sites.

Vegetation survey is undertaken to determine composition of species number and relative density of tree, shrub and other plant species, whereas growth performance is found from height and girth measurements. Nested quadrates of 10×10 m for trees, 5×5 m for shrubs, and 1×1 m for bushes and grasses are used.

Application of Remote Sensing and GIS

To assess changes that have taken place due to project interventions, remotely sensed data and images can be used. Observations on the following indicators were obtained through analysis of such data:

- Additional area covered and storage volume created by conservation structures constructed.
- Quantity of water available in the existing structures in different months of the year.

- Increase in area under wet season and winter season crops.
- Increase in area under forests or perennial vegetation.

The scenario at different times such as pre-project, during the project and post- project period was produced from the remotely sensed data and images. This was supported by an adequate schedule for taking observations at fixed locations for ground-truthing. The images for the target area can be clipped after geo-referencing with the toposheets of the Survey of India. GIS-based maps then could be prepared for different time lines or time frames for comparison.

Flow pattern of the Machan River in Panchmahal District of Gujarat State and increased availability of water were assessed by this tool after a series of 30 water harvesting structures was constructed in the catchments of the river. Two sets of IRS-B images pre- and post-construction period and for the summer and post-summer durations were procured from NRSA and used. The Machan River and its tributaries were found to flow with greater volume of water in the post-construction period. The farmers could lift water from different sections with diesel-driven pump sets and could irrigate their farms in 20 villages. Even when in four consecutive years of drought affected large areas in the adjoining villages, these villages remained unaffected. Effects of increased availability of water on crop yield, vegetation and water levels were not studied (Sadguru Water Development Foundation, 1999).

Secondary Data

Secondary data and information are obtained through reviewing and culling from published and unpublished literature, reports, documents, and maps. Some of these are also worked out by using remote sensing data sources such as digitized CD or thematic maps and GIS.

Components of M&E Study

A study on monitoring and evaluation, especially when entrusted to an external agency, generally comprises the following sub-studies

Baseline Survey

The benchmark survey report provides a pre-project scenario of the area and the community living in it. In most projects in India, including those supported by bilateral and multilateral external agencies, it is released quite some time after the project has been launched and thus becomes a current status report rather than a baseline survey. The limitations of such a survey report are partly overcome by taking adjoining representative areas as control.

Mid-term Evaluation

This is generally carried out after the project has run through about half of its duration and when a reasonable period has elapsed after the completion of the baseline survey. This sub-study attempts to obtain another scenario and when it is compared with that of the baseline survey it provides a trend in the outputs. This reveals the gaps or deviation from the approved plan and offers an opportunity to carry out mid-course corrections.

Final Evaluation

The final evaluation aims at determining the effectiveness of the project implementation. It assesses: the extent and type of outputs vis-à-vis inputs; outcome/effect vs the set of direct or immediate objectives; durable effect vs impact and the sustainability or the continuing effect and consequent impact over the years to come, especially after the project withdraws. It uses the data on the same indicators as have been used in the baseline survey and mid-term evaluation. But it collects more detailed primary data on outputs and outcome to carry out a process-based analysis. It also helps to do some comparative analysis with the results, which could be obtained during the baseline survey as well as during the mid-term evaluation study.

Data Analysis and Data Interpretation

A WSD project has a large basket of results varying greatly in form, magnitude and quality. The data are, therefore, segregated in sets. This helps in reprocessing, collating and analysing within a particular set as well as in association with other sets that lead to appropriate process analysis. Some examples are presented below from some of the generally accepted sets.

Climate

Climate controls development and use of natural resources. Among the given climatic parameters, rainfall and its distribution within a year and over the years profoundly affect the performance of a plant or crop. Over a 10-year period, drought visited the six representative locations of the Siwaliks of Punjab state between 62.5% and 100% of the years. In the critical crop month of July, rainfall at Pathankot varied from a high of 620 mm to a low of 59.3 mm, while for August rainfall ranged from 406.2 mm to 38 mm. The variations or the departures from the normal need to be studied. Similarly, for the winter crop months of January and February, rainfall ranged from 125.6 mm to 4.6 mm and 101.0 mm to 0.0 mm respectively. Trends for seasonal rainfall for wet as well as winter crop seasons exhibited similar wide variations. For micro-level works, which are the common features in WSD, a record of the immediately preceding 10 to 25 years is more relevant than a long record extending to the distant past (Das et al., 1966). This is important with high variability of rainfall over short distances where data from the specific on-site or near-site locations are more useful. Data from such sites are generally of short length. A test for acceptability of data length of short records should be desirable.

Land Use

Distribution of the geographical area under different uses is a significant pointer to watershed degradation. The area under perennial vegetation vis-à-vis areas under crop and other uses reveals land use alienation, as in the details given in Table 3 from a study in the Siwaliks of Punjab.

Soils

Soil characteristics provide information on the land's productivity, its proneness to erosion, moisture status, and other factors. Variations in values of soil organic carbon at different sites and under different uses are given in Table 4. These values showed that soils of project area cannot have reasonable productivity without application of organic manures.

Soil and Water Conservation

Soil and water conservation treatments given to areas of three watersheds of eastern India, which are capable of producing biomass, included land

Table 3 Distribution of MWS under different adequacy classes of green area (percentage of total number of MWS).

District	*< 20% green area – poor*	*20 to 40% green area - fair*	*40 to 60% green area - good*	*> 60% green area – very good*
Gurdaspur	30	20	30	20
Hoshiarpur	50	12.5	12.5	25
Nawan Shahr	-	-	50	50
Ropar	-	-	33	67
Patiala	100	-	-	-

Table 4 Soil organic carbon in MWS of three districts (range in %).

District	*Agriculture*	*Forest/common land*
Patiala	0.480–0.090	0.330–0.030
Ropar	0.750–0.180	0.570–0.120
Nawan Shahr	0.150–0.270	0.120–0.030

treatments, cultivation method, afforestation, and other factors. The package as a whole was designed to make the land safe. The relative picture of the three watersheds reveals the variation in utility of the packages (Table 5).

Intensity of Treatment Package

The high-intensity treatment package for Karkara watershed, which contained three large check dams, created a total of 84.4 ha-m surface storage that could hold 4.82% of incident rainfall. In Karma watershed, small earth-fill dams, gully plugs, and other interventions as a part of medium-intensity package, which were put up across the flow direction and in series and in a networking formation, created 71.5 ha-m of surface storage. The storage created in Karkara watershed has irrigation potential of 450 ha in wet season and 175 ha in winter. The storage created in Karma watershed developed irrigation potential of 360 ha in wet season and 120 ha in winter.

Improved Groundwater Status

The increase in recharge of groundwater and consequent increase in irrigated area due to wells were substantial (Table 6). Increased surface storages and consequent irrigation accompanied reduced well-irrigated area.

Table 5 Conservation protection provided in three watersheds vs three intensity levels of treatments.

Watershed	*Pre-project, % of safe area*	*Post-project, % of safe area*	*Package intensity*
Mahesha	0	0	None
Karma	0	60	Medium
Karkara	0	72	High

Table 6 Improved groundwater availability and increase in irrigated area.

Watershed	*Respondent no.*	*No. of wells*	*Pre-project (ha)*		*Post-project (ha)*	
			Irrigated area	*Total crop land*	*Irrigated area*	*Total crop land*
Karkara	26	21	31.14	62.43	14.02	62.43
Karma	27	29	22.52	62.19	46.67	62.19

Improved Channel Flow

Due to project activities carried out in the first two years, there were perceptible changes in flow duration as well as volume in the streams of some of the MWS in the Siwaliks of Himachal Pradesh (Table 7).

Employment Opportunity

There was substantial gain in employment due to project implementation in both Karkara and Karma watersheds, but the increases varied with the intensity of the treatment package. In Karkara watershed with high-intensity treatments the increase was 258% as against 75% in Karma watershed with medium-intensity treatments. The large increase in Karkara watershed was due to intensification in cultivation and other diversified avenues. In the cultivation sector the variations in employment growth from pre-project to post-project were as follows:

Period	Karkara watershed	Karma watershed
Pre-project	16,410 person-days	11,130 person-days
Post-project	30,990 person-days	21,360 person-days
Percentage increase	88.85	91.93

Due to lower treatment intensity there was little diversification in employment opportunities in Karma watershed and thus the villagers remained engaged in cultivation. Total increase was far below the desired level.

Institutions

The higher-level project inputs in Karkara watershed included a variety of community-based organizations. These are 22 women's self-help groups (SHGs), one watershed development committee, four village agricultural development committees, one inter-village horticultural development

Table 7 Improved flow duration and flow volume due to watershed development activities (obtained by participatory village transect survey).

MWS	*Drainage channel*	*Flow duration, months*		*Increase in flow volume (%)*
		1999	*2001*	
Amrayion	Amraiyon Khad	12	12	10
	Jamli Khad	12	12	20
Bankala	Rukhri Khad	9	10	10
Tikri Bhojpur	Tikri Khad	12	12	10
Koti Kalyanpur	Mohan Nala	12	12	20

committee and another inter-village improved seed and artificial insemination committee, besides user groups for irrigation and fishery. These organizations created their own funds through subscriptions, project contribution for labour provided, sale proceeds, services costs, etc. Many successfully obtained credit for their members from banks through their accounts at a lower rate of interest. They also recovered the loan amounts from their members and repaid the dues to the banks in time.

Economic Status

The post-project family income in three watersheds is given in Table 8.

Table 8 Variation of family income due to intensity of treatments (income in Rs per family per year).

Watershed	*Lowest*	*Highest*	*Average*
Karkara	2,000	132,000	32,000
Karma	2,000	100,000	44,000
Mahesh	5,500	62,000	32,100

In spite of high level of project interventions in Karkara watershed, the average family income is the lowest and almost equal to that of the untreated watershed of Mahesha. This showed that the project interventions and benefits were not equitably distributed. In fact only 4 villages out of 13 received irrigation water created by three relatively large check dams and the lift-irrigation systems. There had been no comparable package for other villages where irrigation potential either did not exist or could not be created.

The distribution of families under different annual income classes revealed that about 50% of the families were still below the poverty line in Karkara watershed, which had received high-intensity treatments. It showed that the level of watershed development was still inadequate and the desired objective of distributed equity not achieved.

Environmental Improvement

Natural regeneration of forest areas close to habitations were studied from the plant composition as well as species diversity index values that have been used in the watersheds of Karkara and Karma (LUCI, 1999). Owing to the proximity of the settlements and high demands for fuelwood the forests were getting depleted and the quality was deteriorating. The village agricultural committees had taken steps to protect the forests from grazing and over-exploitation. A vegetation survey was made to determine the extent of natural regeneration induced through these efforts. The identified sites were Chorah and Sargaon in Karkara watershed and Dhori Chhagaria in Karma watershed. Nested quadrates and line transects were used. Sargaon site had the maximum of 20 species, Dhori Chhagaria 13 and Chorah 10. Among the herbaceous species only five could be seen and the most common one was *Cynodon dactylon*. *Shorea robusta*, the dominant species of the region, was the highest (65%) at Chorah and the lowest (10.5%) at Sargaon. At Dhori Chhagaria it was 30%. *Madhuca indica*, the most revered tree of the region, could be found only at Sargaon and constituted only 5.4% of the total population. Other important species found were *Carissa opeca*, *Diospyros melanoxylon*, *Terminalia tomentosa* and *Butea monosperma*. To determine the differences in quality of composition of forests, Species Diversity Index (SDI) and Similarity Index (SI) values were computed. These were as under:

Site	SDI (%)	SI
Sargaon	29.20	0.47 with Chorah
		0.30 with Dhori Chhagaria
Chorah	3.44	0.52 with Dhori Chhagaria
Dhori Chhagaria	2.35	

The site at Sargaon had a higher SDI because it was further from the villages. At the other two sites SDI values were low because of greater anthropogenic interference due to proximity of the villages. Protection alone cannot raise the SDI significantly within a short period. It would be desirable to accelerate the process of natural regeneration with the dominant species of the region such as *Shorea robusta*, *Syzigium cummini*, *Carissa opeca* (shrub), *Anogeissus latifolia*, *Acacia catechu*, and *Emblica officinalis*, besides *Madhuca indica*, for which a special drive should be mounted.

Increase in green area after project implementation was detected in a number of MWS of the Siwaliks of Himachal Pradesh. These increases were due to land use shifts to trees, horticultural plants or pastures. From the ecological point of view the enlarged green area is a favourable change. Actual increases in 10 MWS can be seen from Table 9.

SUSTAINABILITY

The term *sustainability* is widely used in project planning and also to assess the continuing or recurrent effectiveness of an implemented project. Programmes differ in development or broad as well as in immediate objectives, and each programme has a logical framework linking inputs,

Table 9 Total increase in area under green/perennial cover during mid-term evaluation as compared to that during baseline survey.

Micro-watershed	*Total increase (ha)*	*% increase*
Amriyon-Bakarla	10.70	1.71
Bankala	36.96	8.32
Tikri Bhojpur	22.89	4.71
Koti Kalyanpur	23.41	4.15
Dhang Phalsi	20.00	2.32
Adual	41.00	18.80
Jadla-1	48.80	6.49
Bhadsal	43.34	6.09
Thora Baloon	45.77	5.76
Bari Khad	44.15	9.80
Average	33.70	6.82

outputs and outcomes and thus impact. Therefore, generalized concept and definition of sustainability must yield to specific ones. For WSD programmes, the term needs to be defined in the context of the objectives, target groups and target area, as sustainability has been found to vary with all these three. Endowment of a watershed can be developed and managed variously to serve the needs and aspirations of different target groups. The farmers, artisans and landless would like to ensure availability of food, fodder, fuelwood and water, while urban settlements desire to get water on a continuing basis and industry may look for regular supply of water and a particular raw material. Thus, perceived outputs and consequent outcomes differ and so does the concept of sustainability and the basis on which to assess sustainability (Chakravarty, 1992; LUCI, 1999). This could be seen from two definitions in use as given hereunder.

According to the FAO (1991), "a system, which involves management and conservation of natural resources base and orientation of technological and institutional changes in such a manner as to ensure attainment and continued satisfaction of human needs for the present and future generations, is a sustainable one. Such sustainable development (in agriculture, water, forestry and fisheries sectors) conserves land, water, plants and animal genetic resource in an economically viable and socially acceptable way."

The World Food Programme (WFP) and M.S. Swaminathan Research Foundation (MSSRF, 2001) have observed that sustainable food security will have to be defined as "physical, economic, social and ecological access to balanced diets and safe drinking water, so as to enable every individual to lead a productive and healthy life in perpetuity".

In India, over the years, WSD and management professionals have arrived at the perception (AFC, 2002) that a WSD project, to become sustainable, must strive to achieve a balance among the following processes:

- Production of phytomass vs its consumption and use;
- Regeneration or replenishment of nutrients, water, etc. vs their depletion caused by harvests/use and degradation;
- Creation of livelihoods vs growing work force, especially rural work force; and
- Building capability (technical, managerial and financial) of community-based societies and thus their members vs escalation of rigours in managing the balance as stated above.

The sustainability of a watershed development or management project can thus be defined as restoring and maintaining a dynamic balance between yield of goods and services and continued availability of land area, soil fertility, water and air in quality and quantity as well as inputs from plant and animal communities through a combination of technological and institutional innovations. The task of developing a methodology for assessing the degree of sustainability and providing a rational basis to compare the sustainability of a group of watersheds posed a serious challenge. RODECO Consulting of Germany provided an opportunity to this end, which was used to develop and test such a methodology (LUCI, 1999). The details were published in the proceedings of the International Conference on Land Resource Management held in New Delhi in November 2000 (Das et al., 2000). The methodology uses the following main steps:

- Identification and selection of important parametric or broad areas of concern.
- Selection of individual and composite indicators for each parametric area.
- Data acquisition, after deciding on means to collect data (primary and secondary) as well as identifying those that can be gathered through scientific tools and those that can be transformed from information obtained through participatory mode.
- Developing scales of matrix grades and matrix scores, using a unit-neutral yardstick to compare level of sustainability for different components with diverse units. The scale used is shown here under:

Matrix grade	Matrix score out of 10	Level of sustainability
P - Poor	below 2	Not sustainable
F - Fair	between 2 and 5	Sustainable for a very short term
G - Good	between 5 and 8	Sustainable for a medium term
VG - Very good	between 8 and 10	Sustainable on a long-term basis

To impart a numerical value to grades, a range in percentage of total grades is used, based on Indicator Sustainability Ratings (ISR):

Very good: When total of VG ratings of ISRs alone is 75% and more *or* when total of VG plus G ratings of ISRs is 75% or more

Good: When total of VG and G ratings of ISRs is between 50 and 75%

Fair: When total of VG and G ratings of ISRs is between 25 and 50%

Poor: When total of all VG and G ratings of ISRs is below 25%

- Computing individual Indicator Sustainability Ratings (ISR): The values or matrix score of individual indicators ranged from a high to zero for both Karkara and Karma watershed. This showed that all the interventions were not equally effective.
- Computing Parametric or Component Sustainability Ratings (PSR/CSR): The ratings of eight parametric or component areas were also different. They varied within the watershed as well as between the watersheds.
- Computing Overall Project Sustainability Rating (OPSR): The values of the OPSR with ISR grade ratings are presented in Table 10 and those obtained with ISR score ratings are given in Table 11.

Table 10 OPSR with ISR grade ratings as per conditions or settings defined earlier.

Watershed	*OPSR*	*Reasons*
Karkara	Very good	VG and G ratings together were 83.5%
Karma	Good	VG and G ratings together were 53 %

Table 11 OPSR with ISR score ratings.

Watershed	*Total ISR values*	*Average ISR*	*OPSR*
Karkara	252	7.41	Good
Karma	148	4.35	Fair

The lower ratings with score values are due to the greater ability of the score scale to measure the differences within two grade matrices.

The methodology provided insight into the relevance of intensity of the intervention package to the attainment of sustainability. The higher sustainability attained for Karkara watershed as compared to that for Karma watershed was due to a treatment package of higher intensity. The methodology also revealed the persistent shortfall. The matter of equitable implementation of a package to a greater area may have to be addressed to achieve a higher level of sustainability. Site- and area-specific adaptation will be necessary.

REPLICABILITY

The assessment of various outcomes has largely been based on beneficiary response and participatory appraisals besides those determined by scientific measurements. The sustainability level achieved should give planners confidence to replicate the project module in similar settings of biophysical and socioeconomic attributes. It would thus be desirable to develop a rationale to assess replicability, with a set of important biophysical parameters as well as a socioeconomic scenario. Such a rationale has yet to be developed.

References

Agricultural Finance Corporation Ltd. (AFC). 2002. National Land Resource Management Policy - Report of the Expert Group, New Delhi.

Agnihotri, Y. 1999. Socio-economic indicators and methods of monitoring and evaluation for afforestation and eco-development projects: M&E of afforestation and eco-development projects: Current efforts and future options. pp. 38-43. *In:* D.C. Das and D.V. Nithyanand. [eds.] RC-National Afforestation and Eco-development Board, Agricultural Finance Corporation Ltd., New Delhi.

Chakravarty, S. 1992. Sustainability the Concept and Its Economic Application in the Context of India. pp. 28-31. *In:* Anil Agarwal [ed.] The Price of Forests. CES, New Delhi.

Chopra, K. and G. Kodekodi. 1992. Some issues in evaluation and replication of participatory institution. pp. 262-271. *In:* Anil Agarwal [ed.] The Price of Forests. CES, New Delhi.

Chopra, K. and D.V. Subba Rao. 1996. Evaluation of watershed management in the watersheds of Sahibi Catchment of Rajasthan. Institute of Economic Growth, Delhi.

Consulting Engineering Services (CES) India. 2002a. Final Base Line Survey Report on IWDP (Hill-II), Himachal Pradesh, New Delhi.

Consulting Engineering Services (CES) India. 2002b. Final Base Line Survey and Current Status Report on IWDP (Hill-II), Punjab, New Delhi.

Das, D.C. 1998. Watershed management in India - Experience in implementation and challenges ahead. pp. 743-774 *In:* L.S. Bhushan et al. [eds.] Soil and Water Conservation: Challenges and Opportunities. 8th ISCO Conference, 1994. Oxford & IHB Publishers, New Delhi.

Das, D.C. 1999. M&E of afforestation and eco-development projects - ecological, social, economic and managerial indicators. pp. 16-37. *In:* Monitoring & Evaluation of Afforestation and Eco-development Projects: Current Efforts and Future Options. Agricultural Finance Corporation Ltd., New Delhi.

Das, D.C. and S. Singh. 1981. Small storage works for erosion control and catchment's improvement: mini case studies. pp. 425-450. *In:* R.P.C. Morgan. Soil Conservation - Problem and Prospects. John Wiley & Sons, Chichester.

Das, D.C., G. Honore and K.G. Tejwani. 2000. Determining Sustainability of Watershed Management Projects: Advances in Land Resources Management for 21st Century, Soil Conservation Society of India (SCSI), New Delhi, pp. 503-524.

Das, D.C., B. Raghunath and G. Poornachandran. 1966. Rainfall and climatic associates in relation to soil and water conservation at Ootacamund (Nilgiris), Part I: Acceptability of data, amount and its distribution. J. Indian Soc. Agr. Eng. (ISAE), III(2): 1-14.

Dhyani, B.L. et al. 1997. Socio-economic analysis of a participatory watershed management in Garhwal Himalayas - Fakot Watershed. Central Soil and Water Conservation Research and Training Institute (CSWRTI), Dehra Dun, Bulletin No. WT 35 /D-21.

Gupta, S.K. and M.L. Khybri. 1986. Soil Conservation in India. Jugal Kishore & Co.

Lal, J.B. 1992. Economic value of Indian forest stock. pp. 43-48. *In:* Anil Agarwal [ed.] The Price of Forests. CES, New Delhi.

Land Use Consultants International (LUCI). 1999. Impact and Sustainability of Watershed Management Programme in Bihar. Prepared for RODECO, Germany. Tech paper No. IGBP/WSM/120/99.

Maudgal, S. and M. Kakkar. 1992. Evaluation of forests for impact assessment of development projects. pp. 53-64. *In:* Anil Agarwal [ed.] The Price of Forests. CES, New Delhi.

M.S. Swaminathan Research Foundation (MSSRF) and World Food Programme (WFP). 2001. Food Insecurity Atlas of Rural India. MSSRF, Chennai.

Ram, B. and B.L. Dhyani. 1998. Socioeconomic aspects WSM programme in India. pp. 1649-1662. *In:* L.S. Bhushan et al. [eds.] Soil and Water Conservation: Challenges and Opportunities, Vol II. 8th ISCO Conference, 1994. Oxford & IHB Publishers, New Delhi.

Sadguru Water and Development Foundation (SWDF). 1999. Annual Report M.M. SWDF, Dahod, Gujarat.

World Bank. 1999. Regional environmental assessment for the proposed integrated watershed development projects (Hill-II). Prepared by Lea Associates South Asia Pvt. Ltd., New Delhi.

3

Monitoring and Evaluation of Soil Conservation and Watershed Development Projects: An Approach to Sustainable Development

Janya Sang-Arun[1], Machito Mihara[2], Eiji Yamaji[1], and Samran Sombatpanit[3]

[1]The University of Tokyo, Tokyo, Japan. E-mail: janyasangarun@yahoo.com, janyasan@gmail.com and yamaji@k.u-tokyo.ac.jp

[2]Tokyo University of Agriculture, Tokyo, Japan. E-mail: m-mihara@nodai.ac.jp

[3]World Association of Soil and Water Conservation, 67/141 Amonphant 9, Sena 1, Bangkok 10230, Thailand. E-mail: sombatpanit@yahoo.com

ABSTRACT

The main purpose of soil conservation and watershed development is to better the quality of life of citizens by increasing land productivity and income in harmony with the ecosystem. However, the results of such development often conflict with the intentions. Hence, there is a need to monitor and evaluate the situation and impact of soil conservation and watershed development projects from various points of view to reach the objective of sustainable development.

This chapter proposes a guideline and criteria for selecting practical monitoring and evaluation (M&E) methods for soil conservation and watershed development projects on the basis of project objectives, natural resources and ecology, socioeconomics, public participation and public hearing, local perception and acceptance, and sustainability of adoption of technology.

Because of budget constraints on many projects, M&E is not thorough and many projects report only positive results. Furthermore, most projects evaluate success on the basis of quantitative outcomes, with less awareness of qualitative factors. To minimize costs, we propose ways to select parameters for M&E and we suggest choosing high impact parameters and paying attention to negative impacts. The parameters, methodology and schedule should be decided before the project is begun. Additionally, participatory M&E is recommended for enhancing people's perception of the project needs, which may strongly affect the adoption or non-adoption of project innovation. Also, the locals should monitor and evaluate the project's achievements and its effect on their lives and natural resources.

INTRODUCTION

Recently, soil conservation and watershed development projects have been significant issues for conserving land productivity and improving the quality of life. However, each development project has negative and positive impacts on human beings and the environment. There is a need to monitor and evaluate the situation and impact of soil conservation and watershed development projects from various points of view to reach the objective of sustainable development.

Not every project may be able to set up a monitoring and evaluation (M&E) system, because of budget limitations and high technical requirement, especially small projects in developing countries. Hence, we propose a guideline and criteria for selecting practical M&E parameters of soil conservation and watershed development projects on the basis of project objectives, natural resources and ecology, socioeconomics, public participation and public hearing, local perception and acceptance, and sustainability of adoption of technology.

SOIL CONSERVATION AND WATERSHED DEVELOPMENT

A watershed is an area from which all water discharges into a bounded river system, making it an attractive unit for technical efforts to manage water and soil resources for production and conservation (Kerr, 2002; Kerr et al., 2000). The watershed development criteria may differ according to existing problems, social requirements, and implementer's interests. However, the general activities are irrigation and water storage, soil conservation and erosion control, enhancing agricultural production, creating off-farm employment, conserving or rehabilitating natural resources and the ecosystem, and the construction of roads and other public facilities.

Soil conservation is a part of watershed development. Basically it aims to maintain soil fertility (Young, 1989) and to sustain maximum crop productivity from a given area of land by minimizing the effects of agricultural practice on soil erosion, water quality, and sedimentation and by minimizing crop damage by wind or water erosion (Morgan, 1995). Inevitably, soil degradation is a biophysical process exacerbated by socioeconomic and political factors (Lal, 2001; Hellin and Haigh, 2002). Top-down planning with little participation (Erenstein, 2003) or little emphasis on intensive soil conservation and erosion control (Lai, 1992) leads to low adoption and insufficient erosion control. Therefore, advanced soil conservation projects in the context of better land husbandry, sustainable rural livelihoods and people's participation are essential for increasing adoption and adaptation of soil conservation technology in farmlands (Pretty and Shah, 1997; McDonald and Brown, 2000; Johnson et al., 2001; Hellin and Haigh, 2002; Dorward et al., 2003; Erenstein, 2003).

The main purpose of soil conservation and watershed development is to better the quality of life of citizens by increasing land productivity and income in harmony with the ecosystem. However, the results often conflict. For example, the Qingshishan watershed project in China (Jianbo et al., 2002) achieved the initial objectives of increasing net income, water use, employment, and land use, but the rates of losses of nutrients, pesticide and soil have increased after the establishment of the project. In addition, the development in the upper streams disrupted water use in the downstream areas. For example, seven dams constructed for power generation and irrigation along the Mekong River (Nakayama, 2002) decreased the diversity of fish, fishery and water use of the lower part and between the dams. Also, intensive cash-cropping in the watershed areas in northern Thailand has reduced downstream water supply in the dry season and increased the level of sedimentation in the lowland (Scoccimarro et al., 1999). Furthermore, many watershed development projects around the world have failed to take into account the needs, constraints and practices of the local people (Johnson et al., 2001). Therefore, monitoring and evaluation is necessary to understand soil conservation and watershed development projects in the fullest way, to achieve sustainable development or at least to minimize undesirable results.

MONITORING AND EVALUATION

There are many specific definitions of monitoring and evaluation. Monitoring is observation and assessment of the activities, results, and impacts of a project in the context of implementation schedules, primarily to inform or warn all stakeholders of project performance. Evaluation refers to periodic assessment (for example, before-project, mid-term and end-of-project) of the relevance, performance, efficiency and impact of the project in the context of its stated objectives using information from the monitoring and presenting the overall performance of the project (Morgan and Ng, 1990; Vine et al., 2000; Crawford and Bryce, 2003; Gyorkos, 2003). However, many scientists prefer to use M&E together, as they are intimately linked (Crawford and Bryce, 2003).

The purposes of M&E are to observe and provide information for donors, implementers, citizens and other interest groups on the decision making for the better resolution of undesirable results. Besides this, the donors also evaluate the performance of the implementers and the project achievement. Some project implementers, therefore, are hesitant to be evaluated and try to present only positive results, even when they know some unfavourable outcomes have occurred during project implementation.

WHO SHOULD MONITOR AND EVALUATE

Everyone related to the project, including donors, implementers, NGOs, governors, stakeholder partnerships, and the citizens in the project area and nearby, should do M&E either formally or informally. Effective M&E can minimize the negative impacts of project activities on the environment, on sustainable use of natural resources, and on the well-being of citizens. Formal M&E can be carried out through investigation, public hearing, questionnaire survey and other means. The parameters and methodology are decided by the major stakeholders and summarized in the formal report. It is usually conducted by donors, implementers, government offices, NGOs and stakeholder partnership. Informal M&E involves observation in the field and/or follow-up of news and announcements. For example, farmers may observe the effect of canal dredging on fish and other aquatic diversity by counting the numbers of fish and aquatic foods that they catch after and before the project. They can then inform the implementers what is going on and lobby for modifications in the project. If the locals conduct informal M&E, the implementers will concern themselves more about the side effects of their projects.

DIMENSIONS OF M&E

To meet the goal of sustainable development, M&E should address (1) achievement of project objectives, (2) project impact against natural resources and ecology, (3) project impact on socioeconomic conditions, (4) participatory and public hearing, (5) perception and acceptance of the local people, and (6) sustainability of adoption of technology. Impact pertains to both negative and positive outcomes. Although these six dimensions have been widely applied for M&E, they are often applied individually depending on specialization of each group (e.g. Lai, 1992; Scoccimarro et al., 1999; He et al., 2000; Reynolds et al., 2000; Vine et al., 2001).

Achievement of Project Objectives

The objectives and activities of each project are generally monitored and evaluated by quantitative methods. Cost-benefit analysis (Kerr and Chung, 2001) and cost-effectiveness analysis (Murray et al., 2000) are widely used by donors and implementers. Typically, qualitative analysis is ignored except in seminars and workshops.

Project Impact on Natural Resources and Ecology

Though every watershed development project purposes to improve the present situation, it may worsen the ecological system and biodiversity of the watershed. Sometimes it also affects public use of resources and human health in the long run. Many countries require the implementers to conduct an environmental impact assessment (EIA) before starting large projects that may affect the environment, public use of resources, and quality of life of the locals. Unfortunately, many small- and medium-sized projects carry out neither EIA nor M&E.

Project Impact on Socioeconomic Conditions

Income is so far the most important issue for the locals. Inevitably, every development project affects their socioeconomic conditions, causing them to lose their income or giving them no direct benefit. Additionally, changes in natural resources and ecological conditions affect the resource use of citizens in and around the project area. A river, for example, is a source of food, culture, transportation, trading and habitat for the people living around it, and a watershed project will affect all these aspects of their lives. Social scientists and economists pay more attention to beneficiaries and non-beneficiaries, social change, income, and equity of advantage, factors that are less often considered by typical natural scientists, engineers and project implementers.

People's Participation and Public Hearing

There are many levels of people's participation in a project (Table 1). People's participation in setting criteria, identifying priority constraints,

Table 1 Levels of people participation in the projects (Adapted from Pretty and Shah, 1997).

Level	*Level of participation*	*Activities*
0	No participation	People know nothing, are ignored, or are refused information
1	Passive participation	People receive information but make no response
2	Participation in information dissemination	People give information through interviews only
3	Participation by consultation	People listen or follow advice but make no comment
4	Participation for material incentives	People support project activities for its incentives without understanding
5	Functional participation	People form or join groups to support the project
6	Interactive participation	People give opinions, share experiences, ask questions
7	Self-mobilization	People play roles in decision-making, give ideas and logic, try to improve and solve their problems, try to learn and look for the right solution

evaluating possible solutions, and M&E are likely to achieve what coercion and subsidies cannot toward more successful or sustainable development (Johnson et al., 2001). Participatory evaluation can facilitate mutual learning, build local capacity of decision making and community-center development, and enable participants to evaluate their own needs, analyze their own priorities and objectives, and undertake action-oriented planning to solve their own problems (Bradley et al., 2002). In addition, participatory M&E may enhance acceptability, adoptability, and sustainability of project innovation. This is because the real persons who take care of the project area are the local people, while implementers support it only for a short duration. The implementers can promote participatory M&E through public hearing, group discussion, questionnaire surveys, and combined measures.

Public hearing is normally conducted in the planning stage or before the projects are begun (Leach et al., 2002). Its purposes are to announce the project contents and to gather comments from local citizens, because the project activities may cause undesirable effects on the locals. Generally, public hearing is a part of EIA before initiation of large-scale development projects in many countries and is becoming a part of people's participation in rural development projects. However, we consider that the perspective of the locals is the sign of success in sustainable development, and therefore public hearing should be conducted several times during the life of project; for example, before starting the project, during the project, at the end of the project, and, if possible, some time after the end of the project.

Perception and Acceptance of the Local People

Perception and acceptance of the local people affect the success of projects (Bielders et al., 2001; Jones, 2002; Batz et al., 2003; Dorward et al., 2003). A top-down style of management ignores the real stakeholder of soil conservation and/or watershed development activities. Some implementers decide what the problems are and what should be developed, and farmers are considered irrational and ignorant. In such cases, people's participation means people helping to implement plans that are considered good for them (Shaxson, 1992).

Such implementers succeed in project achievement in quantity but not in quality. The adopted technologies are often ignored after the project's end. In addition, lack of technical knowledge of certain practices (e.g. soil conservation measures) promoted in the project site leads to misuse of the technologies themselves. For example, farmers in northern Thailand constructed various types of bench terraces different from the recommended one and kept them bare. The terraces became prone to erosion by rainfall. Rill and gully erosion became dominant (Fig. 1) and it is doubtful whether these bench terraces are beneficial for soil conservation.

Sustainability of Adoption of Technology

Sustainability of adoption of technology is very important for evaluation of the project's success. The locals should be able to manage and apply the promoted technology by themselves after the end of the project. A monitoring system should include beneficiary-level parameters on the relevance and acceptability of specific project outputs and services to farmers and the ways in which these are integrated into the agricultural and land use system (Lai, 1992).

This means that the number of farmers who adopted the promoted techniques is not the only key measure of the success of the project, but also the perception, awareness and perspective of the locals on their development needs, including the possibility that they may apply or manage the promoted technology by themselves after the end of the project.

The appropriate technology in each location is also important. The implementers may select a number of effective technologies for each project, but limitations such as their cost, labour requirement, and

Fig. 1 Rills on bench terraces in Pang Prarachatan village, northern Thailand.

techniques for constructing and managing them are major obstacles to their sustained use.

HOW TO SELECT PARAMETERS FOR M&E

Monitoring and evaluation parameters should be decided at the initial stage of project planning. Figure 2 shows the process of parameter selection, M&E, and remedial action. The process for selection of parameters is explained as follows:

- First, investigate the details of the projects, activities, scale, period and budget.
- Second, list the impacts that may occur if the project is carried out. The list should cover all the dimensions of M&E mentioned earlier.
- Third, forecast the direction and weight of the impacts if the project is carried out. Weighting can be done by classifying impacts as high, medium and low or by ranking them with numbered scores. Table 2 demonstrates a list of parameters according to the six dimensions of M&E for soil conservation and watershed development projects. Table 3 shows an example of a checklist,

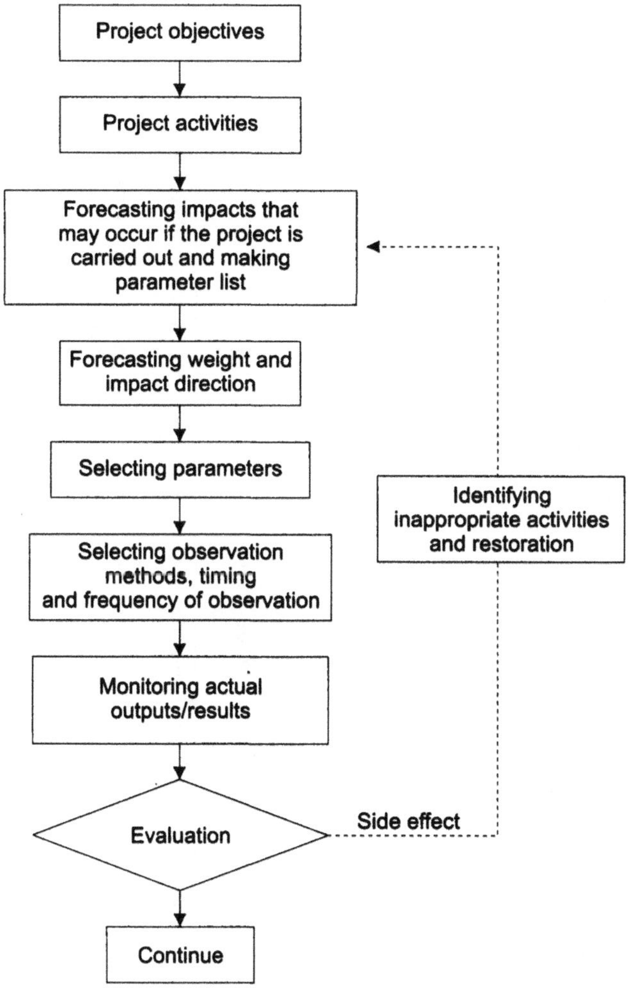

Fig. 2 Process of parameter selection, monitoring, evaluation and remedial action.

weighting and direction of impacts that may occur within and outside the project area for canal dredging in Thailand. Note that forecasting, weighting and direction of impacts may differ with the project period, scale, location, land use system, present situation or problem in the project site, background of evaluators and other factors (Kammerbauer et al., 2001).

Fourth, select monitoring parameters that are weighted as severe impacts. The budget may not permit all of these parameters to be monitored; however, the negative impacts should be monitored and evaluated, while positive impacts can be ignored.

Table 2 Sample parameters of monitoring and evaluation.

Project activity:		
Impact dimension	*Parameters*	*Remark/question*
Project achievement	• Objectives or purposes of this activity	Success or not? What percentage of participants have succeeded?
Natural resources and ecology	• Soil: soil loss, nutrient loss, agrochemicals, pollutants, fertility, soil biota, soil physical properties • Surface water: water resource, water quality, sedimentation, eutrophication, aquatic biodiversity, pollutants • Ground water: ground water table, water quality, pollutants • Air: pollutants • Climate: air temperature, rainfall, season, weather • Forest: forest area, tree, wildlife, rare species, cave, waterfall, attractions	How will the project affect the environment? Is the effect on environmental quality positive or negative? Should the project continue or not? How can the negative impact on the environment be minimized?
Socioeconomic conditions	• Population: beneficiary (advantage), non-beneficiary (disadvantage) • Education • Public health • Transportation • Water supply • Electric supply • Land use system • Agricultural activity	How will the project affect socioeconomic conditions? Is the effect on social and economic conditions positive or negative? Should the project continue or not?

(Table 2 Contd.)

(Table 2 Contd.)

Socioeconomic conditions	• Industrial activity • Trading • Tourism • Occupation • Income • Tradition and culture	How can the negative impact on social and economic conditions be minimized?
People's participation and public hearing	• Agreement on project involvement • Level of participation (see *Table 1*) • Condition of participation: voluntary or regulatory • Activities of participation	Do the locals agree with project involvement? How many people attended the project? What is the participation level? Can it reach self-mobilization in the future?
Perception of local people	• Perception of project objectives and activities • Perception of introduced technology • Perception of present environmental situation • Perception of impact of project on human being and environment	Do the locals understand the project objectives and the introduced technology? What percentage of them can understand?
Sustainability of adoption of technology	• Cost of equipment and construction • Investment • Technical requirement for construction and management • Labor requirement • Incentives • Long-term subsidies • Socioeconomic impact • Local perception • Local participation	Can the locals manage or conduct the recommended technology by themselves after the end of the project or not?

Table 3 Sample weighted checklist for M&E decision of natural resources and ecology and socioeconomic conditions.

Project activity: Waterway dredging for irrigation and flood control				
Impact dimension	*Parameters*	*On-site impact*	*Outside impact*	*Remarks*
Natural resources and ecology	• Soil:			• Less soil and nutrient loss because floods may be decreased • Less pollutant that may contaminate in excess water
	- Soil loss	–	–	
	- Nutrient loss	–	–	
	- Agrochemicals	–	–	
	- Pollutants	+	–	
	- Fertility	–	+	
	- Soil biota	+	0	
	- Soil physical properties	+	0	
	• Surface water:			
	- Water resource	++++	++/–	• Increase in waterway capacity
	- Water quality	–	–	• High amount of suspended solids in the water, high bank erosion due to high velocity of runoff
	- Sedimentation	–	–	
	- Flood	++	++	
	- Eutrophication	+	+/–	
	- Aquatic biodiversity	–	–	
	- Pollutants	+	+	
	• Groundwater:			
	- Groundwater table	+	+	• Increase in percolation
	- Water quality	0	0	• No effect on water quality and pollutant if less chemical contamination in surface water
	- Pollutants	0	0	
	• Air: pollutant	0	0	

(Table 3 Contd.)

(Table 3 Contd.)

	• Climate:			
	- Air temperature	–	–	• Clearing of plants and trees along ridge of the waterway
	- Rainfall	–	–	
	- Season	–	–	
	• Forest:			
	- Forest area	–	0	• Clearing of plants and trees along the water way
	- Wildlife	–	+	
	- Rare species	–	+	
	- Attractions	–	0	
Socioeconomic conditions	• Population:			
	- Beneficiary	++	+	• Irrigation water, aquatic food, natural food resource
	- Non-beneficiary	–	+	
	• Education	0	0	
	• Public health	0	0	
	• Transportation	0	0	• If no boat transportation in the project area
	• Water supply	0	0	• If no water supply service in the project area
	• Electric supply	0	0	
	• Agricultural activity	++	+	
	• Industrial activity	0	0	• If no industrial activity in the project area
	• Trading	+	+	
	• Tourism	+/–	0	
	• Occupation	++	+	
	• Income	++/–	+/–	
	• Tradition and culture	–	–	

(+) for positive impact, (-) for negative, and (0) for no impact; the weight range is very low, low, medium, high and very high

HOW TO MONITOR

The measuring method, timing and frequency need to be decided. There are many measuring methods such as field survey, general observation and measurements by technical analysis on farm or in laboratory.

For large-scale projects, technical analysis with high accuracy of measurement techniques should be applied in the monitoring process. For M&E of small projects, field observation and public hearing or questionnaire surveys may be sufficient. However, the parameters, methodology and schedule should be decided before the project starts. Otherwise it may affect the perception and participation of the locals. The locals can help the implementers to monitor the impacts of the project through low-technology field observations such as turbidity of water during/after rainfall, fish or animal diversity, flooding, drought, temperature, rainfall events, rills and gullies, or landslides. Local participation in the monitoring process may increase understanding between the locals and the implementers and may assist sustainability of the project achievement.

Timing of sampling and analysis is important to indicate the severity of project impact on natural resources and ecology. For example, the amount of soil loss or water quality due to the effects of soil erosion should be monitored during rainfall events. It may not be possible to get pollution results after the time has passed, as the soil and pollutants will have already run off or settled in river bottoms. High frequency of sampling and analysis is useful for the accuracy of evaluation but expensive. The schedule and frequency of each parameter may therefore be determined by its severity and significance, including financial significance, for M&E of the project.

HOW TO EVALUATE

The evaluation should cover at least seven questions:

1. Does the project succeed in its objectives or not?
2. What is the difference between before and after (or with and without) the project?
3. Did the project bring about improvement without side effects?
4. Should this project continue or not?
5. What should be improved or restored for better results from ongoing activities?
6. Will the success be sustained or not?
7. Do the locals perceive and realize the project needs or not?

Sometimes, the impact of a project cannot be measured immediately after it ends, as with impact on water quality or underground water level. Research information or experience from other projects may be necessary for evaluation.

REPORT AND RESPONSE

All of the results observed by project staff should be reported and evaluated without bias. They should be reported to not only the donors but also the locals. The results of M&E are useful for the people's perception and perspective. Additionally, they may increase the number of participants.

The M&E will be more meaningful if both donors and implementers understand its objectives and its results as well as the locals. This process will create challenges for future projects on sustainable development.

CONCLUSION

Monitoring and evaluation parameters may differ for each project according to its objectives, activities and budget. However, it is recommended that six dimensions be monitored and evaluated, particularly for negative impacts: achievements with respect to objectives, impact on natural resources and ecology, impact on socioeconomic conditions, people's participation and public hearing, local perception and acceptance, and sustainability of adoption of technology. Additionally, participatory M&E is recommended for enhancing people's perception of the project needs, which may strongly affect the adoption or non-adoption of project innovation. Also, the locals should monitor and evaluate the project achievement and its effect on their lives and natural resources. The M&E is meaningful if donors, implementers and locals understand its objectives and use the results to create challenges for future works on sustainable development.

References

Batz, F., W. Janssen and K.J. Peters. 2003. Predicting technology adoption to improve research priority-setting. Agr. Econ. 28: 151-164.

Bielders, C.L., S. Alvey and N. Cronyn. 2001. Wind erosion: the perspective of grass-roots communities in the Sahel. Land Degrad. Dev. 12: 57-70.

Bradley, J.E., M.V. Mayfield, M.P. Mehta and A. Rukonge. 2002. Participatory evaluation of reproductive health care quality in developing countries. Soc. Sci. Med. 55: 269-282.

Crawford, P. and P. Bryce. 2003. Project monitoring and evaluation: a method for enhancing the efficiency and effectiveness of aid project implementation. Int. J. Proj. Manag. 21: 363-373.

Dorward, P., M. Galpin and D. Shepherd. 2003. Participatory farm management methods for assessing the suitability of potential innovations. A case study on green manuring options for tomato producers in Ghana. Agr. Syst. 75: 97-117.

Erenstein, O. 2003. Smallholder conservation farming in the tropics and sub-tropics: a guide to the development and dissemination of mulching with crop residues and cover crops. Agr., Ecosyst. Environ. 100: 17-37.

Gyorkos, T.W. 2003. Monitoring and evaluation of large scale helminth control programmes. Acta Tropica 86: 275-282.

He, C., S.B. Malcolm, K.A. Dahlberg and B. Fu. 2000. A conceptual framework for integrating hydrological and biological indicators into watershed management. Landscape and Urban Plann 49: 25-34.

Hellin, J. and M.J. Haigh. 2002. Better land husbandry in Honduras: towards the new paradigm in conserving soil, water and productivity. Land Degrad. Dev. 13: 233-250.

Jianbo, L., W. Zhaoqian and F.W.T. Penning de Vries. 2002. Application of interactive multiple goal programming for red soil watershed development: a case study of Qingshishan watershed. Agr. Syst. 73: 313-324.

Johnson, N., M.R. Ravnborg, O. Westermann and K. Probst. 2001. User participation in watershed management and research. Water Pol. 3: 507-520.

Jones, S. 2002. A framework for understanding on-farm environmental degradation and constraints to the adoption of soil conservation measures: case studies from highland Tanzania and Thailand. World Dev. 30(9): 1607-1620.

Kammerbauer, J., B. Cordoba, R. Escolán, S. Flores, V. Ramirez and J. Zeledón. 2001. Identification of development indicators in tropical mountainous regions and some implications for natural resource policy designs: an integrated community case study. Ecol. Econ. 36: 45-60.

Kerr, J., G. Pangare, V.L. Pangare and P.J. George. 2000. An evaluation of dryland watershed development projects in India. EPTD Discussion Paper No. 68, 130 pp.

Kerr, J. 2002. Watershed development, environmental services and poverty alleviation in India. World Dev. 30(8): 1387-1400.

Kerr, J. and K. Chung. 2001. Evaluating watershed management projects. Water Pol. 3: 537-554.

Lai, K.C. 1992. Monitoring and evaluation of soil conservation projects. pp. 133-149. *In:* S. Arsyad, I. Amien, T. Sheng and W. Moldenhauer. [eds.] Conservation Policies for Sustainable Hillslope Farming. SWCS and WASWC.

Lal., R. 2001. Soil degradation by erosion. Land Degrad. Dev. 12: 519-539.

Leach, W.D., N.W. Pelkey and P.A. Sabatier. 2002. Stakeholder partnerships as collaborative policymaking: evaluation criteria applied to watershed management in California and Washington. J. Pol. Anal. Manag. 21(4): 645-670.

McDonald, M. and K. Brown. 2000. Soil and water conservation projects and rural livelihoods: options for design and research to enhance adoption and adaptation. Land Degrad. Dev. 11: 343-361.

Morgan, G.S. and R.C. Ng. 1990. A framework for planning, monitoring, and evaluating watershed conservation project. pp. 157-172. *In:* J.B. Doolette and W.B. Magrath. [eds.] Watershed Development in Asia: Strategies and Technologies. World Bank Technical Paper No. 127.

Morgan, R.P.C. 1995. Soil Erosion and Conservation, 2nd ed. Longman Group Limited, England, 198 pp.

Murray, C.J.L., D.B. Evans, A. Acharya and R.M.P.M. Baltussen. 2000. Development of WHO guidelines on generalized cost-effectiveness analysis. Health Econ. 9: 235-251.

Nakayama, M. 2002. Institutional aspects of international water management: lessons from the Mekong River Basin. pp. 69-78. *In:* L. Jansky, M. Nakayama and J.I. Uitto. [eds.] Lakes and Reservoirs as International Water Systems: Towards World Lake Vision. United Nations University, Tokyo.

Pretty, J.N. and P. Shah. 1997. Making soil and water conservation sustainable: from coercion and control to partnerships and participation. Land Degrad. Dev. 8: 39-58.

Reynolds, K.M., M. Jensen, J. Andreasen and I. Goodman. 2000. Knowledge-based assessment of watershed condition. Comput. Electron. Agr. 27: 315-333.

Scoccimarro, M., A. Walker, C. Dietrich, S. Schreider, T. Jakeman and H. Ross. 1999. A framework for integrated catchment assessment in northern Thailand. Environ. Model. Software 14: 567-577.

Shaxson, T.F. 1992. Crossing some watersheds in conservation thinking. pp. 81-89. *In:* K. Tato and H. Hurni. [eds.] Soil Conservation for Survival. SWCS and WASWC.

Vine, E.L., J.A. Sathaye and W.R. Makundi. 2000. Forestry projects for climate change mitigation: an overview of guidelines and issues for monitoring, evaluation, reporting, verification and certification. Environ. Sci. Pol. 3: 99-113.

Vine, E.L., J.A. Sathaye and W.R. Makundi. 2001. An overview of guidelines and issues for the monitoring, evaluation, reporting, verification, and certification of forestry projects for climate change mitigation. Global Environ. Change 11: 203-216.

Young, A. 1989. Agroforestry for Soil Conservation. CABI, Wallingford, UK, 276 pp.

4

Impact Monitoring of Soil and Water Conservation: Taking a Wider Look

Karl Herweg
Centre for Development and Environment, Berne University, Berne, Switzerland.
E-mail: karl.herweg@cde.unibe.ch

ABSTRACT

Despite the huge amount of information available on soil and water conservation (SWC), there seems to be a considerable gap in knowledge about the impact of SWC technologies, such as the effectiveness of on-farm technologies in controlling soil erosion, their impact on human and natural resources, cost-benefit ratios, or the level of integration into prevailing farming systems. This paper introduces a methodology for impact monitoring and assessment developed by an international expert group over the past few years. More than anything else, impact monitoring requires a significant change of mind. This involves looking beyond one's own profession and even one's own mandate. The methodology presented is focused on sustainable land management, which puts SWC into a wider thematic framework and thus helps to identify a broader range of the side-effects and impacts of SWC activities.

TAKING A WIDER PERSPECTIVE

Experience gained in the WOCAT programme (World Overview of Conservation Approaches and Technologies) shows that available information and knowledge about soil and water conservation is quite comprehensive but fragmented. Individual actors such as researchers, experts, planners and farmers have highly specific know-how in limited areas of research or practice. Interestingly, a huge knowledge gap seems to exist with respect to the impact of soil and water conservation (SWC) technologies in particular, such as the effectiveness of on-farm technologies in controlling soil erosion, their impact on human and natural resources, cost-benefit ratios, or the level of integration into prevailing farming systems (Liniger and Schwilch, 2002; Liniger et al., 2002).

The reasons for this may be numerous. But development programmes generally seem to emphasize performance rather than impact. For example, an SWC programme may be responsible for the design and implementation of SWC measures (performance) but have no mandate to monitor consequences in the mid- or long-term (outcome/impact). Usually, programme planning tools merely take a long-term perspective into consideration when formulating overall goals, objectives and purposes. Programme monitoring, in contrast, covers mostly short-term performance only, i.e. checking whether or not a programme has achieved the expected results, such as the number and length of terraces implemented. Another issue that prevents people from seriously dealing with impact monitoring is the fact that there are always unintended impacts, both positive and negative. And who wants to admit negative effects when this could lead to a loss of funding?

The term "impact" refers to the mid- to long-term implications of a programme in its particular context. The context can be described as the biophysical, sociocultural, economic, institutional and political milieu or environment. It consists of several levels, from the micro-level (local level) to the macro-level (national policy, economy, etc.), and involves different stakeholders, such as local land users, women groups, extension workers, trainers, teachers, health specialists, economists, and policymakers. The numerous elements and stakeholders in such a context and their interrelationships are the reason that a programme has multiple effects, requiring a wider thematic perspective. The impact of an SWC programme cannot be seriously assessed by taking only soil loss into account and ignoring, for example, economic and social aspects. Therefore, a wider perspective such as sustainable land management (SLM) is recommended. The fact that changes in land management may become visible only after several years or even decades also suggests a wider perspective in terms of time. The question is, how can programmes with a rather narrow temporal and financial scope systematically engage in impact monitoring?

DEVELOPMENT OF AN INSTRUMENT FOR IMPACT MONITORING AND ASSESSMENT IN SLM

This chapter concentrates on an instrument for impact monitoring and assessment (IMA) in SLM that was developed on the basis of requests and needs of several development agencies. An international group of experts drafted an initial version by 1998 that was distributed in three languages (Herweg et al., 1998; 1999a,b). Over a period of three years this version was applied in and commented on by several programmes, workshops and consultants. Their views were incorporated in the fully revised present version (Herweg and Steiner, 2002, 2003). This document is intended as a contribution to an on-going discussion of quality management of development programmes.

Impact monitoring and assessment is designed to provide programme staff and consultants with building blocks for developing their own programme-specific monitoring system as part of a programme's self-evaluation process. It does not offer standard solutions; its procedure and tools must be adapted to individual circumstances. It aims at plausible indications of a programme's impact - not scientific proof. Thus, IMA is an instrument of reflection and learning that helps assess impacts during a programme's lifetime to better adapt its activities to a changing reality or context.

SIX STEPS IN IMPACT MONITORING AND ASSESSMENT

Experience gained during the development phase of this instrument has shown that a shift of paradigm from performance (efficiency or doing things right) to impact (effectiveness or doing the right things) is essential. Creating a desired impact requires that a programme context be sufficiently understood. However, financial constraints usually do not permit extended preliminary investigations. In order to save time and costs, the six steps in IMA briefly described below are closely linked to existing procedures of programme cycle management and are not meant to be an additional, independent tool. They take up common management aspects such as problem analysis, indicator selection and evaluation, but add a longer-term and broader impact perspective. Since IMA focuses on SLM, its methodology can be very useful for consideration in SWC programmes. The examples given below were developed during impact monitoring workshops between 1999 and 2001.

Step 1: Involvement of Stakeholders and Information Management

A development programme or project is in a position to trigger changes, but only stakeholders can actually bring about change. If, for example, the implementation of SWC technologies is the "output" of an SWC programme, farmers who apply these outputs will soon observe positive and/or negative outcomes, such as increased or decreased production or technical problems. The learning process connected with this, and related changes in mind and attitude, can either initiate or hinder further social processes such as dissemination. Stakeholders are not simply adopting or rejecting a technology; they are adapting, developing and integrating SWC, creating and observing impacts at the same time. A programme that supports and systematizes such a process by impact monitoring and assessment is in need of a transparent tool that actively involves stakeholders, helps collect and disseminate information that is actually relevant for them, and can be understandably communicated through appropriate means and media. Therefore, impact monitoring starts with a stakeholder analysis and the development of suitable information management.

Step 2: Review of Problem Analysis

Many development programmes start with the formulation of a core problem and try to derive activities to respond to it. But a programme context is complex; its main elements are highly interconnected, and not every connection is known. Any intervention can create multiple side-effects. Different (partly hidden) agendas of stakeholder groups complicate the situation, which means that the precise outcome of an intervention is basically unpredictable. Consequently, a programme should take a wider perspective from the beginning, starting with - or complementing earlier problem analysis by - a broader systems analysis before focusing on a specific component such as SWC. Table 1 presents so-called fields of observation that can help identify the relevant elements in an SLM context, as well as meaningful impact indicators. Development specialists probably have to accept that they cannot precisely plan and control how the context will change. Their role is to provide an impulse, monitor and assess effects and reactions, design subsequent impulses, and so forth. A well-tested tool, the participatory systems analysis, is described in detail by Herweg and Steiner (2002).

Table 1 Fields of observation in SLM.

Level	*Dimensions of sustainability*			
	Institutional	*Socio-cultural*	*Economic*	*Ecological*
Household (including farm plot level)	• Education and knowledge • Access to natural resources • Household strategies • …		• Household income, assets and consumption • Labour and workload • Land management and farming system • …	• State of natural resources • …
Community	• Local leadership • Local institutions • Producer and self-help organizations • …	• Gender issues • Conflict management • Innovation • …	• Markets, prices and credit • Public property • …	• Land use • Water resources • …
		• Social and economic disparities • …		
District	• Education, training and extension • Land and water rights, tenure • …	• Change in social values • …	• Employment opportunities/migration • Infrastructure • …	• Land cover • Off-site effects • …

Step 3: Formulation of Impact Hypotheses

Impact hypotheses are formulated automatically when analysing a programme context (the SLM "system"), or when selecting elements and discussing their interrelations. For most of us it may not be easy to describe, for example, exactly what a more sustainable form of land management would look like. In addition, it may take several years to actually see a context changing towards SLM, which may be beyond the lifespan of a common development programme. If the goal is vague and the scope is very broad, a final impact is difficult to formulate. But it is possible to divide the "future" into manageable steps by setting up impact chains. An impact chain consists of a sequence of smaller effects leading towards SLM (positive impact) and/or away from it (negative impact). This division enables a programme to learn its lessons and apply corrections along the way and not after the end of the programme!

An impact chain can start with the programme's outputs, for example, implemented SWC technologies. The second part of the chain covers *use* of the outputs. Are land users working with the new technology and, if yes, how many of them are? It has to be kept in mind that the use of incentives produces a high application rate but may easily hide the true impact. Therefore, the third part refers to the *usefulness* of the output. Table 2 presents an example of an impact chain. In applying SWC technologies, a land user experiences both benefits and drawbacks (positive and negative effects or outcomes). In case of predominantly positive effects, a process of further developing and integrating a technology into a farming system may begin. Negative effects may lead to modifications or even abandonment of the technology. Impact monitoring helps to determine where the output is useful and where not, to identify both positive and negative effects, and to draw appropriate conclusions. The important point is to admit that negative impacts or side-effects are the rule, not the exception. They serve as lessons learned in a process of technology development. The formulation of both positive and negative impact hypotheses at the planning stage helps ensure better preparation when reality turns out not to correspond with the plan.

Step 4: Selection of Impact Indicators

Impact indicators are not only understandable representations of a more complex reality but also an important means of communication between different stakeholders (Tables 3 and 4). Indicator selection should not be reduced to a one-day theoretical exercise conducted by experts. Instead, sufficient time should be allocated to identify a reasonable selection of impact indicators that is acceptable to the stakeholders involved.

Table 2 Example of an impact chain and corresponding impact indicators.

Links in the impact chain	*Impact chain (positive and negative implications)*	*Possible impact indicators*
Output	A new production and conservation technology	-
Use of output	Most (very few) farmers in the programme area apply new production and conservation technologies (applicability) and adapt them to their specific situations (adaptability)	• % of farmers adapting new technologies without incentives
Benefits/drawbacks (effects/outcomes)	Crop production increases (decreases), pests and diseases are minimized (increase), soil degradation decreases (increases). Improved agricultural production is (not) marketable, household income increases (decreases), and women's economic status is strengthened (weakened)	• Crop yield • Occurrence of pests and diseases • Soil erosion
Impacts	Men and women decide jointly (men decide) how to re-invest household income; farmers experiment more (less) than before; soil fertility improves (decreases); more (fewer) boys and girls attend school	• Household income • Women's labour income • Household decision-making • % of farmers experimenting with cropping practices • Soil fertility status • Boys and girls with school leaving certificate

Table 3 Selected impact indicators related to SLM fields of observation.

SLM fields of observation at household level	*Impact indicators*
Education and knowledge	% of school children / No. of school drop-outs (separate for boys and girls), No. of people with school leaving certificate
Access to natural resources	No. and size of plots managed by women and men, management of communal land
Household strategies	Household structure, labour division, changes in perceptions and behaviour, innovations
Household income, assets and consumption	Household income, male and female earnings, gross margins, clothing, housing, nutrition, purchasing power, spending power, months of food security, re-investment in new farm implements, seeds, etc.
Labour and workload	Labour division, labour income
Land management and farming system	Labour income, change in farming system, adapted farming practices, abandoned technologies, application rate of conservation-effective practices
State of natural resources	Soil fertility status, soil erosion, salinity, compaction, water availability and water quality, biodiversity, plant growth, plant cover, pests and diseases, No. and quality of animals
SLM fields of observation at community level	*Impact indicators*
Local leadership	Access to natural resources by women/ men, actions taken when local by-laws are neglected
Local institutions	Active participation, survival rates of trees, conservation structures maintained without incentive, representation of social strata
Producer and self-help organizations	No. of farmers' associations, representation of social strata
Gender issues	% of women in decision-making institutions and meetings, % of women with land title; gender-specific access to credit, workload, income
Conflict management	Conflicts over natural resources, taboos with regulatory character, binding local agreements
Social and economic disparities	Wealth, status of minorities, clothing, housing, % of landless people

(Table 3 Contd.)

(Table 3 Contd.)

Innovation	No. of innovative technologies, social status of innovators
Markets, prices and credit	Distance to markets, new shops and business, No. of credits, interest rates
Land use	% of cropland, pasture, forest / bush land and other, visible signs of resource degradation, deforestation rate, cultivation of marginal land, overgrazing, abandonment of cropland
Water resources	No. of people suffering from water-borne diseases; No. of conflicts over water resources, water colour, months when springs and rivers have water
SLM fields of observation at district level	*Impact indicators*
Education, training and extension	District radio programmes with environmental messages, farmers' and schoolchildren's environmental awareness
Land and water rights, tenure	Environmental laws, regulations, land title, land price, local taboos with regulatory character, enforcement of regulations
Change in social values	Crime, conflicts between generations; social status of farmers
Employment opportunities / migration	Unemployment rate, vacancies, in- and out-migration, No. of female household heads
Infrastructure	Access to markets, schools, services, credit, scholars per family, frequency, price and reliability of transport, frequency of power cuts
Land cover	% of crop, pasture, forest land
Off-site effects	Flash floods, sedimentation of dams, water quality, destruction of roads and bridges

Step 5: Development and Application of Impact Monitoring Methods

Development programmes without a research component cannot afford a time-consuming monitoring exercise. Nonetheless, data of a certain quality must be maintained to allow a meaningful assessment. A suitable compromise is required between the highest possible quality and the lowest possible input. During the test of the first version of IMA (Herweg et al., 1998), most programmes responded in confused and negative fashion when confronted with a larger number of cost-effective field methods and the choices left for them. Therefore, the final version of IMA presents three methods only, but with a higher probability of application.

Table 4 Examples of linking generic and local indicators for better communication.

Generic indicators	*Corresponding local indicators*
Soil erosion in t ha^{-1}	Increased seeding rate: Seeds are washed away as a consequence of soil erosion and need to be re-sown.
Organic matter content, cation exchange capacity, nutrient content (soil fertility indicators)	Indicator plants: These point to locations where soil fertility is high, where the nutrient status of the soil has recovered during a fallow period, where the groundwater table is high or waterlogging occurs frequently, etc.
Human nutrition	Fat/slim cats and dogs: In villages where the human population does not have enough to eat, domestic animals such as dogs and cats will be slim.
Increased household income	Men have two or more wives: In some Muslim areas this is a sign of economic well-being.

Used in triangulation, they can produce a comprehensive picture of an SLM context, provided that the user assures data of acceptable quality. The three methods are interviews and discussions, photo-monitoring, and observations made during participatory transect walks.

- *Interviews and discussions* with local stakeholders are the basis for IMA. The information obtained can be very detailed but will be guided by individual perceptions and the different (often hidden) agendas of the stakeholders. Although all kinds of visible and invisible changes might be discussed, socioeconomic aspects may dominate. A cross-check of the information, particularly invisible (e.g., social) changes, can be made through interviews with other stakeholders. Visible improvements or deterioration can be cross-checked with photo-monitoring and participatory transect walks.
- *Photo-monitoring* provides an overview of visible changes in the programme context, which may be predominantly related to biophysical and economic issues. But photos require interpretation and further investigation of the background. This can be done through interviews and discussions, as well as during participatory transect walks, depending on which aspects need further clarification.
- *Observations* made and discussed during a *participatory transect walk* provide a detailed view, especially of biophysical issues, although social and economic issues can also be addressed. A transect walk highlights the spatial interrelations of, for example, soil degradation and nutrient, water and energy flows. Discussions often start with visible aspects but can ultimately include links with invisible aspects. A transect walk is an excellent opportunity to identify local impact indicators. The information can be cross-checked with interviews and photo-monitoring.

Step 6: Impact Assessment

Sustainable changes in a context are usually the result of social processes such as learning, innovation, and integration. These processes are influenced by many factors, such as external policies, economic developments, and internal social mechanisms of change. It is quite difficult for a programme - which can be considered an external factor - to say in the end what exactly its own contribution was. But it is still possible to:

- assess changes in a context together with stakeholders involved (Table 5, Fig. 1),
- ask whether the programme in question actually did trigger changes, and
- assess whether these changes can lead towards or away from the goal (SLM).

The regular assessment of changes in the programme context (Fig. 1) should be part of an on-going discussion about possible impacts and actual developments together with stakeholders. It deepens the understanding of the context dynamics, reveals the weak parts, and at the same time indicates what activities to carry out next.

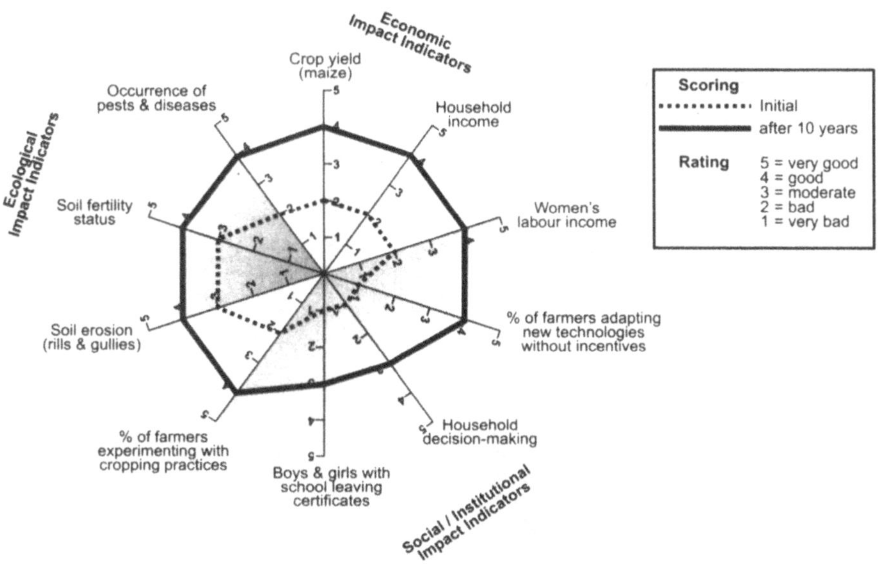

Fig. 1 Participatory assessment of changes in a mid- to long-term programme context (examples).

Table 5 Preparing impact assessment benchmarks (reference values) for several impact indicators should largely reflect an agreement between a programme and local stakeholders.

	Rating*				
	5 Very good	*4 Good*	*3 Moderate*	*2 Bad*	*1 Very bad*
Short-term indicators					
Crop yield (maize)	> 3 t ha^{-1}	> 2 - 3 ha^{-1}	> 1.5 – 2 ha^{-1}	1- 1.5 ha^{-1}	< 1 ha^{-1}
Household income	>20% increase	> 10-20% increase	1-10% increase	stagnating	decreasing
Women's labour income	>20% increase	> 10-20% increase	1-10% increase	stagnating	decreasing
% of farmers adapting new technologies without incentives	> 60%	> 40-60%	> 20-40%	10-20%	< 10%
Occurrence of pests and diseases	no	rarely, little evidence	sometimes, but can be controlled	control is often difficult	high, every year
Soil erosion (rills and gullies)	no signs of erosion	smoothened soil surface, but no rills	sometimes, few rills	most years, many rills	every year, rills and gullies
Mid- to long-term indicators					
Household decision-making	jointly in most households		jointly in a few households	by men in most households	
% of farmers experimenting with cropping practices	regular modifications by > 70%	regular modifications by > 50-70%	regular modifications by > 30-50%	irregular modifications by 5-30%	< 5%
% boys and girls with school leaving certificate	> 80	> 60-80	> 40-60	30-40	<30
Soil fertility status**	deep, dark topsoil, high earthworm activity, high root density		moderately deep and dark topsoil, earthworm activity, root density	light soil colour, yellow and red plant leaves, no earthworms, low root density	

* The rating is highly site-specific and requires intensive discussion with stakeholders.
** Rating of soil fertility status requires consultation with soil specialists.

CONCLUDING REMARKS

Impact monitoring requires, more than anything else, a significant change of mind. It forces the staff of a development programme to consider a longer time perspective and a wider thematic context, such as SLM instead of SWC. This involves looking beyond one's own profession and even one's own mandate. In addition, impact monitoring should make it difficult to paint rosy scenarios and omit mention of negative effects. Instead, it takes account of negative impacts as part of a learning process, but that is not intended to assign blame. Only an open, critical, and sometimes controversial debate with stakeholders about use and usefulness of a programme's outputs may finally lead to optimization of the management of land resources.

Fig. 2 How will I feel when being confronted with my own mistakes?

References

Herweg, K., J. Slaats and K. Steiner. 1998. Sustainable land management - guidelines for impact monitoring. Working documents for public discussion. Workbook 79 pp., toolkit 128 pp. Berne, Switzerland.

Herweg, K., J. Slaats and K. Steiner. 1999a. Manejo sostenible de la tierra - Lineamientos para el monitoreo del impacto. Documentos de trabajo para discusión pública. Manual 79 pp., utilería 128 pp. Berne, Switzerland.

Herweg, K., J. Slaats and K. Steiner. 1999b. Gestion durable des terroirs - Guide pour le suivi des impacts. Documents de travail soumis au débat. Manuel 79 pp., boîte à outil 128 pp. Berne, Switzerland.

Herweg, K. and K. Steiner. 2002. Impact Monitoring and Assessment. Instruments for use in rural development projects with a focus on sustainable land management. Vol. 1: Procedure (48 pp.). Vol. 2: Toolbox (44 pp.). Berne, Switzerland.

http://www.cde.unibe.ch/programmes/mandates/pdf/imavol1en.pdf

http://www.cde.unibe.ch/programmes/mandates/pdf/imavol2en.pdf

Herweg, K. and K. Steiner. 2003. Monitoreo y valoración del impacto. Instrumentos a usar en proyectos de desarollo rural con un enfoque en el manejo sostenible de la tierra. Vol. 1: Procedimiento, (48 pp.). Vol. 2: Utilería (44 pp.). Berne, Switzerland.

Liniger, H.-P. and G. Schwilch. 2002. Better decision making based on local knowledge - WOCAT method for sustainable soil and water management. Mountain Res. Dev. J. 1:22

Liniger H.-P., G. van Lynden and G. Schwilch. 2002. Documenting field knowledge for better land management decisions - Experiences with WOCAT tools in local, national and global programs. Proc. ISCO Conf. 2002, Vol. I, pp. 259-267, Beijing.

5

M&E as Learning: Rethinking the Dominant Paradigm

Jim Woodhill

Head, Social and Economic Department, International Agriculture Centre, Wageningen University and Research Centre, The Netherlands. E-mail: jim.woodhill@wur.nl

ABSTRACT

This chapter argues that for monitoring and evaluation (M&E) to make a useful contribution to improving the impact of soil and water conservation work there must be a much greater focus on learning. A learning paradigm challenges the quantitative indicator-based and externally driven approaches that have characterized M&E in the development field. The chapter proposes five key functions for M&E: accountability, supporting operational and strategic management, knowledge creation and empowerment. From this perspective on the functions of M&E, current M&E trends and debates are examined, which leads to the identification of the key building blocks for a learning-oriented M&E paradigm. The chapter concludes by outlining the elements of a learning system that embodies such a paradigm. The argument of the chapter is not to throw away indicators (both quantitative and qualitative) or to compromise the collection and analysis of good data. Solid learning requires solid information. Rather, this chapter asks those in development initiatives to place the indicator and information management aspects of M&E in a broader context of team and organizational learning. The challenge to be faced is to use effective reflective processes that can capture and use actors' wealth of tacit knowledge that is all too often ignored.

INTRODUCTION

Monitoring and evaluation (M&E) is high on the development agenda. Yet there remains a vast gap between theory and practice. Almost universally donors, development organizations, project managers, and development practitioners want to see better M&E. But this does not come easily. Why is this the case and what is going wrong?

This chapter offers the building blocks of an alternative M&E paradigm[1] that aligns more closely with practice and the realities of how people create knowledge, make sense of their situations, and adapt to change. Such a paradigm focuses on individual, group, and organizational learning, a perspective that has been absent in classical numerical and indicator-driven approaches to M&E. These building blocks emerge from a critical look at M&E in the broader development agenda. This provides important background for the soil and water conservation (SWC) work described in this book that often takes place within the context of development cooperation systems and procedures. The intention is to provide a context for the more specific SWC-related aspects of M&E discussed in other chapters.

To varying degrees, most development practitioners now agree that M&E should incorporate more participatory approaches, that learning lessons is important, that the focus should be on providing management information, that outcomes and impacts need greater emphasis, that M&E must be linked with planning, and that accountability to beneficiaries, partners and donors is critical. All these elements could make M&E more "useful"[2] (Patton, 1997) and can be considered elements of a new M&E paradigm. However, outdated assumptions and practices continue to hamper the development of such learning-oriented innovations to M&E. This chapter argues that the current gap between theory and practice can only be resolved by shifting the perspective from indicator- and data-driven M&E systems to learning-oriented systems.

The idea of organizational learning and the value of facilitating learning within communities, project teams, and professional groups has become well recognized (Argyris and Schön, 1978; Bawden, 1992; Senge, 1992). Sadly, learning remains an ambiguous concept, one that is deemed

[1] The perspectives presented in this chapter are based on the author's reflection on some 10 years' experience with M&E in the following contexts: M&E consultant for the International Fund for Agricultural Development (IFAD) and co-author of IFAD's manual on M&E – Managing for Impact; developing performance reporting and learning approaches in PLAN International; regional planning, M&E facilitation for the World Conservation Union; and participatory M&E issues in watershed management and Landcare in Australia.

[2] See Patton (1997) in relation to the concept of use-focused evaluation.

as simply "training" by many in development work. The everyday image of learning is much coloured by classroom experiences, with teachers expounding facts and students expected to remember and regurgitate those facts.

The idea of learning that underpins the paradigm of M&E outlined in this chapter is quite different. Learning is viewed not only as the accumulation of knowledge or skills but also the ability to constantly improve the efficacy of action. The implications of such a learning perspective for M&E will be outlined in the chapter.

Much development work, including soil and water conservation, has traditionally been supported via time-bound, output-focused projects. However, growing doubts about the effectiveness of projects as such have fed interest in more flexible programme approaches and the provision of support to build the self-reliance and enabling capacity of key institutions and organizations. Until recently, the theory and practice of M&E in development has been shaped almost exclusively by a concern with projects. This chapter is concerned with M&E in a wider context as it relates to not only projects but also to programmes, organizational performance, and processes of institutional change. Reflecting this broader concern, the chapter will use "development initiatives" as an inclusive term to cover M&E at a project, programme or organizational level.

The argument in the chapter for a learning systems approach to M&E is divided into three main sections. First, a foundation is laid by establishing the key functions of M&E. This leads, secondly, into a critical look at emerging issues and debates within the M&E field. From this critique of current theory and practice the chapter then outlines eight building blocks for an alternative M&E paradigm. The chapter concludes by discussing design implications for learning-oriented M&E systems.

THE KEY FUNCTIONS OF M&E

The functions of M&E systems are often taken for granted and not carefully examined. As a foundation for the discussion about alternative approaches to M&E, this section defines the terms being used and proposes five key functions of M&E systems.

Many M&E experts like to make a very clear distinction between monitoring and evaluation. This author instead views them as two overlapping spheres of activity and information. "Monitoring" does focus more on the regular collection of data, while evaluation involves making judgements about the data. In theory, monitoring is viewed as a regular activity while evaluation is a more periodic occurrence. But even in everyday life, monitoring and evaluation are closely interlinked. As an example, when driving and monitoring the speed indicator, it is necessary to simultaneously evaluate the appropriateness of our speed relative to the road and traffic conditions. Leaving evaluation for later would be downright dangerous. This example illustrates that where monitoring stops and evaluation begins is rather less clear than M&E theory often claims.

The separation of monitoring from evaluation has been partly driven by the classical approach to development projects, in which evaluation was undertaken every now and then by external experts, while monitoring was the task of project implementers. It is exactly this scenario that has resulted in an inability of many development initiatives to learn effectively, as it disconnects information collection from the sense-making that precedes improved action.

In summary then, monitoring and evaluation is viewed in this chapter as an integrated process of continual gathering and assessment of information to make judgements about progress towards particular goals and objectives, as well as to identify unintended positive or negative consequences of action. As will be discussed later, M&E must also provide insight into why success or failure has occurred.

The term "M&E system" refers to the complete set of interlinked activities that must be undertaken in a coordinated way to plan for M&E, gather and analyse information, report, and support decision-making and the implementation of improvements.

The alternative paradigm of M&E outlined in this chapter presupposes that any M&E system needs to fulfil the following five purposes. This is not simply an assumption but one borne out by the hands-on practice of M&E.

1. *Accountability* - demonstrating to donors, beneficiaries, and implementing partners that expenditure, actions, and results are as agreed or are as can reasonably be expected in a given situation.
2. *Supporting operational management* - providing the basic management information needed to direct, coordinate and control the human, financial, and physical resources required to achieve any given objective.
3. *Supporting strategic management* - providing the information for and facilitating the processes required to set and adjust goals, objectives, and strategies and to improve quality and performance.
4. *Knowledge creation* - generating new insights that contribute to the established knowledge base in a given field.

5. *Empowerment* - building the capacity, self-reliance, and confidence of beneficiaries and implementing staff and partners to effectively guide, manage, and implement development initiatives.

Within development, *accountability* and, in particular, reporting to donors has tended to drive most M&E efforts. Such reporting has often been seen by the implementers as a tedious administrative task that has to be done but that contributes little to the quality of their efforts or achievements. Furthermore, reporting requirements have tended to focus on the input and activity level and be descriptive rather than analytical about performance. This has inevitably fed a focus on quantitative indicators rather than qualitative explanations. While the need to be accountable is clearly important, the way M&E is conceived to meet this function is rather different from what is required for the remaining four functions. However, the broadening of the idea of accountability to include "downward accountability" aimed at beneficiaries is bringing about some changes in the mechanisms of M&E for this function (Guijt, 2004).

It would seem common sense that M&E should be able to provide the necessary information for operational as well as strategic *management*. In reality, this is often not the case. Most development initiatives appear to have sufficient monitoring (although often of an informal nature) to manage the operational side of basic activity implementation and financial management. Rarer are systems that enable development organizations to make a critical analysis of progress towards outcomes and impacts in a participatory and learning-oriented way with beneficiaries, staff, and partners.

Strategic management involves asking questions such as: Is the initiative really working towards the correct objectives? Are failures occurring because of wrong assumptions (incorrect theory of action[3]) or because of problems with implementation? How can problems be overcome and successes built on? Such questions cannot be answered by a few quantitative indicators but require in-depth discussion and engagement between different actors in a development initiative or organization.

It is at this point that the boundaries between M&E and management also begin to merge. An important, and sometimes unpleasant, lesson for M&E specialists is that M&E cannot drive management. There needs to be a demand from management for the type of M&E that will enable performance to be assessed and improvements to be made. Unfortunately,

[3]Theory of action refers to the set of underlying assumptions (theories) about cause and effect relationships that justify the actions to be taken in order to achieve goals and objectives.

a common assumption often made by M&E system developers is that improving M&E will lead to improved management and performance. This is most definitely not a guaranteed causal connection. In part due to the image of M&E as number counting and dull reporting, many managers do not engage closely with M&E systems or issues and do not consider M&E useful for supporting their management responsibilities. This disengagement becomes self-fulfilling as the lack of management orientation during the M&E design stage will certainly make it ineffective in terms of that function.

The fourth function of M&E is *knowledge generation*. All human actions are based on a set of underlying assumptions or theories about how the world works. These assumptions or theories may be explicit but are also often just implicit everyday understandings about what does and does not work. When, for example, a watershed management programme is designed, it hopefully draws on up-to-date knowledge about watershed management. As the programme proceeds, analysis of what is and is not working and investigation of why this is so may challenge or confirm existing theories and assumptions. In so doing, these insights may contribute to new theory. In this way, M&E in the form of action research can contribute to the established knowledge base.

All SWC initiatives, or indeed sustainable development more generally, take place within contextually specific environmental and socio-political phenomena and processes. This requires those involved to adapt theoretical ideas about SWC to suit their situation and to innovate continually. Not surprisingly, solutions for the complex challenges they face very often emerge from the trial-and-error of experience. Consequently, structured reflection, documentation, and communication about the experiences of a particular development initiative in relation to existing theory become critical components of society's overall knowledge process. The importance of this aspect of M&E is likely to grow in importance. However, as will be discussed below, there is still much to learn about how to generate useful lessons learned.

The fifth function, and yet perhaps the most overlooked, is empowerment. This means *empowering* all stakeholders, whether beneficiaries, managers, staff, or implementing partners, to make a constructive contribution to optimizing the impact of the development initiative. As is well known, knowledge is power; involving or not involving different stakeholder groups in generating, analysing, and making decisions about the knowledge associated with a development initiative can be extremely empowering or disempowering.

A CRITICAL LOOK AT EMERGING ISSUES FOR M&E THEORY AND PRACTICE

This section critically examines six issues that are central to the current theory and practice of M&E. It begins by examining the logical framework approach, which has perhaps been the key force in shaping development planning and M&E over the last several decades. The current concern with accountability for impact is then discussed before moving on to the dilemmas of a quantitative indicator-driven approach to M&E. Subsequently, questions are raised about the effectiveness of M&E practices that claim to be participatory or to be learning lessons. Finally, the thorny issue of adequately resourcing M&E is raised.

The Eternal Logframe

The logical framework approach (or "logframe") is central to the story of M&E in development and has fed much fierce debate about advantages and disadvantages (Gasper, 2000). The logframe entered into development practice from about 1970 on. Over time, and now present under various guises and evolutions, it has become close to a universal tool for development planning. On the surface, the logical framework approach embodies much good common sense. It involves being clear about objectives and how they will be achieved, making explicit the underlying assumptions about cause-and-effect relationships, identifying potential risks, and establishing how progress will be monitored. Who feels the need to argue about this?

However, in practice, the logical framework approach also introduced significant difficulties for those planning and implementing development initiatives.

1. *Lack of flexibility*: In theory, a logical framework can be modified and updated regularly. However, once a development initiative has been enshrined in a logframe format and funding has been agreed on this basis, development administrators wield it as an inflexible instrument. Further, it may be the case that while the broad goals and objectives of an initiative can be agreed on ahead of time, it is not possible or sensible to focus on defining specific outputs and activities, as demanded by the approach.
2. *Lack of attention to relationships*: As any development practitioner well knows, it is the relationships between different actors and the way these relationships are facilitated and supported that ultimately determines what will be achieved. The logical framework's focus on output delivery means that often too little

attention is given to the processes and relationships that underpin the achievement of development objectives. The outcome mapping methodology developed by the International Development Research Centre has been developed to respond to this issue (Earl et al., 2001).

3. *Problem-based planning*: The logframe approach begins with clearly defining problems and then works out solutions to these problems. Alternative approaches to change emphasize the idea of creating a positive vision to work towards rather than simply responding to current problems. Further, experience shows that solving one problem often creates a new problem; the logframe approach is not well suited to iterative problem solving.
4. *Insufficient attention to outcomes*: For larger-scale development initiatives, the classic four-level logical framework offers insufficient insight into the crucial "outcomes" level, critical to understanding the link between delivering outputs and realizing impact.
5. *Oversimplification of M&E*: The logframe implies that M&E is simply a matter of establishing a set of quantitative indicators (means of verification) and associated data collection mechanisms. In reality, much more detail and different aspects need to be considered if an M&E system is to be effective.
6. *Inappropriateness at programme and organizational levels*: The logical framework presupposes a set of specific objectives and a set of clear linear cause and effect relationships to achieve these objectives. While this model may be appropriate for certain aspects of projects, at the programme and organizational level there is mostly a more complex and less linear development path. For programmes and organizations there are often cross-cutting objectives best illustrated using a matrix approach rather than a linear hierarchy. For example, an organization may be interested in its gender or policy advocacy work in relation to a number of content areas such as watershed management planning and local economic development.

While the core ideas behind the logical framework approach can be used in flexible and creative ways, this is very rarely the practice and even the basic mechanical steps are often poorly implemented. Consequently the dominance of its use and poor application has become a significant constraint to more creative and grounded thinking about M&E and the way development initiatives are managed.

The Demand for Accountability and Impact

In all countries, the consequences of free market ideology and policy have led to pressure on public expenditure. The result is much greater scrutiny over the use of public funds for development and environmental programmes. Furthermore, growing public and political scepticism about the results from the last 50 years of international development cooperation (whether justified or not) is forcing development agencies to demand greater accountability and greater evidence of impact for each euro spent. The number of development organizations competing for both public and private funding has also dramatically increased, making accountability an important aspect of being competitive in bidding for funding.

However, it is not only upward accountability that is important. Some development organizations are now putting much more emphasis on transparency and accountability towards the people they aim to serve and their implementing partners. ActionAid International is one of the better-known examples (David and Mancini, 2004). During annual reflections, expenditure is openly shared with partners and local people and the question "Was it worth it?" is discussed openly as the basis for mutually agreed cost reallocation. This process has become a powerful symbol of ActionAid International's effort at transparency and has improved relations along the entire aid chain.

Increasingly, donors want to know – and development agencies want to demonstrate – the ultimate results of investments made. "How have people's lives changed for the better?" or "How has the environment actually been improved?" are recurring questions. This quest for insights about impact, while understandable, brings with it four challenges for M&E practitioners.

First, there is much confusion around what "impact" means and what it is the donors are actually requesting and expecting. Second, impact is often (but not always) a long-term result that occurs after a development initiative has ended. There is often no follow-up funding or mechanism to track this impact. (It should also be realized that this issue is often used inappropriately as an excuse for not even considering the impact dimension.) Third, attributing impact to a particular organization or intervention is often extremely difficult, if not impossible, given all the other actors and factors that also influence the situation (Roche, 1999). Fourth, as one moves from assessing inputs, outputs and outcomes and eventually tracking impacts, it becomes increasingly difficult, if not impossible, to define simple, meaningful, and easily measurable indicators. Usually a more complex story of a range of interacting factors must be told to explain impact in a meaningful manner. For example, it is easy to monitor how many soil conservation structures have been put in place by a community; it is more difficult to assess the result on yields and eventually the overall livelihood benefit for the community.

These issues give rise to a fundamental paradox. For accountability to the wider public, politicians, or the media, simple, highly summarized, and ideally numerically formulated information is demanded. Yet, the nature and complexity of much environment and development work makes it extremely difficult, if not impossible, to produce meaningful information in this form.

Quantitative Indicators Only, Please

Not everything that is important can be measured. The classic mantra for M&E has been to develop simple, measurable, achievable, reliable and time-bound (SMART) indicators. The drive for setting up M&E systems based only on easily measurable quantitative indicators has perhaps been one of the key reasons for the failure of M&E systems to be operationalized or to contribute useful information for the management of development initiatives. What keep-it-simple advocates overlook in setting up M&E systems is the importance of explaining why there is success or failure of expected results. Quantitative indicators can often tell what is happening but fail to answer the question why. That question is fundamental if appropriate improvements are to be identified and implemented. To understand why certain changes are and are not occurring requires a level of critical analysis that depends upon qualitative information and well-facilitated dialogue.

Monitoring and evaluation has also tended to focus on the collection of predetermined indicators and information. However, the reality of any change process is that unforeseen changes and problems will inevitably emerge. Consequently, M&E systems based predominantly on predetermined indicators will logically fail to give the information necessary for responsive management (Guijt, forthcoming).

A false dichotomy is often made between numerical information being objective and reliable and qualitative information being subjective and less valid (Davies, 1996; Roche, 1999). As any experienced development practitioner knows, donors who ask for a report full of numbers will get the numbers. However, whether these numbers have any bearing on reality is a different question. In essence, much more sophistication is needed in understanding the role of qualitative and quantitative information in M&E and the methods required to ensure the quality and reliability of either type of information. Indeed, ensuring the reliability of quantitative information is often highly dependent on having good people management processes in place that enable dialogue and debate and cross-checking with qualitative information.

Participation

The participation of primary and other stakeholders in the formulation of development initiatives has become a widely accepted "good practice", underpinned to a large extent by the Participatory Rural Appraisal (PRA) methodology. The value and importance of participation has now flown over to the M&E field with much interest in participatory monitoring and participatory impact assessment, which draw on many of the visual PRA tools (Estrella et al., 2000; IFAD et al., 2001).

However, as with participation in planning, a big gap is evident between the rhetoric and the practice. A lack of capacity to appropriately and effectively use participatory tools and methods often leads to poor implementation and hence poor outcomes from supposedly participatory processes. Further, a naiveté or deliberate avoidance of power issues has led to much criticism of participatory processes (Leeuwis, 2000; Cooke and Kothari, 2001; Cornwall, 2004; Groves and Hinton, 2004).

Meaningful participation implies much involvement in and control over the development process by those benefiting from or directly involved in implementing development initiatives. This challenges the top-down planning and upward accountability-driven M&E that characterizes much development administration. The call for greater participation in M&E has fundamental implications for both theory and practice and is a key rationale for a more learning-oriented paradigm.

The Discovery of Lessons Learned

Development initiatives are increasingly focusing on capturing lessons learned or identifying "good" and "best" practices and producing these as knowledge outputs. In some cases, these efforts have led to useful insights. On the whole, however, lessons are of poor quality (see Box 1) (Patton, 2001; Snowden, 2003).

Such lessons-learned efforts fail for several reasons (Guijt and Woodhill, 2004). To start with, those involved are rarely clear about who the lesson is relevant to or who needs to learn what. Furthermore, they fail to make a solid connection between the existing knowledge and theory base — and any knowledge gaps — and the "new" lesson learned. Hence, it is often unclear whether the lesson is really "new" or whether it confirms or contradicts existing theories and practices. In many cases, lessons learned are also often either too specific or too generalized to be considered a useful contribution to the existing knowledge base. A further

Box 1. Poor quality lessons (Guijt and Woodhill, 2004)

- The lesson learned does not have a generalized principle that can be applied in other situations. It is simply a description of an observation, or a recommendation that lacks justification.
- The lesson has not been related to the assumptions or hypotheses on which the programme or project has been based and so lacks a meaningful context.
- The lesson is an untested or inadequately justified assumption or hypothesis about what might happen if something is done differently. In other words it would be foolish to rely on the lesson unless it is first tested.
- The lesson is either too general or too specific to be useful.
- The lesson has not been related to existing knowledge, hence it is unclear whether it represents a repetition of existing understanding or offers a fresh insight.

problem occurs because the extent to which the lesson or good practice has been tested and validated over time or in different contexts is unclear, making the validity of its wider relevance questionable (Patton, 2001). Finally, and perhaps most important, instead of a constructivist perspective on knowledge (discussed later in this chapter), a "commodity" model of knowledge is often assumed with the belief that, if a lesson learned is documented, it can be transmitted to others and they will use it. This ignores the complex and dynamic way in which practitioners engage with theory and practice and the processes by which they learn how to improve what they do.

While it is clearly a positive step to see those involved with M&E perceiving the "knowledge" function of their work, this trend will only yield real benefits if much more attention is given to understanding underlying knowledge and learning processes.

Everything for Nothing

The M&E of projects and programmes is often poor simply because the financial resources and human capacity needs have been dramatically underestimated. The situation is easily illustrated by comparing the resources committed to financial accounting to those committed to monitoring the outputs, outcomes and impacts. In most projects or organizations you will find well-qualified accountants, bookkeepers and financial managers backed up by accounting software and adequate computer facilities. By comparison, more often than not the capacities and resources for monitoring the deliverables and impacts will be significantly less. Yet, monitoring the results and understanding reasons for success or failure is without doubt a more complex and demanding task than keeping track of finances, and in the end even more important in terms of overall performance.

No one would want to fall into the trap of arguing that more resources are going to M&E than are getting to "the ground". However, the failure rate of development initiatives is high in any event (arguably at least in part because of more M&E). While there is no detailed research on the subject, it has become generally accepted that a 5-10% investment in M&E is reasonable. Such investment in M&E can, in most situations, be easily recouped through M&E improving the impact of development initiatives.

THE BASIS FOR AN ALTERNATIVE PARADIGM

Given these pervasive and pernicious issues with the conceptualization and implementation of M&E, a fundamental shift is required to ensure that M&E becomes a learning process by which development actors gain insights into how they can improve their performance. Despite the islands of M&E innovation, the dominant paradigm about M&E is only changing very gradually. The following eight points provide a basis for an alternative M&E paradigm.

Learning from a Constructivist Perspective

How do humans make sense of the world around them? How does one know what is true, fair, reliable or plausible? How is the relationship between science and politics to be understood? Such philosophical questions might seem a world away from the practicalities of setting up effective M&E systems. However, if M&E is to be understood as "learning", then a basic understanding on the philosophy of knowing becomes essential. This is particularly so given the dominant objectivist and positivist influence of classical scientific thought on the field of M&E.

A positivist scientific outlook sees a world of "reality" independent from human perceptions that awaits discovery through rigorous scientific observation, experimentation, and analysis (Miller, 1985). Scientific method, with its perceived objectivity, is consequently seen as the route to truth and valid knowledge (the white-coated scientists in television commercials are testament to how deep this idea runs in western society). The positivist scientific methodology has proved highly effective for understanding biophysical phenomena and for technological development. However, major theoretical and practical difficulties emerge when the methodology is applied to social phenomena or to the sort of complex, politically infused and interdisciplinary problems that characterize the quest for sustainable development (Capra, 1982; Guba, 1990a; Woodhill and Röling, 1998).

A constructivist philosophy argues that "reality" as humans experience it is constructed through our social interaction. In other words, what is experienced as "real" is, at least in part, a social construction influenced by history, culture, and language (Berger and Luckman, 1991; Guba, 1990b; Maturana and Varela, 1987). Consequently, existing theories and assumptions determine what is perceived and the way we make sense of the world that surrounds us. A constructivist perspective has far-reaching consequences for social "science" research and intervention strategies in social contexts (Guba and Lincoln, 1989; Guba 1990a).

In practice, a constructivist perspective means focusing on how adults learn and on how groups, organizations, and communities create shared understanding and meaning. An important foundation for the design and facilitation of such learning processes is the model of experiential learning developed by Kolb (1984) that drew on earlier work by Lewin (1948), Dewey (1933), and Piaget (1970). Much of the current thinking on learning-oriented approaches to development is based on this body of theory.

In essence, Kolb argues that there are four dimensions to experiential learning: having an experience, reflecting on that experience, conceptualizing from the experience, and then testing out new ideas or concepts that lead to a new experience. Paying attention to these four dimensions of experiential learning has proved enormously helpful to facilitating the processes that enable individuals, organizations, or communities to improve their performance and respond to change.

Recognizing Dynamic Environments and Uncertainty

The context for soil and water conservation is inevitably one of rapid social and environmental change and considerable uncertainty. To build a bridge or a dam, it may be possible and is even desirable to have a clear, upfront, step-by-step plan that can be followed and monitored with preset indicators. Any kind of integrated natural resource management endeavour is a totally different story due to the layered, long-term and multi-stakeholder nature of the work. As has been well articulated, such work requires a very flexible and adaptive approach to management (Gunderson et al., 1995; Dovers and Mobbs, 1997; Ghimire and Pimbert, 1997; Defoer et al., 1998; Hinchcliffe et al., 1999; Lee, 1999; Roe et al., 1999; Borrini-Feyerabend et al., 2000; Jiggins and Röling, 2000).

Accepting the reality of dynamic environments and uncertainty has dramatic implications for the way in which planning, management, and M&E processes are conceived. Most significantly, management must be highly responsive and adaptive. This means regularly checking that goals and objectives remain relevant and constantly adjusting and refining implementation strategies in response to changed circumstances and new insights. Classically, many development initiatives are contracted to an "implementation" team with contractual payments based on the delivery of pre-determined outputs. Much of the aid administration system is still structured around this model. If the reality of dynamic environments and uncertainty is accepted, then by definition, this classical model of development intervention is a recipe for failure and dashed expectations and must give way to more adaptive models.

Moving from External Design and Evaluation to Internal Learning

Combining a constructivist perspective with the consequences of dynamic environments and uncertainty implies shifting from external design and evaluation of development initiatives to effective processes of internal learning. Therefore, those implementing and benefiting from an initiative will carry greater responsibility for its strategic guidance. This calls for far-reaching changes to the way in which development initiatives are designed, managed, contracted and, most significantly, monitored and evaluated. The external expert-oriented processes of development initiative design and evaluation must cede to ongoing internal learning process with key stakeholders. While external experts can add much value to development initiative formulation and evaluation activities, it is ultimately those most directly involved who are in the best position to improve development performance.

Managing for Impact

As any M&E specialist quickly finds out, the ability to implement and effectively use the potential of M&E results depends to a very large extent on management style, interest and capacity. Accepting the above points about dynamic environments, internal learning, and a concern with realizing impact logically leads to the idea of managing for impact (IFAD, 2002) or what others refer to as managing for results. This represents a significantly different paradigm about management than that which underpins many development initiatives.

In part driven by the logical framework model, M&E logic has prescribed a model that assumes that managers can only be held accountable and responsible for the delivery of specified outputs that are directly within their control. This model offers few incentives for managers to be concerned with the realization of higher-level objectives.

A managing-for-impact model argues quite the opposite by stating that managers should be (held) responsible for guiding an intervention towards the achievement of its higher-level objectives. Note the difference between being responsible for such guidance and holding management directly accountable for the achievement of impacts. Clearly in many development situations, the level of uncertainty is such that direct accountability would be unfair and impractical. However, managers should have a deep understanding of whether the intervention strategy is proving to be effective and be highly responsive to emerging difficulties. Such impact-oriented and responsive management is, of course, not just a function of the management of a development initiative but, as also mentioned above, is strongly influenced by how donor funding is administered.

Two aspects of managing for impact require particular attention from a learning-oriented perspective. One is recognizing that in most development contexts what can be achieved has to do with the coordination, integration, and commitment of a range of different actors (Earl et al., 2001). Consequently, the challenge of managing for impact does not concern just delivering outputs within managers' control but rather influencing the relationships and the actions of others that may well be beyond direct management control. The second aspect is for management to be able to distinguish between problems that are due to a failure of effective implementation and those problems that are due to a failure of theory (assumptions) in the intervention strategy (Margoluis and Salafsky, 1998). Dealing with these two quite different problems requires unique management responses.

A managing-for-impact model can be viewed as four interlinked elements (IFAD, 2002).

1. *Guiding the strategy* – taking a strategic perspective whether an initiative is heading towards its goals (impacts) and reacting quickly to adjust the strategy or even the objectives in response to changed circumstances or failure.
2. *Ensuring effective operations* – managing the day-to-day coordination of financial, physical, and human resources to ensure the actions and outputs required by the current strategy are being effectively and efficiently achieved.
3. *Creating a learning environment* – establishing a culture and set of relationships with all those involved in an initiative that will build

trust, stimulate critical questioning and innovation, and gain commitment and ownership.

4. *Establishing information-gathering and management mechanisms* – ensuring that the systems are in place to provide the information needed to guide the strategy, ensure effective operations, and encourage learning.

At the heart of this model are the "people processes" that enable the necessary information to be gathered, good decisions to be taken, and individuals and organizations to give their best. Potential learning events that can contribute to this are partner meetings, participatory planning workshops, annual reviews, staff performance appraisals, informal discussions, social gatherings, rewarding of good performance, and participatory impact assessment. Despite the vast knowledge about effective non-hierarchical management techniques and extensive array of participatory, learning-oriented methods and tools that can be used (International Agriculture Centre ([IAC], 2005), only a fraction of this knowledge and the available processes are employed in most development initiatives.

Types and Sources of Information for Learning and Management

As already introduced, M&E can only be useful if it answers the question *why* has there been success or failure. Many donors recognize this and are rejecting activity reporting, instead asking for results and impact reporting. Taking this one step further into the arena of improved next steps requires addressing the questions of *so what* are the implications for the initiative and *now what* will be done about the situation.

Answering such questions requires using diverse information types and sources, many more than are present in conventional M&E efforts. Six aspects of information are discussed below.

1. Formalized and Informal Knowledge

Monitoring and evaluation systems often revolve around information that can be formally measured, summarized, and reported. Paradoxically, managers, while certainly using formalized information when they have access to it, make much use of informal information in their decision-making. They pick up this informal information in daily interactions with staff, partners and clients. No family has a formalized M&E system (or a logical framework, for that matter) but yet a constant flow of informal information enables decisions to be taken and family life to go on. The valuable insights held by most involved in a development initiative are often left locked up and do not get a chance to inform evaluation and influence decision-making. Monitoring and evaluation needs to invest in understanding the role of informal information, and then building on existing practices to ensure a smooth and useful flow of such information.

2. Qualitative and Quantitative Information

Much has been already been said in this chapter on the complementary nature of qualitative and quantitative information. Both kinds are critical. An indicator-driven approach to M&E often drives systems in the direction of quantitative information, but it is often the qualitative information that is required for explanation, analysis, and sound decision-making. It also needs to be recognized that this can be a false distinction as qualitative information can often be summarized quantitatively. For example, a qualitative question can be asked of women about how SWC training has changed their faming practices. The percentage of women giving the same responses can be reported, but the qualitative nature of the question does not preclude reporting on unexpected responses.

3. Content and Process

There has been a tendency for M&E systems to focus on the realization of objectives (results) and ignore the processes by which such objectives can be achieved. Managers of an SWC initiative clearly need to know how many of what different types of measures are being constructed and where. However, only with insights about the implementation and stakeholder engagement processes can they also help to spot and address problems. Processes precede results. Thus, clarity about what constitutes a good process and putting in place ways of assuring the quality of such processes are critically important.

4. Levels in the Objective Hierarchy

Monitoring and evaluation systems, albeit with varying terms, distinguish between inputs, activities, outputs, outcomes, and impacts. This leads to an oversimplified image of a four-layered results chain. In reality, a change process passes through more levels (of identifiable cause-and-effect relationships) as well as non-linear sideways interactions. Thus, rather than thinking of development as a linear chain of cause-effect linkages, an image of a nexus of interwoven events is closer to the truth. Irrespective of the image, the key point is that monitoring must happen at all levels of a hierarchy or in all "corners" of the nexus. Within the most commonly used "hierarchy" logic, managing for the "outcome" level is of particular importance as this forms the strategic link between short-term operational perspective and long-term impact level.

5. Descriptive and Explanatory Information

Effective M&E is iterative. There should be some indicators that show whether things are going as expected and these will be descriptive. However, once an indicator shows a problem, the M&E system needs to investigate and move into a more explanatory mode. This is only done when there is a need. This is just like an oil pressure light for the engine of the car. If the light is off there is no need to investigate further. However, as soon as the light indicates a problem there is a need to explain why. A common complaint about reporting is that it remains at a very descriptive level about what has been achieved and gives little attention to explaining the reasons for success, failure, and changes in strategy.

6. Objective and Subjective

Some of the information needed to manage for impact will be objective (something that everyone can agree is most likely correct), for example, the expenditure based on audited accounts or the area of deforestation calculated from aerial photographs. Other information will be more subjective, such as the opinions of different stakeholders on the reasons why local people are encroaching into a protected area. Knowing about the subjective views of different people and organizations is just as important for management as the objective information. It is these opinions that are leading people to act in a particular way and so understanding their opinions is critical to the development process.

Integrating Action Learning (Research) into Development Initiatives

Earlier in the chapter, the emerging practice of capturing lessons learned was subjected to some scrutiny. This section suggests how the problems raised earlier can be overcome. An important starting point is to be clear about the potential of a particular initiative to contribute to knowledge generation. Some development initiatives may be simply implementing well-established and theoretically sound practices, in which case little added value would be gained from further reflection.

However, many development initiatives contain an experimental element or at least are based on one or more assumptions about which there is not entire clarity or consensus. In this case, valuable lessons can be learned to improve the efficiency and effectiveness of implementation of similar endeavours. For example, a large-scale watershed management programme may have much to offer the existing theory and practice of natural resources conflict management, water resource policy, the politics of multi-stakeholder engagement, or hydrology.

If there is agreement about this potential, then clear learning or research objectives can be set. This requires an extra dimension to a "normal" M&E system. In particular, it requires even more attention to the explanation of observed changes. Importantly, if an initiative aims to contribute new insights to existing theory and practice, then it must be established and resourced in a way that makes this possible. This may include working in partnership with a research institute, giving staff time to write research articles, and encouraging participation in seminars and conferences.

The Politics of Critical Reflection

Most, if not all, development initiatives involve a network of diverse stakeholders with varying interests and varying types and levels of power. They include, for example, the primary stakeholders (beneficiaries), clients, development NGOs, and government agencies. This makes a development initiative deeply political, as with any other social process. Within organizations there are power dynamics between the work floor and management, while power struggles between organizations are par for the course. Meanwhile, in communities, power differences are played out in all corners: between leaders and others, women and men, young and old, rich and poor.

An M&E process that engages stakeholders in critical reflection and brings greater transparency to the actions and performance of different groups can threaten existing political relations and power dynamics. Critical reflection is often not welcomed by some (Klouda 2004). Individuals or organizations may fear that their position and credibility will be affected by transparency about their performance or excessive frankness about the performance of others. Individuals in a management position may well feel too personally insecure to feel comfortable exposing themselves to criticism, or to the potential consequences of loss of face or contract termination. The consequences to the individual or group doing the critical thinking must be acceptable. This means that there must be no fear of retribution from others, whether the immediate boss, the authorities in the ministry to which the development initiative belongs, an international NGO that provides the funding to its local counterpart, or peers. This crucial factor must not be underestimated in any development initiative that is embedded in a broader set of institutions and relationships (Guijt, forthcoming).

For M&E to be effective under these circumstances, those who are to introduce learning elements must have a solid understanding of the power dynamics and politics of the situation. This understanding will help them introduce the process in ways that can build trust or to accept the limits of what is possible in contexts where open and transparent M&E processes are not (yet) politically feasible.

In relation to power, M&E can itself become a critical tool for empowerment - or disempowerment. For example, participatory impact assessment can be developed in a way that holds donors and implementing agencies and NGOs accountable to those they should be benefiting. Insightful large-scale experiences now exist with participatory auditing of public accounts (Lucas et al., 2005).

Capacities, Incentives and Resources

An M&E officer has often conventionally been someone who can collect, synthesize, and report data. The picture of M&E being presented above calls for a very different set of skills and abilities. A "new paradigm" M&E officer needs to be a skilled process facilitator who can build trust and who is sensitive to the politics of the situation. Such officers need a good grounding in participatory methods and tools and qualitative approaches as well as the more classical M&E skills of being able to develop good indicators, monitoring methods, and data collection and synthesis processes. Currently there is little recognition of the need for this set of skills in M&E professionals.

The capacity issue pertains not only to M&E specialists but also to managers or leaders of projects, organizations and communities. Many in positions of authority find it difficult to even conceive of what a learning-oriented M&E system might look like, much less have the capacity to bring it about in their organization, project or community.

To put in place any effective M&E system requires a careful look at the incentive structures at all levels. What are the incentives for a manager to be more open and to admit mistakes? What are the incentives for field workers to report failure that might reflect poorly on their performance? What are the incentives for a development NGO to report on genuine lessons learned and problems they have had to their donor rather than giving only the good news? An M&E process that can lead to learning and constructive improvements requires an incentive structure and a culture that reward innovation and openness about failure; it also requires norms and procedures that ensure the transparency of performance.

It hardly needs saying that effective M&E can only be realized with an appropriate level of investment in capacity building, information management, facilitation, and time for monitoring and reflection processes.

CONCLUSION — DESIGNING M&E AS LEARNING SYSTEMS

Classically, M&E systems in development projects, programmes and organizations are designed with a focus on what data needs to be collected and processed in order to report (mostly to donors) on a set of predetermined indicators. This chapter challenges the paradigm that underpins what has proven to be inadequate M&E practice. This technical information and external accountability-oriented approach needs to be replaced by an actor-specific learning approach. Such an approach focuses on the learning processes that enable different individuals and groups to continually improve their performance, while, importantly, recognizing that they are working in highly dynamic and uncertain contexts. The challenge then is to design effective learning systems that can underpin management behaviours and strategies aimed at optimizing impact, rather than simply delivering predetermined outputs.

A learning system is characterized by the following:

- Clear analysis of the stakeholders involved, their information and learning needs, and their power relations.
- Creation of a set of norms and values and level of trust that makes transparency of performance and open dialogue about success and failure possible.
- Design and facilitation of the necessary interactive learning processes that make critical reflection on performance possible.
- Establishment of clear performance and learning questions (including qualitative and quantitative indicators where appropriate) that deal with the *what, why, so what* and *now what* aspects of M&E.
- Collection, analysis and presentation of information in a way that triggers interest and learning from those involved.

A learning systems approach recognizes that much learning is already occurring, often in informal ways, and that the individuals involved in any situation usually have considerable knowledge about what is happening. The challenge is to enhance these informal processes and to capture and use the wealth of tacit knowledge through effective reflective processes, supplemented by formal processes that optimize the learning. This chapter does not argue that indicators (both quantitative and qualitative) should be thrown away or that the collection and analysis of good data should be compromised. Solid learning requires solid information. Rather, this chapter asks those in development initiatives to place the indicator and information management aspects of M&E in a broader context of team and organizational learning. This requires a significant paradigm shift, the building blocks of which have been outlined in this chapter.

References

Argyris, C. and D.A. Schön. 1978. Organizational Learning: A Theory of Action Perspective. Addison-Wesley, Massachusetts.

Bawden, R.J. 1992. Systems approaches to agricultural development: The Hawkesbury experience. Agr. Syst. 40: 159-176.

Berger, P. and T. Luckman. 1991. The Social Construction of Reality: A Treatise in the Sociology of Knowledge. Penguin, London.

Borrini-Feyerabend, G., M.T. Farvar, J.C. Nguinguiri and V. Ndangang. 2000. Co-management of Natural Resources: Organising, Negotiating and Learning-by-Doing, 95 pp. GTZ and IUCN, Kasparek Verlag, Heidelberg, Germany.

Capra, F. 1982. The Turning Point: Science Society and the Rising Culture. Collins, Glasgow.

Cooke, B. and U. Kothari. 2001. Participation: the New Tyranny? Zed Books, London.

Cornwall, A. 2004. New democratic spaces? The politics and dynamics of institutionalised participation. IDS Bull. 35: 1-10.

David, R. and A. Mancini. 2004. Going against the Flow. The Struggle to Make Organisational Systems Part of the Solution Rather than Part of the Problem. IDS, Brighton.

Davies, R. 1996. An evolutionary approach to facilitating organisational learning: An experiment by the Christian Commission for Development in Bangladesh. Unpublished report, Centre for Development Studies, Swansea, Wales, UK.

Defoer, T., S. Kante and T. Hilhorst. 1998. A Participatory Action Research process to improve soil fertility management. pp. 1083-1092. *In:* H.-P. Blume et al. [eds.] Towards Sustainable Land Use. Furthering Cooperation between People and Institutions. Proc. 9th International Soil Conservation Organisation Conference, 1996. Vol. II. Catena Verlag, Bonn, Germany.

Dewey, J. 1933. How We Think. Heath, New York.

Dovers, S.R. and C.D. Mobbs. 1997. An alluring prospect? Ecology and the requirements of adaptive management. pp. 39-52. *In:* N. Klomp and I. Lunt. [eds.] Frontiers in Ecology: Building the Links. Elsevier Science, Oxford.

Earl, S., F. Carden and T. Smutylo. 2001. Outcome Mapping: Building Learning and Reflection into Development Programs. International Development Research Centre - Evaluation Unit, Ottawa.

Estrella, M., J. Blauert, D. Campilan, J. Gaventa, J. Gonsalves, I. Guijt, D. Johnson and R. Ricafort. [eds.] 2000. Learning from Change: Issues and Experiences in Participatory Monitoring and Evaluation. Intermediate Technologies Publications, London.

Gasper, D. 2000. Evaluating the logical framework approach. Towards learning-oriented development evaluation. Public Adm. Dev. 20: 17-28.

Ghimire, K.B. and M.P. Pimbert. 1997. Social change and conservation: An overview of issues and concepts. *In:* K.B. Ghimire and M.P. Pimbert. [eds.] Social Change and Conservation. Earthscan Publications, London.

Groves, L. and R. Hinton. 2004. Inclusive Aid: Changing Power and Relationships in International Development. Earthscan, London.

Guba, E. and Y. Lincoln. 1989. Fourth Generation Evaluation. Sage Publications Ltd., Newbury Park, California.

Guba, E.G. 1990a. The Paradigm Dialog. Sage, London.

Guba, E.G. 1990b. The alternative paradigm dialog. pp.17-27. *In:* E.G. Guba. [ed.] The Paradigm Dialog. Sage, London.

Guijt, I. 2004. ALPS in Action: A Review of the Shift in ActionAid towards a New Accountability, Learning and Planning System. ActionAid, London.

Guijt, I. Forthcoming. Strengthening and Critical Link in Adaptive Collaborative Management: The Potential of Monitoring Learning from Collaborative Monitoring: Triggering Adaptation in ACM. CIFOR, Bogor.

Guijt, I. and J. Woodhill. 2004. 'Lessons Learned' as the experiential knowledge base in development organisations: Critical reflections paper presented to European Evaluation Society Sixth International Conference, Berlin.

Gunderson, L.H., C.S. Holling and S.S. Light. [eds.] 1995. Barriers and Bridges to the Renewal of Ecosystems and Institutions. Columbia University Press, New York.

Hinchcliffe, F., J. Thompson, J. Pretty, I. Guijt and P. Shah. [eds.] 1999. Fertile Ground: The Impact of Participatory Watershed Development. Intermediate Technology Publications Ltd, London.

IFAD. 2002. Managing for Impact in Rural Development: A Guide for Project M&E. IFAD, Rome.

IFAD, ANGOC and IIRR. 2001. Enhancing Ownership and Sustainability: A Resource Book on Participation. IFAD, ANGOC and IIRR, Manila.

International Agriculture Centre (IAC). 2005. MSP Resource Portal: Building Your Capacity to Facilitate Multi-stakeholder Processes and Social Learning [Online]. Available from International Agriculture Centre www.iac.wur.nl/msp (posted 2005).

Jiggins, J. and N. Röling. 2000. Adaptive management: Potential and limitations for ecological governance. Int. J. Agr. Resour., Govern. Ecol. 1.

Klouda, T. 2004. Thinking Critically, Speaking Critically. Unpublished paper. [Online] http://www.tonyklouda.pwp.blueyonder.co.uk/.

Kolb, D.A. 1984. Experiential Learning: Experience as the Source of Learning and Development. Prentice-Hall, Englewood Cliffs, New Jersey.

Lee, K.N. 1999. Appraising adaptive management. Conserv. Ecol. 3: 3-13 [online].

Leeuwis, C. 2000. Reconceptualizing participation for sustainable rural development: Towards a negotiation approach. Dev. Change 31: 931-59.

Lewin, K. 1948. Resolving Social Conflicts: Selected Papers on Group Dynamics. Harper and Row, New York.

Lucas, H., D. Evans, K. Pasteur and R. LloydPasteur. 2005. Research on the Current State of PRS Monitoring Systems. IDS, Brighton.

Margoluis, R. and N. Salafsky. 1998. Measures of Success: Designing, Managing and Monitoring Conservation and Development Projects. Island Press, Washington, D.C.

Maturana, H.R. and F.J. Varela. 1987. The Tree of Knowledge – The Biological Roots of Human Understanding. Shambala, Boston.

Miller, A. 1985. Technological thinking: Its impact on environmental management. Environ. Manag. 9: 179-90.

Patton, M.Q. 1997. Utilization-Focused Evaluation: The New Century Text. Sage Publications Inc., California.

Patton, M. 2001. Evaluation, knowledge management, best practices, and high quality lessons learned. Am. J. Eval. 22: 329-36.

Piaget, J. 1970. Genetic Epistemology. Columbia University Press, New York.

Roche, C. 1999. Impact Assessment for Development Agencies: Learning to Value Change. Oxfam Publishing, Oxford.

Roe, E., M.V. Eeten and P. Gratzinger. 1999. Threshold-based Resource Management: the Framework, Case Study and Application, and Their Implications. Report to the Rockefeller Foundation. University of California, Berkeley.

Senge, P.M. 1992. The Fifth Discipline: The Art and Practice of the Learning Organisation. Random House, Sydney.

Snowden, D.J. 2003. Managing for Serendipity. Or why we should lay off "best practice in knowledge management". First published in ARK Knowledge Management.

Woodhill, J. and N.G. Röling. 1998. The second wing of the eagle: How soft science can help us to learn our way to more sustainable futures. pp. 46-71. *In:* N.G. Röling and M.A.E. Wagemakers. [eds.] Facilitating Sustainable Agriculture: Participatory Learning and Adaptive Management in Times of Environmental Uncertainty. Cambridge University Press, Cambridge.

Part 2

M&E in Practice

6

Monitoring and Evaluation for Efficient Sustainable Development Projects

Edgar Hernandez
Universidad de Los Andes, Mérida, Venezuela. E-mail: edgarah@icnet.com.ve

ABSTRACT

There is much evidence of a global environmental crisis damaging human well-being, notably in more vulnerable ecosystems such as the Andes. The majority of plans and programs in soil and water conservation, integrated rural development, and watershed management have had less significant results than expected. Decision-makers are seeking new strategies and new methods to meet these challenges. Sustainable development is such a new approach, emphasizing continuous adaptability and learning and sensitive to the interests of an ample range of stakeholders. It is an interdisciplinary approach to understand and to solve problems associated with the complex interrelations of human society, the natural, economic, political and institutional systems. This approach demands an appropriate monitoring and evaluation component. A method is proposed to facilitate the design of activities to measure indicators, as well as to evaluate the effects and impacts generated by a project for sustainable development in a community or in a watershed. A detailed case study from Venezuela is used to apply the proposed method in the following steps:

- Problem analysis by participatory assessment
- Characterization of the context of the area, its elements and interrelations

- Selection of problems, indicators, and establishment of the baseline
- Goals of the project, proposed actions and treatments
- Predicting short-, medium- and long-term impacts
- Design of a measurement plan
- Measurement and documentation of change, data processing
- Impact assessment and interpretation of changes in the context
- Information dissemination

INTRODUCTION

Soil deterioration throughout the world is creating a worrisome scene because the actual productive ground surface of 2.0 ha per person could change to 0.5 ha per person in 2020 because of the increase of the population and land mismanagement (Brooks and Eckman, 2000). In addition, most reports about natural resources projects indicate that soil and water degradation continues and soil productivity decreases because the sustainable soil management projects have been insufficient and/or ineffective (Herweg, 1998). Perhaps this is due to the complexity of the land and water use problems. They are products of many dominant factors, some political and institutional, which have not been approached appropriately.

Also, projects have not been designed with viable objectives, nor with the participation of all stakeholders. Many lack modern management methods such as monitoring and evaluation (M&E) of project achievement. There is not yet a conviction of the importance of M&E or a culture of process evaluation. This activity does not have assigned financial resources. Changes of civic employees affect continuity in the implementation and evaluation of projects. In addition, methods and results have not been sufficiently disseminated.

There is thus an urgent requirement for leaders and governments to execute programs with greater creativity and effectiveness, to stop the deterioration of the soil-water-forest system, and to mitigate the impoverishment of communities. For example, the Andean mountains are fragile ecosystems with geologic and tectonic instability. Their slopes often have inclination of greater than 50%. The original vegetation cover is forest and moor, with a pleasant climate that makes it propitious for cultivation of temperate zone crops with high yields. These mountains have for centuries been the abode of hard-working people. Recently they have become important tourist spots. They are also the sources that supply water to the cities located at the foothills and to irrigation systems. In spite of their natural fragility, they are productive ecosystems with their own economy of great regional strategic value.

However, population growth, lack of effective development programs, and changing natural conditions are increasing their fragility. Fertility and agricultural productivity are diminishing. Soil erosion, sedimentation, deforestation, loss of biological diversity, water pollution, and the impoverishment of the population are increasing. In addition, disasters caused by land movements, torrents, and floods appear to have become more frequent. In order to face this situation, plans and programs for soil and water conservation, integrated rural development, and watershed management have been executed. However, most have had less significant results than expected. This worries governments, leaders, and national and international organizations that promote the development. They are looking for new strategies and new methods.

THE SUSTAINABLE DEVELOPMENT APPROACH

Since 1960, after the Brundtland report and the Rio Conference, there has appeared a new, attractive approach, broad and holistic. It tries to look for greater effectiveness in order to harmonize use of natural resources with conservation. This is known as sustainable development. This concept recalls the philosophy of sustainable yield developed by European foresters in the 19th century who, in a situation of diminishing wood supply, wanted to harvest wood and yet preserve the forests (Eckman et al., 2000).

Sustainable development has been defined as:

- Development that satisfies present requirements without jeopardizing the capacity of future generations to satisfy their own requirements (WCED, 1987).
- A development process that looks for the production of environmental goods and services in order to keep or increase the well-being of the present population while protecting the natural resources base and the environment on which will depend the well-being of future generations (Eckman et al., 2000).

The development process can be maintained by itself, without decreasing existing resources. This new approach has been widely disseminated. It is hoped that it will help mitigate the deterioration of natural resources and promote well-being at the level of communities, watersheds, and regions. It is a more integral and ecological approach but could become a "high political" concept that is transformed into a practice and operative strategy (Eckman et al., 2000). In this sense, two things have to be applied (Herweg, 1998):

- Participatory assessment and planning
- Monitoring and evaluation of project achievements

Sustainable development is a process, not a final state. It is an organized action that does not end, that continuously looks for improvement in the operation of the system. It has continuous adaptability and learning and is sensitive to the interests of an ample range of stakeholders, who demand stability of policies translated into a financing and stable operation, using an interdisciplinary approach to understand and to solve problems associated with the complex interrelations of human society with natural, economic, political and institutional systems. This approach demands synergy and an M&E component (Eckman et al., 2000). The intention of this document is to demonstrate the usefulness of M&E, to motivate those in charge of sustainable development projects to develop their own instruments of M&E.

MONITORING AND EVALUATION OF PROJECT ACHIEVEMENTS

Definitions

Monitoring is the systematic measurement, throughout the life of a project, of a set of indicators that can be verified objectively. These indicators are related to the area's problems, project goals, efficiency, and impacts. In addition, they can be associated with the processes that happen in the "context" of the area. Evaluation is the cross-examination, analysis and interpretation of the results measured in the monitoring. It contrasts the measurements with the prognostic values that were hoped for during the accomplishment of the project. Evaluation is a mean of showing the merits of the activities implemented and providing feedback on the project phases that have already been implemented so that they may be improved upon in future projects. Monitoring and evaluation are two sequentially interrelated actions (Hernandez, 1995). Evaluation aims at the analysis of the factors that have facilitated the attainment of the objectives and the limitations that have prevented the desirable results, with the respective recommendations of adjustments. It is an activity that generates information to help stakeholders to take actions (Rivera and Herrera, 1998)

Evolution

Like any scientific or technological approach, M&E is the result of the continuous search for improvement by means of accumulation of knowledge and experiences. It has been an evolution tied with the progress of the theory of development. Rivera and Herrera (1998) claim that M&E appeared for the first time in United Nations projects during the

1950s and soon extended to financial and cooperative organizations and governments.

The preoccupation was to improve the efficiency of the projects. Initially, the interest was only to control expenses and to measure the amount and quality of physical goals, ensuring better use of the invested resources (Rivera and Herrera, 1998). These evaluations were made at the end of the project (terminal evaluation) or some years later (ex-post evaluation). For these reasons, there was no possibility of correcting deviations or faults that appeared during the operation of the project. This deficiency was later corrected by conducting M&E during the life of the project, that is, continuous M&E. Still, the main focus has been on expenses and physical goals. This type of evaluation neglected to analyze the processes by means of which those benefits were generated, and the elements internal and external to the project area. For such reasons, "M&E of processes" was created in the 1990s. It has been an interesting, complex proposal. Its use has not yet been extended (Rivera and Herrera, 1998).

The indicators were at first designed, measured, and processed only by technical personnel, without the participation of the affected population. To correct this unilateral vision, participatory M&E, which supports the organized intervention of the beneficiaries and social actors, was put into practice (Davis-Case, 1989). Currently, we have observed another approach of M&E that includes the surroundings as a whole and the evaluation of external elements: political, social, economic and environmental. Such elements often determine the performance of the project, so they must be monitored and be evaluated with an "integral systemic approach".

In all this evolutionary development, international organizations have played a very important role, for example, the United Nations, UNICEF, FAO, and financial organizations such as the World Bank and the Inter-American Development Bank. The M&E component was increasingly enforced in public projects, especially those financed by international banks or donors (IADB, 1997).

Justification of M&E

Monitoring and evaluation is highly recommended for the effective development of projects. When it is properly used, the beneficiaries, operative personnel and managers can take informed decisions (Red latinoamericana de cooperacion tecnica en manejo de cuencas hidrograficas, 2003). It is useful to see M&E from 18 different dimensions (IADB, 1997):

At the beginning of the project:

- It induces efforts to gather data to establish the baseline data on watersheds and communities' problems in a quantitative form.
- It helps to stimulate participation.

During implementation:

- It enforces reflection on the project as it is implemented.
- It ensures use of resources as planned.
- It offers effectiveness, efficiency and transparency to the project.
- It allows quantitative expression of the initial effects of actions and treatments.
- It provides early indication of the appearance of problems.
- It facilitates early adjustments to correct deviations.
- The people affected by the project participate in and understand better the project and its context.
- It improves their participation in the execution.
- It helps protect the investment process.

At the end:

- It allows quantitative expression of the impacts, which are products of actions and treatments.
- It forces comparison of results with the baseline.
- It facilitates the search for complementary financial resources based on proven results.
- It produces data to improve designs and strategies taking into account learned experience and lessons.
- It provides data on the effectiveness of the original designs.
- It fulfills the accountability requirements of donors or national or international banks.
- It facilitates the spreading of good practices.

Mistakes to Be Avoided

Several mistakes should be avoided (Hernandez, 1995), such as gathering too much data and information that is of little help in the project management. Sometimes, there is too much interest in obtaining information related more to implementation than to impact, or greater interest in statistical analysis and less in accurately quantifying changes. Sometimes the monitoring is designed in a complex manner like a scientific experiment with high investment. Frequently, the stakeholders do not participate with enthusiasm. Too much importance can be given to the design phase and too little to implementation. Also, the results of the evaluation have not been used properly.

Design of an M&E Achievements System

On the basis of experiences reported by different authors (Davis-Case, 1989; Sthapit and Shrestha, 1989; Hernandez, 1995; Big Thompson Watershed Monitoring Program, 2000; Herweg and Steiner, 2002), a method is proposed to facilitate the design of activities to measure indicators, as well as to evaluate the effects and impacts generated by a project of sustainable development in a community or in a watershed.

Conditions

The proposed method must fulfill several conditions:

- Very clear objectives of the M&E system.
- Real conviction by the directive personnel of the utility of an M&E system.
- Incorporation of M&E in all phases of the project, especially in its initial period.
- Capability of the operative personnel to design and to apply a suitable M&E method.
- Encouragement of beneficiaries to participate in the design and application of the method.

A solid plan for M&E must be completed in the design phase of a project. The process of planning for M&E involves detailed scrutiny of each stage of a project - design, start-up, implementation, completion - and a detailed analysis of M&E requirements at each of those stages (IADB, 1997).

During the design phase, the primary M&E concern is to help design a project that can be evaluated later. Throughout the implementation of a project, mechanisms are needed to track its performance — its ongoing relevance, effectiveness, and efficiency. At the end of a project, the M&E function is responsible for measuring and assessing a project's results and impacts, and contributing lessons learned for the future (IADB, 1997).

The Method

The method proposed to design an M&E system uses nine steps, three of which belong to the traditional cycle of project formulation (Table 1).

Table 1 Steps to design an M&E system (modified from Herweg and Steiner, 2002).

Steps	*Questions to be answered*
1. Problem analysis by participatory assessment	What are the most important problems according to the people?
2. Characterization of the area (sound understanding of the project context, its elements and their interrelations)	What are the most important elements of the project context? What are other important problems identified from the external point of view? How are the elements interlinked? What role do they play in the context? Is the context moving towards or away from sustainability?
3. Selection of problem indicators,establishment of the baseline	What set of indicators describes the problems? What are the initial values of these indicators?
4. Goals of the project, proposed actions and treatments	What will be the goals of the project? What will be the proper actions and treatments to solve the problems? What actions can a project give towards more sustainable development?
5. Predicting short-, medium- and long-term impacts, selection of impact indicators	What positive or negative impacts will occur in association with the proposed actions and treatments? What set of indicators will show whether changes help to achieve the goals? What are the initial values of these indicators? How can the number of indicators be reduced to a minimum?
6. Design of a measurement plan	Where, when, how often should the proposed indicators be measured? Which methods are applicable within the means and capacities of the project?
7. Measurement and documentation of change, data processing	How can the impact indicators and the context be monitored and documented? How can impact assessment be prepared?
8. Impact assessment, interpretation of changes in the context	How did the indicators change? What impacts did appear? How did the context change from the stakeholders' point of view? What is the link between the impacts and the goals of the project? Do the lessons learned indicate that the project has stimulated important social processes? What proposed action or treatment should be strengthened or adjusted? How should it be done?
9. Information dissemination	How will information be presented? Who should get the report? How often?

A SAMPLE APPLICATION

In 1996, a work group of the University of the Andes was requested by Mérida State Government to design a project to promote the sustainable development of the community of San Jose de Limones, and to reduce the deforestation in a mountain forest in the western region of Venezuela (UFORGA, 1998). We applied the planning method outlined in Table 1, giving attention to M&E of achievements.

1. Problem Analysis by Participatory Assessment

We identified the problems based on the results of semi-structured interviews, farmer's records, field transects, and group meetings with active participation of the entire community (Table 2). We looked for consensus, understanding of the problem, and exchange of views among the stakeholders.

The stakeholders tended to place greater weight on problems related to marketing, agricultural production (in this case banana and coffee), housing quality, and lack of public services. They did not point out ecological or organizational issues.

2. Characterization of the Context of the Area, Their Elements and Interrelations

The assessment was complemented with specialized studies of soil, topography, and physiopathology. These studies led to the description of other problems and an understanding of the functioning of the area in the

Table 2 Problems identified by participatory assessment in San Jose de Limones, Mérida State (UFORGA, 1998).

Problems or causes	*Priority*
Deficient marketing	1
Far-away schools	2
Far-away dispensaries	2
Absence of loans	2
Lack of electricity	2
Lack of technical assistance	2
High incidence of plagues	2
Absence of training for housewives	2
Bad conditions of houses	2
Deficient roads	3
Problems in water distribution	3
Lack of property titles	3

Table 3 Problems identified by specialists in San Jose de Limones, Mérida State (UFORGA, 1998).

Other problems and causes
Decreasing soil fertility
Erosion
Deforestation
Intense hunting of *Agoutis lapa*
Decreasing crop productivity
Low income
No non-agricultural income
Absence of organizations
Lack of leadership
Low education level
Social conflicts
Conflict with National Park authorities

wider framework of a mountain ecosystem. The detected additional problems, along with their representative elements, are presented in Table 3.

These elements, along with those of the participatory assessment, allowed us to construct a tree of cause-effect relations (Fig. 1). This graph clarifies the context of the project and the interrelations between causes and consequences. We identified low income and low quality of life as the main basic problems.

The main negative impacts of this context are an unmotivated population and deforestation in search of more land to cultivate, with the consequent alteration of the hydrologic regime, loss of biological diversity, and indiscriminate hunting of *Agoutis lapa*. This exhaustion contributed to the emigration of young people towards the cities.

This system was moving away from sustainability. The main causes (Fig. 1) were divided into two groups: the lack of suitable technical attendance and the absence of organization, leadership, and participation.

3. Selection of Problems and Indicators and Establishment of the Baseline

The design involves identifying the variables related to the problems at the initial stage of the project. They must be objectively verifiable and directly linked to the problems. The selected indicators to represent the problems and their initial value are expressed in Table 4.

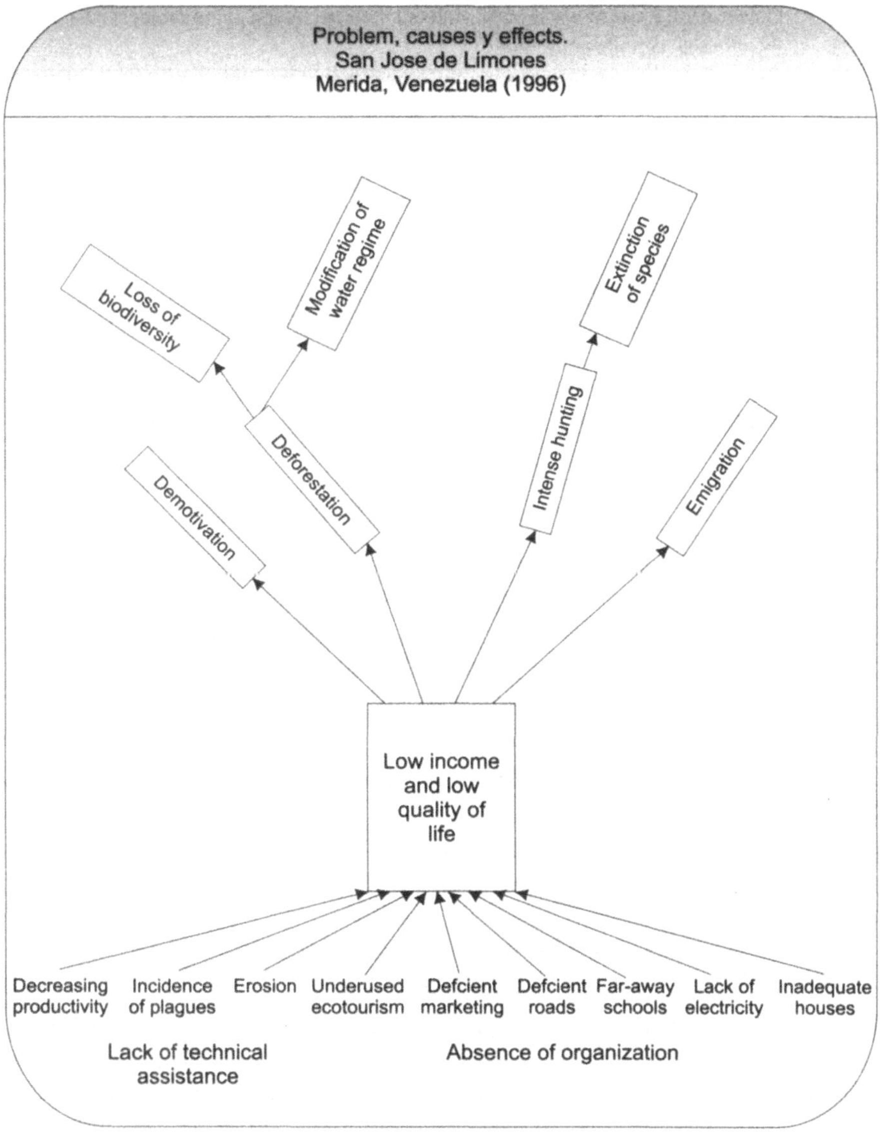

Fig. 1 Problems, causes and effects, San Jose de Limones, Merida, Venezuela (1996).

4. Goals of the Project, Proposed Actions and Treatments

The main objective usually stems directly from the need to solve one or several problems or to take advantage of concrete development opportunities and should be clearly stated in the project. In this project, the main objective was to establish a model of sustainable development

Table 4 Indicators of the problems. Initial values at the beginning of the project in San Jose de Limones (UFORGA, 1998).

Indicators	*Initial value (1996)*
Crop price at farm/price at market	0.3
School distance	3 km
Dispensary distance	4 km
Farm loans	0
Houses with electricity	0
Farms attended by extension service	0
Farms affected by plagues	0
Housewives training	0
Bad houses	15
Roads deteriorated	5 km
Houses with treated water	0
Land titles	0

conducted by the beneficiaries themselves in order to improve their income and quality of life and to halt deforestation. The activities proposed to achieve the stated objectives were the following:

- Establishment of a Sustainable Development Association to solve problems related to water, electricity, school, roads, loans, etc.
- Education and leadership building
- Assistance services linked to crop production system, marketing, and habitat
- Land tenure and ownership titles
- Headwaters forest protection and ecotourism

5. Predicting Short-, Medium- and Long-term Impacts and Selection of Impact Indicators

The indicators have to be related to the estimated impacts produced by the actions. They should measure progress toward the benefits and enable us to quantify the achievements. One set of benefits will be related to solving the problems and another set will be connected with the development of new potentialities. The indicators selected in our case are shown in Table 5.

6. Design of a Measurement Plan

The design of a measurement plan seeks to answer seven questions:

Where do we monitor the indicators?

Which methods do we use?

Table 5 Predicted values of impact indicators. Project San Jose de Limones. Mérida, Venezuela (UFORGA, 1998).

Indicators	*Initial value (1996)*	*Future values (estimated, 2002)*
Problems solved by the association	0	12
Distance to school, km	8	1
Farms attended by extension service, % total area	0	100
Banana productivity, kg/ha/yr	41,000	50,000
Coffee productivity Qq/ha/yr	10	15
Houses with electricity	0	12
Houses improved	0	12
Roads improved, km	1	5
Net family income, % avg increase	-	30
Crop price at farm/price at market	0.3	0.8
Forest protection area, % of headwaters	60	100
Land titles, % of families	16	100

What kind of precision do we require?
How often do we monitor?
For how long?
Who monitors?
How much will the system cost?

7. Measurement and Documentation of Change

The outcomes of the data processing are indicated in Table 6.

Table 6 Values of impact indicators. Project San Jose de Limones, Mérida, Venezuela (UFORGA, 1998).

Indicator	*Future values (estimated, 2002)*	*Actual values (2002)*
Problems solved by the association	12	8
Distance to school, km	1	0.1
Farms attended by extension service, % total area	100	30
Banana productivity, kg/ha/yr	50,000	38,000
Coffee productivity Qq/ha/yr	15	8
Houses with electricity	12	18
Houses improved	12	10
Roads improved, km	5	4
Net family income, % avg increase	30	10
Crop price at farm/price at market	0.8	0.55
Forest protection area, % of headwaters	100	75
Land titles, % of families	100	16

8. Impact Assessment and Interpretation of Changes in the Context

The data measured during the monitoring after six years was analyzed. It is shown in Table 7 according to the form proposed by Herweg and Steiner (2002). The table shows values for each indicator classified in five categories (very good, good, moderate, bad and very bad). With this table, it is possible to change the absolute units of the values to relative ones. This change allows comparison of the performance of all indicators.

Impact assessment consists of elaborating a table with values for each indicator classified in five categories (very good, good, moderate, bad and very bad). With this table, it is possible to change the absolute units of the values to relative ones. This change allows comparison of the performance of all indicators.

It is observed in Fig. 2 that after six years the M&E system shows important impacts and benefits in habitat, organization and roadways. It is good to point out that the deforestation process has stopped. Nevertheless, crop yields show worrisome performance because of

Table 7 Classification of the indicators. Project San Jose de Limones, Mérida, Venezuela.

Indicator	5 *Very good*	4 *Good*	3 *Moderate*	2 *Bad*	1 *Very bad*
Problems solved by the association	> 11	10-7	6-4	1-3	0
Distance to school, km	< 0.5	0.5-1	1.1-2	2.1-3	> 3
Farm attended by the extension service, % total area	> 60	40.1-60	20.1-40	10-20	< 10
Banana productivity, % avg increase	> 20	10.1-20	1-10	stagnating	decreasing
Coffee productivity, % avg increase	> 20	10.1-20	1-10	stagnating	decreasing
Houses with electricity, %	> 80	80-60.1	60-40.1	40-20	< 20
Houses improved, %	> 96	80.1-96	50.1-80	10-50	< 10
Roads improved, %	> 96	40.1-96	10.1-40	1-10	< 1
Net family income, % avg increase	> 20	10.1-20	1-10	stagnating	decreasing
Crop price at farm/price at market	> 0.9	0.9-0.71	0.7-0.51	0.5-0.3	< 0.3
Forest protection area, % of headwaters	> 96	80.1-96	50.1-80	10-50	< 10
Land titles, % of families	> 96	80.1-96	50.1-80	10-50	< 10

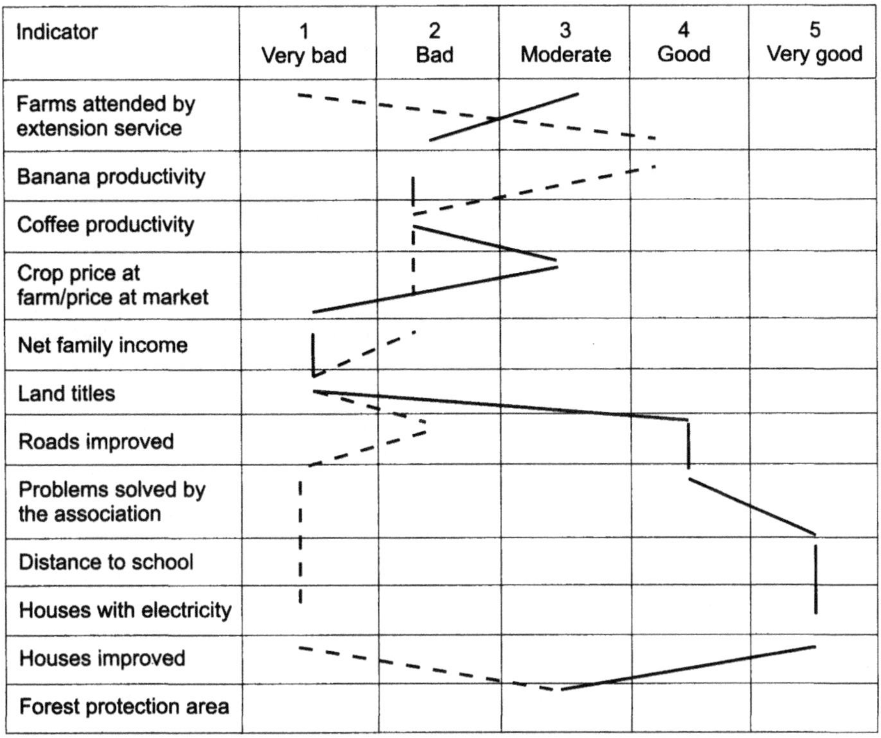

Fig. 2 Impact profile after six years of the project in San Jose de Limones, Mérida, Venezuela.

continuous plagues and difficulties in applying the fertilization program. The previous conclusions point to a readjustment of agricultural extension activities in the project. The difficult economic situation of the region affects the development of the project in the economic component.

9. Information Dissemination

The present report of evaluation must be analyzed broadly with the settlers and the authorities at local, regional and national level so that readjustments can be made in the areas in which the indicators were below the expected values.

CONCLUDING REMARKS

It is hoped that with the discussion of the previous example, and with the theoretical and methodological framework presented in this book, project directors and technical personnel, as well as stakeholders, will incorporate the component of M&E of achievements in projects of sustainable development.

References

Big Thompson Watershed Monitoring Program. 2000. Big Thompson Watershed Monitoring Program. Fort Collins, Colorado, USA.

Brooks, K. and K. Eckman. 2000. Global perspective of watershed management. Conference proceedings: Land Stewardship in the 21st Century – The Contributions of Watershed Management. Tucson, Arizona, USA.

Davis-Case, D. 1989. Community forestry. Participatory assessment, monitoring and evaluation. Community Forestry Note 2. FAO, Rome, Italy.

Eckman, K., H. Gregersen and A. Lundgren. 2000. Watershed management and sustainable development: Lesson learned and future directions. Conference proceedings: Land Stewardship in the 21st Century – The Contributions of Watershed Management. Tucson, Arizona, USA.

Hernandez, E. 1995. Monitoring and evaluation of watershed management project achievements. Conservation Guide 24. FAO, Rome, Italy.

Herweg, K. 1998. Sustainable land management for guidelines impact monitoring. [online] http://srdis.ciesin.columbia.edu/pdf/slm.pdf

Herweg, K. and K. Steiner. 2002. Impact Monitoring and Assessment. Vol. 1: Procedure. Vol. 2: Toolbox. Berne, Switzerland.

Inter-American Bank of Development (IADB). 1997. Planning for Monitoring and Evaluation. [online] http://www.iadb.org/cont/evo/EngBook/anexii.htm

Red latinoamericana de cooperación técnica en manejo de cuencas hidrográficas. 2003. Tercer Congreso Latinoamericano de manejo de cuencas. INRENA, FAO. Arequipa, Peru.

Rivera, R. and H. Herrera. 1998. Manual de seguimiento y evaluación de proyectos de desarrollo rural. CIARA, FIDA, CAF. Caracas, Venezuela.

Sthapit, K.M. and J. Shrestha. 1989. Conceptual framework for monitoring and evaluation system for soil conservation and watershed management activities in Nepal. Department of Soil Conservation and Watershed Management, Government of Nepal, FAO. Kathmandu, Nepal.

UFORGA. 1998. Aldea Ecológica San José de Limones. Municipio Andrés Bello. Estado Mérida. Universidad de los Andes, Consejo de publicaciones. Mérida, Venezuela.

World Commission on Environment and Development (WCED). 1987. What Is Sustainable Development? [online] http://www.dep.org.uk/cities/WhatisSD1.htm

7

Using a Watershed Focus for Monitoring and Evaluation of SWC in Nicaragua

Robert Walle[1] and Sarah W. Workman[2]

[1]Senior Water Resource Specialist, CPESC 2134, P.O. Box 93, Tegucigalpa Honduras, C.A. (Zamorano-DSEA Social and Environmental Development). E-mail: robertow@ibw.com.ni

[2]312 Hoke Smith Bldg., University of Georgia, Athens, Georgia 30602-4356, USA. E-mail: sworkman@uga.edu

ABSTRACT

In response to Hurricane Mitch, which occurred in 1998, donors financed reconstruction initiatives in Central America. With descriptions of the ProCuencas project watershed approach, we present the methods by which the active monitoring and evaluation components evolved during agricultural reactivation, watershed restoration, and conservation management activities in Nicaragua. Technical training in methods and the subsequent installation of soil and water conservation practices with farmers were primary objectives. A vision of deliverable products guided the set-up of a monitoring plan before project activities began. After critical watersheds were selected, participatory planning events and simple practice standards guided conservation interventions, while GIS databases simplified monitoring and planning connectivity on a landscape scale. Continuous monitoring of project activities allowed incremental evaluation and feedback on intermediate results to respond to the needs of the affected populations and integrate activities into a watershed program to meet strategic objectives. Evaluation of the larger watershed recovery during Mitch Reconstruction showed that targeted reforestation along riparian areas was both the easiest to accomplish and the most beneficial to people and the environment. It was necessary to have an impact on the water resources being used by the beneficiaries as well as attend to stream corridor restoration. Considering all the indicators together integrated the restoration effort and reinforced the potential for lasting impact. Project results emphasize how community participation in soil and water conservation can be increased and how quality of conservation impacts thus improves.

INTRODUCTION

Interest in soil and water conservation (SWC), its use in watershed management, and agricultural restoration became essential after Hurricane Mitch hit Central America in late October 1998. Responding to this disaster, bilateral aid from the United States financed reconstruction initiatives in Central America. The underlying principles of reconstruction were to foster transparency, accountability and community participation. One of the number of governmental and non-governmental institutions called upon for assistance, the Pan-American Agricultural College, Zamorano, undertook agricultural reactivation, watershed restoration, and conservation management in Nicaragua. Technical training in methods and the subsequent installation of SWC practices with farmers were primary objectives in the results framework.

Since SWC was part of its training activities in the field, Zamorano already had experience in implementing conservation practices proven effective in controlling erosion and sedimentation, slowing runoff, and promoting infiltration under local conditions in Central America. With an established regional presence, Zamorano was prepared to rapidly and successfully apply the practices in this special situation. Continuous monitoring of project activities allowed incremental evaluation and feedback on intermediate results to respond to the needs of the affected populations and, rather than implement individual or isolated actions across a landscape, to integrate activities into a watershed program to meet strategic objectives.

Data from monitoring activities informed colleagues and donor coordinators about success and progress of field activities and opportunities for synergistic partnership actions. Donor representatives monitored monthly field reports and expenditures, then participated directly with field staff to evaluate quarterly results. Monitoring and evaluation were of utmost concern to both implementer and donor to ensure financial accountability, progress toward project goals, and application of lessons learned in previous watershed management and SWC projects in the region.

PROJECT OBJECTIVES AND ACHIEVEMENT INDICATORS

The ProCuencas team at Zamorano reviewed the final products to be delivered to the donor before determining the plan for monitoring progress. The vision of deliverable products guided set-up of a monitoring plan before project activities began and allowed for considered revision after each quarterly report was submitted. Donor involvement identified how the project would be evaluated and permitted Zamorano to set up its in-field monitoring and financial systems to address indicators and reporting needs.

Based upon the contract with the donor to mitigate vulnerability to flooding and other disasters and also restore economic livelihood of farmers and micro-entrepreneurs in areas affected by Hurricane Mitch, the project objectives were the following:

- Promote adoption of environmentally sustainable agriculture practices in hurricane-affected areas, measured in number of benefiting farm families;
- Reclaim or rehabilitate agricultural lands, including community water resources, measured in number of benefiting farm families;
- Stabilize priority watersheds with SWC practices, measured in surface area covered and length of water courses;
- Mitigation of damage to productive infrastructure within the watershed, measured in number of hectares.

The indicators measured had predicted goal levels and, although estimates of quality were not required, many became evident during the two-year limit set for completion of activities and expenditures. Information about numbers of participants reflected gender and household composition, while adoption data was used for internal programming and decision-making to guide activity site selection. Local partners obtained additional technical assistance from US government agencies that helped fill key gaps in personnel capabilities during the project time frame.

WATERSHED APPROACH

A problem with many SWC efforts later repackaged as watershed projects is that individual and dispersed actions in affected areas do not mitigate further damage from normal rainfall events (Perez and Tschinkel, 2003), much less damage suffered during a hurricane. This is primarily because they are agronomically instead of hydrologically oriented. A priority watershed focus allowed the ProCuencas project to connect areas where farm families adopted sustainable agricultural practices or SWC techniques, households benefiting from reclamation and restoration of agricultural lands, or communities benefiting from repair of farm-to-market roads. Considering all the indicators together integrated the restoration effort and reinforced the potential for lasting impact.

Because there were diverse objectives within areas stabilized by improved practices, as well as agricultural activities, viewing intermediate results in the watershed context facilitated monitoring and increased connections between different project activities. Monitoring the activities within each watershed made it possible to track impact on a wider scale and plan for connectivity both within the watershed and between watersheds within the larger landscape. Figure 1 represents project components and how the ProCuencas project operated.

PROGRESSION OF PROJECT

Realistically, it was not possible to attend to all of the farms in a typical watershed given the extensive hurricane damage, negative effects of previous deforestation, and erosion from many hillside farms. To ensure project success, the following critical areas of the affected watersheds were identified to better use the available time and resources:

- Areas in need of disaster mitigation to protect infrastructure such as power lines and roads from further damage;
- Farmland under annual crops on highly erodible soils near surface water;
- Water sources, riparian areas, and uplands where continuity of conservation work was possible.

Recently repaired culverts and certain susceptible road cuts requiring stabilization were identified by technicians, prioritized in project plans, and targeted for planting activities. In addition, since most eroded soil is deposited down slope and never reaches surface water, and it was not possible to reforest a sufficient area of the watersheds to effect a favorable change, much of the project work concentrated on farmland surrounding critical areas.

Applying SWC Practices On-farm

From previous experience in the project area, the project manager had in hand technically successful SWC practices that were acceptable to farmers. What was needed were specific instructions to field technicians on practice installation, materials necessary, proper layout, and how to document the location. On-site training also included how to observe and document effects and work with farmers on how to monitor and maintain areas during the life of the effort with a vision for longevity.

The simple practice standards available included the following:

- Appropriate selection of plants adapted to location or need;
- Correct contour determination and design criteria; and
- Location and layout for maximum effect.

Monitoring involved verifying that the SWC practices conformed to the simple practice standard. Each technician had a checklist form to include for each site in the bi-weekly work report. The project maintained a photo library for project sites that resulted in an extensive digital collection to demonstrate progression of activities.

Plant Selection (Vetiver, Trees, Combinations)

Plant species that would be appropriate were well known in the project area. The need for rapid establishment of live barriers that would be effective in reducing erosion, retaining soil, slowing runoff, and promoting infiltration led to the use of vetiver grass, *Vetiveria zizanioides*, a proven practice in Central America. Technicians had access to vetiver propagated in previous efforts, but due to the increased farmer demand for the conservation technique, the donor endorsed purchase of additional vetiver with plant material funds. *Pennisetum* species were used under certain conditions where tillage could control expansion of the barrier into cropland or where fodder was needed.

The project established and managed tree seedling nurseries in each of the priority watersheds. Species selection focused on trees native to the watershed region and adapted to the conditions of the planting habitat, though limited numbers of seed sources of some species made them less available (e.g. native fruits). Technical assistance from US Department of Agriculture scientists greatly enhanced the forestry sector of the reconstruction program (Workman, 2001). Improved nursery techniques, seed collection, compost and soil mixes, and selection for site-species compatibility all increased species diversity, seedling vigor, and plant survival. The predominant tree species (scientific and local common name) used for specific habitats are listed in Table 1.

Within the five national Departments (Chinandega, Leon, Matagalpa, Estelí, Nueva Segovia) of the project watersheds, in 106 communities during the first year of the project, there were approximately 1,400 nurseries producing from 44,000 to 575,000 tree seedlings of up to 25

Table 1 Species selected for the forestry sector of the reconstruction program.

Stream corridor restoration	
Salix humboldtiana, Sauce de rio	*Andira inermis*, Almendro de rio
Pithecellobium saman, Genízaro	*Enterolobium cyclocarpum*, Guanacaste
Hymenaea courbaril, Guapinol	*Albizia guachapele*, Gavilán
Upland plantings	
Swietenia humilis, Caoba del Pacífico	*Cedrela odorata*, Cedro real
Tabebuia rosea, Roble	*Cordia alliadora*, Laurel
Rapid growth	
Caesalpinia velutina, Mandagual	*Gliricidia sepium*, Madero negro
Restoration and cultural beauty	
Calycophyllum candidissimum, Madroño	*Guaiacum sanctum*, Guayacán
	Various indigenous fruit trees

species. The second year, restoration activities used a greater diversity of species and over five times the number of seedlings produced in the previous season.

Correct Contour and Design

Contour marking was a joint activity with farmers and technicians actively participating on farmers' fields. Sharp turns in the barriers that made tillage difficult and hindered adoption of the recommended practices were corrected to improve the contour. The correction of the contour was important to assure ease of tillage operations, whether by animal draft or machinery. Monitoring activities checked contour level layout and design and reported any adjustments made to prevent pooling or rivulets.

Location and Layout of Practices On-farm for Maximum Effect

While complete protection of all cultivated hillsides was desirable, it was necessary to first protect surface water from contamination and stabilize critical areas. Farmland closest to the rivers was of greatest priority. Vegetative practices were located where their effect would be maximized and locations nearest surface waters were prioritized. Precise farm maps were made with simple hand-held GPS receivers once initial conservation practices were completed. The precise maps served as basic templates for participatory planning events and helped farmers within the watershed create a spatially accurate representation of their work. Entering locations into GIS databases simplified monitoring and maps could then be joined and represented at a larger scale for monitoring and planning connectivity of landscape activities.

Off-farm Connectivity

While farm families were primary beneficiaries of project activities, areas outside individual farms were also key to achieving larger watershed goals. Where off-farm areas formed part of restoration works, their inclusion and connections with on-farm SWC completed a larger watershed picture. Areas along rivers and near community water supplies were targeted for this effort. While conserving soil and water on farms is the cornerstone of many projects, it was necessary to have an impact on the water resources being used by the beneficiaries and to attend to stream corridor restoration.

Riparian areas devoid of vegetation, whether from sedimentation, streambank erosion, or the general environmental degradation common in Central America, were considered priorities for project activities in Nicaragua. The two site selection criteria used for restoration were width and connectivity. *Width* determined if there was sufficient area for the riparian vegetation to promote sediment deposition and to protect critical areas from further damage. Generally, this was considered to be 30 m to comply with current Nicaraguan law. *Connectivity* was the key to improving larger areas of the watershed. Multiple and strategically located plantings provided a greater landscape effect than single plantings on a large scale. Periodic plantings over time were less labour intensive and not only gave greater resilience, through better survival from diverse planting dates, but also allowed for design modifications in areas of break-through erosion (rills) or problem spots.

Uncontrolled grazing on public areas, much along the lines of the tragedy of the commons, was and remains the most significant threat to establishment of vegetation. The main activity was finding areas lacking vegetation and then connecting them to areas of still acceptable cover. Instead of a hundred or more separate areas where SWC work had been accomplished, connecting the natural features of the watershed with the on-farm conservation work formed larger areas of effective protection.

MONITORING AND INTERNAL EVALUATION

Record Keeping in the Field

Since the project was weighted toward agricultural activities in the beginning, technicians had to record what SWC practice was applied and where the practice was located. This included names and basic family structure on the farms (age, gender, number of children). A geographic reference point and a brief description of the work done were included as well. Taking at least one GPS coordinate, in the case of smaller interventions, or multiple points to define the perimeter along the borders of river buffers and of larger areas where the intervention was reported in terms of area, also marked the location of the work. When unified in a GIS-based inventory these individual points showed areas of greater concentration of project activity and helped guide further actions in the watersheds.

Individual field technicians were required to keep a record of their daily activities including what practice was applied, where on the farm, and what materials were used. Basic record keeping of vehicle mileage and per diem expenses were reported for technicians for reimbursement of expenses; these details were required at the time they received their pay, so that information was kept up to date. Field technicians estimated that 20% of their time was devoted to information gathering and documentation for the monitoring efforts.

Documenting Land Use Changes

It was determined that large reforestation efforts and plantings would be replaced by specific plantings in critical areas of the stream corridors in the priority watersheds. Monitoring of these plantings was done by frequent visits along the river sections and reporting on progress. Using field maps based on the GPS points from individual technicians' records helped certify areas as completed, find new areas for planting, and connect practices in the watershed.

To achieve a greater watershed focus, links with existing patches of riparian forest were established. Vulnerable land under annual cropping was changed to perennial cropping according to farmer preference. Instead of a general promotion of perennial crops, an effort was made to target critical areas under annual cropping and replace them with perennial crops. Initially this included grafted fruit seedlings, improved market varieties of coffee, and cashew. Decline in coffee prices in 2000 led to the limitation of areas to 0.7 ha of coffee per farmer, and an intensification of efforts on other crops. Where annual cropping existed on sloping parcels near surface water, enhancing the productive capacity of the buffers as well as the protective capacity with perennial species supported family livelihoods as well as conservation goals (Workman et al., 2002). Perennial crops such as cashew and grafted fruit trees (mango, avocado and citrus) in the form of riparian buffers were prioritized over general plantings with farmers for integrated watershed management plans.

Monitoring Effectiveness of Conservation Practices with Farmers

Farmers were trained in the beneficial aspects of installing soil conservation practices. The steep slopes and friable soils of hillside farms in Nicaragua are susceptible to erosion. Installing live barriers, particularly of vetiver grass, allowed rapid establishment of barriers that quickly accumulate eroded soil from upper slope positions. In the rainy season following establishment, farmers spotted the retained soil easily and, comparing it with the upper slope positions, saw the benefits of soil conservation. Control of erosion between barriers was evident by significantly less rill erosion. Soil accumulation and increased fertility gradients measured elsewhere in Central America (Walle and Sims, 1998, 1999) became evident to farmers comparing properties of the soil retained with soil where recent erosion had taken place. These observations were demonstrated successfully in repeated farmer-to-farmer training events and adoption among peers and neighbors promoted consolidation of erosion control practices.

During group training events for watershed restoration and participation in conservation activities, community participants expressed a strong desire for increased project focus on water resources. Participants in watershed training events specified that what was most important to them was the water they drank, so SWC issues for these water sources had to be addressed.

Additionally, during an internal evaluation conducted midway in the project, participants expressed their desire for more work specifically on potable water. Project repair of damage to exposed pipes and cleaning out wells that had been flooded reestablished water systems damaged during Hurricane Mitch. At the time there were, and there still are, longstanding problems associated with water in Nicaragua. Many wells were merely holes in the ground and lacked basic sanitary protection. Contamination from entrance of runoff water remains common. Greater community benefit can be obtained from aqueduct construction and shortening the distance women and children go to fetch water.

There were noticeably different perceptions regarding the application of watershed management as presented to the participants. The project required increased watershed protection in specific areas. Most men were interested in agricultural applications and women were primarily interested in water resources. To provide incentives for off-farm and away-from-home watershed and conservation work, vegetative materials (budwood for fruit tree grafts and vetiver grass) were contributed for off-farm participation (mostly to men). Incentives of well restoration, water storage, and construction of spring boxes or well casings for sanitation were welcome in all communities (mostly women-driven). Monitoring showed that people were drinking irrigation water because it was of a higher quality than the water provided for drinking.

Barbed wire was a frequent item requested for the rehabilitation activities. Examination showed a tendency for it to be requested by ranchers for use on their farms. It was determined that it would be easier to securely fence in water sources for protection and let the community take care of the larger problem of free-roaming cattle. This proved a more cost-effective use of project resources to achieve the goal of water source protection. By the end of the ProCuencas project, water resources became the most important focus of the watershed restoration project.

Evaluation and Use of the Results from Monitoring

Feedback through evaluation of technicians and personnel moved the project through incremental completion of the objectives in the watersheds. Due to the nature of the contract, individual technicians had to accurately record their work to record how the project was meeting its contracted goals. Simply counting the number of trees planted, area rehabilitated, and households involved could not give an accurate picture of what was needed for continued benefits from the project efforts. Monitoring not only the progress toward certain goals but also other socially important factors helped the project adapt its efforts to serve the people affected by Hurricane Mitch.

The results of the field reports allowed management to evaluate technical progress in the watershed, report progress, and plan for expansion of successful activities and modifications in the project. Based on results from the mid-term monitoring and internal evaluation and, most importantly, beneficiary suggestions, an increased focus was placed on rural community potable water systems. This included water source protection, improvements in distribution lines, basic maintenance, operation, and sanitation. The process also allowed for adjustment of rewards and incentives for technicians based on performance and evaluations of beneficiaries.

Connecting the conservation practices of different technicians was needed to cover larger stream corridor and watershed features for increased impact. It was not only the number of things done by individual technicians that made up the watershed focus, but also how well they were connected together throughout the watershed. Monitoring individual technicians' progress and achievements allowed an overview of how activities could be brought together. For example, monitoring results showed that two technicians in the same sub-watershed were both completing goals, but their work areas had begun to diverge. To achieve a larger section of the stream corridor with an adequate level of protection, a management decision that both should work on the same stream corridor simply required a bit of work plan modification and adjustment in their work zones.

When the actual work done by individual technicians was represented on the GIS, it was easy to see where connections could be made. The use of a GIS to represent the overall project was a popular tool with technicians and helped evaluators review project activities. We used GIS to correlate and present field data in map formats and database summaries to evaluators during their quarterly visits.

Donor-assigned personnel, for both technical and financial aspects of the project, conducted external evaluations throughout the Hurricane Mitch Reconstruction Project each calendar quarter. During the Hurricane Mitch restoration project in Nicaragua, monitoring was done at the main program office level (administrative backstopping unit), project level (cost-effectiveness of investment, outcomes, community impacts), and field office level (accountability, efficiency, quality of technical implementation).

EXTERNAL TECHNICAL AND FINANCIAL EVALUATION

The donor was determined to avoid all forms of corruption to ensure efficient use of resources and maximum benefit to the recipients of USAID funds. Within the donor institution a cognizant technical officer maintained close communication with the project administration and was the point of contact between the donor and the project manager. The position required both technical knowledge and familiarity with donor rules and regulations. The officer coordinated field inspections and supervised field visits by auditors of the General Accounting Office. All monitoring results from project activities were available to the officer so that he could visit sites anytime or request site visits with project technicians and management staff.

Both the General Accounting Office and Deloitte Touche accounting consultants audited financial transactions and account management. Individual items in the financial records were examined and tracked, from purchase to field implementation. Audit personnel visited project sites they chose to verify use of materials and community counterpart contributions to the efforts. An example of a field visit would be to a community water project to see the materials (e.g. cement, PVC pipe, faucets) in the field and to interview beneficiaries about the activity. These procedures brought together individual record-keeping items with their beneficiaries in the field situation. This also gave the evaluation team members a realistic view of what was behind the numbers and reinforced the fact that local people were the fundamental actors and beneficiaries, helping to modify initial plans to include community water resources in the SWC activities.

Natural resource conservation specialists from the USDA evaluated the appropriateness of some practices, the location in the watershed, and the functionality once installed. Suitability and function of the structures were primary concerns for watershed and conservation work. Where watershed restoration works were done, they were supervised through a small grants process that approved work before intervention and required technical monitoring and later evaluation to ensure efficacy. For greater community participation, vegetative conservation practices or "bio-engineering" techniques were employed where technically possible and, when necessary, in conjunction with larger constructed works such as Bendway weirs. Joint planting of willow cuttings (*Salix humboldtiana*) into rock protection improved vegetation establishment along the stream corridor.

On-site evaluation of the stream restoration work helped greatly in orienting the rehabilitation work into a watershed context. Results from early monitoring activities showed the community's high level of interest in their local water resources, especially their drinking water. We found that provision of materials for three wells in the community served as a successful incentive for the community labour needed to complete the stream restoration efforts in the riverside communities.

Small grants recipients also benefited through USDA technical assistance from the Natural Resources Conservation Service, the US Forest Service, and the Cooperative State Research, Education and Extension Service. When called upon, the USDA personnel helped design and later evaluate the work done and, in view of their technical evaluations, adjustments were made to successfully close out the project. Before intervention, the fundamental question to measure restoration progress was, "Is this a priority work?" After intervention, the question was, "Did the work achieve the desired effect?"

Evaluations were considered useful by local technicians who benefited by interaction with Natural Resources Conservation Service specialists. These active and participatory evaluations were instrumental in increasing local capacity in watershed work such as stream corridor restoration where previously only basic SWC practices were used. After intervention, community members frequently responded in participatory evaluations that the river area was now a nicer place to be. Continued and regular interaction with the community helped broaden the participation and indicate who would be likely to undertake future activities.

FOLLOW-UP AFTER THE MITCH RECONSTRUCTION PROJECT

Lessons learned from monitoring and evaluation results of the ProCuencas Nicaragua activities were incorporated into the design of the two follow-up projects. Pertinent results of previous monitoring and evaluation pointed out the need to encompass even *more* community water systems and increase or continue promising reforestation activities. Accurate record keeping allowed a rapid start-up of activities, with communities for water sources already diagnosed and areas for riparian reforestation already identified. As a result of intensive monitoring and evaluation, positive changes in project activities were made and subsequent evaluation resulted in an increase in Zamorano capacity to provide existing expertise to major donor projects at a higher level of financial and field commitment in community water resources.

Community Water Systems — Potable Water for Miraflor Nature Reserve

Evaluations of the ProCuencas project showed that the greatest impact was received from water projects. The Pan-American Health Organization's interest in municipal water systems led it to fund a small grant to coordinate watershed protection under its program of primary environmental assistance. With the Pan-American Health Organization, a primary focus of community water systems was backed up with watershed management to protect the water sources. This is a different focus from that in the previous project, where community water systems supported watershed activities.

Where the provision of access to potable water was the primary concern of the Pan-American Health Organization, experience in integrated watershed management allowed Zamorano to provide overall service from the beginning. Protecting water sources from the outset of activities and a continual focus on water resources were the guiding forces. Connectivity was again a driving factor, given the condition of the upper Estelí River Watershed and the clear need of residents within the Miraflor Nature Reserve, northeast of the city of Estelí.

Reforestation, Soil Health and Conservation Activities

Through a collaborative exchange grant from the USDA, we undertook enhancement of vegetative conservation practices with mycorrhizal inoculation of tree seedlings. This was a direct follow-up on previous work in the watershed context for strategic reforestation and watershed stabilization in the Estelí River Watershed. Additional collaboration with the National Autonomous University in Leon, UNAN, and with Zamorano in Honduras promoted work with a variety of tree species, and specifically with Musaceae and fruit trees that were prominent in Mitch restoration activities.

Large attempts at reforestation have been largely unsuccessful in dryland areas of Central America (Arcia, 2001; Varmola and Carle, 2002). The effects have been difficult to perceive, there has been high mortality, and there has been little positive change in the landscape. Evaluation of the larger watershed recovery during Mitch Reconstruction showed that targeted reforestation along riparian areas was both the easiest to accomplish and the most beneficial to people and the environment. It was evident that reforestation along stream corridors supported conservation, provided visual impact, and was easily understood and appreciated by the watershed population. The goal of the Estelí follow-up project with mycorrhizal inoculation was simply to get quality trees established under difficult edaphic conditions in critical areas. The principal objective was to increase seedling survival and adaptation to stressful growing conditions. Previous monitoring activities allowed rapid site selection within the watershed. We anticipate that the results of the work will provide evidence and demonstrations that can be useful to future reforestation efforts in the region.

CONCLUSION

Attending to the millennium goals of reducing in half the number of people without access to drinking water and sanitation around the globe will require a greater focus on potable water resources. These resources are intimately linked to SWC activities. Because soil erosion and land degradation continue to threaten food security, development actions must continue to address both soil and water quality issues. Where donors want greater impact, there will be increasing tendency to focus on water supply projects.

With active monitoring and evaluation components, community participation in SWC can be increased and the quality of conservation impacts improved, as the ProCuencas project demonstrated in Western Nicaragua following Hurricane Mitch. Achievement indicators as outlined in a logical framework monitoring plan, application of SWC on-farm with farmers along with training farmers in maintenance practices, and keeping records of how activities connect for design modification or amplification were all useful methods towards watershed conservation. Monitoring the direct effects of SWC practices and incorporating them into larger watershed actions helped ensure that these important project activities continue and will improve the livelihood of residents in Nicaragua and Central America.

References

Arcia, D. 2001. La situacion forestal y las propuestas de accion del Grupo Intergubernamental de Bosques en los paises de Centroamerica [The forestry situation and action proposals of the Intergovernmental Group on Forests in the countries of Central America]. Revista Forestal Centroamericana 33: 27-32.

Horton, M. and P. Freire. 1991. We Make the Road by Walking: Conversations on Education and Social Change. Temple University Press, Texas.

Perez, C. and H. Tschinkel. 2003. Improving watershed management in developing countries: A framework for prioritizing sites and practices. Agricultural Research and Extension Network, UK Department for International Development (DFID). ODI AgRen Paper No. 129, 20 pp.

Varmola, M.I. and J.B. Carle. 2002. The importance of hardwood plantations in the tropics and sub-tropics. Int. Forest. Rev. 4(2): 110-121, 165-167.

Walle, R.J. and B.G. Sims. 1998. Natural terrace formation through vegetative barriers on hillside farms in Honduras. Am. J. Alternative Agr., 13(2): 81-84.

Walle, R.J. and B.G. Sims. 1999. Fertility gradients in naturally formed terraces on Honduran hillside farms. Agron. J., 91: 350-53.

Workman, S.W. [ed.] 2001. Watershed approach to soil and water conservation: Hurricanes Mitch and Georges Reconstruction Projects in Central America and the Caribbean. SWCS Conference, Indianapolis, Indiana, July 2001. Synthesis of International Activities Session (Powerpoint series) on-line at http://www.swcs.org/.

Workman, S., C. Rodriquez and R. Chavez. 2002. Reforestation activities for watershed restoration in Nicaragua. p. 61. *In:* Proceedings, Working Forests in the Tropics: Conservation through Sustainable Management, Feb. 2002. University of Florida, Gainesville. http://conference.ifas.ufl.edu/tropics/abstracts.pdf.

8

Monitoring and Evaluation of Watershed Projects in India – Field Application and Data Analysis

K.K. Gupta and D.C. Das

Members, Soil Conservation Society of India. B-19, Parijat Apartments, Pitampura (West), New Delhi 110034, India. E-mail: dinesh_ranu2003@yahoo.com

ABSTRACT

A people-centred, bottom-up approach to watershed interventions is described in this chapter. It is demand driven and flexible in adjusting to local resources and skill. In the monitoring and evaluation for watershed development projects too, people are placed at centre stage to gather information at village, household, and specific site or beneficiary level, using participatory modules and semi-structured participatory rural appraisal for village profiles, household surveys and village transect surveys. However, direct measurements are necessary for quantified data on important indicators. Participatory models with on-site people and grassroots-level functionaries largely leave out the impact on vital biophysical linkages and remain inadequate to provide quantified and graded outputs as well as outcome that are essential for close scrutiny by a third party for taking investment decisions or assessing potential for replication. It has thus become necessary to make the participatory team a mixed one including beneficiaries and outside experts. The data collected with specially designed schedules provided quantified results that helped draw important inferences. These deliberations established a rationale for participatory concepts, approach, and principles developed in India for impact assessment of watershed development projects spread over large areas.

INTRODUCTION

The concept and principles of monitoring and evaluation of watershed development and management (WSD/WSM) projects, methodologies for sampling area units and sites, identification of parametric areas and sub-areas, selection of indicators and tools, and techniques for data acquisition are given in a separate chapter in this book. In the past decade, focus on people-centred approaches has increased and beneficiaries have been given opportunities to plan, implement and monitor projects. This bottom-up approach is based on felt needs, driven by demand, and flexible to adjust to local resources and capacity of target households. Therefore, capacity building of village communities, village institutions, women, and disadvantaged groups has become important.

But the balance between on-site and off-site compulsions and aspirations as well as sustainability of an ecosystem, especially for recurrent availability of water, depends on upstream and downstream linkages and also the continuity of processes of regeneration and replenishment. This calls for use of scientific principles and tools. Thus, even in the ongoing transformation, the need for participatory modules involving on-site beneficiaries and experts from outside is inescapable. The question is how to develop such modules, which should be time- as well as cost-effective. This chapter presents the breakthrough achieved in this direction for projects in operation over large areas with diverse situations such as the lower Himalayas in northern India and plateaus of the east and west of the country and being implemented by government and corporate agencies as well as NGOs. This chapter concentrates on evaluation or impact assessment.

METHODS

For impact evaluation and assessment of sustainability, data series of multiple indicators are required at different time frames. These series offer opportunities to compare the status or scenario before and after or with and without the project. The data must be applicable to major spatial variations and against different socioeconomic backdrops. Further appropriate interpretation of project outputs would require process analysis; thus, data gathered are of two types, primary and secondary. Primary data comes from available records, reports, and other published literature collected by way of a village profile developed through participatory rural appraisal (PRA) and specially prepared schedules and checklists.

A participatory village transect survey is used to gather site-specific data and direct beneficiary responses with the checklists, schedules, and questionnaires (Appendix I, after Epilogue in this volume). Household interactive surveys are held with structured schedules and questionnaires. These schedules are different for different components such as agriculture, horticulture, livestock development, and socioeconomics (Appendix II).

Data gathered through direct field survey or measurements comprise depth of well water, discharge of small streams, composition, number and growth of trees, shrubs, grasses, and collection of samples of soil, leaf litter, and grasses.

RESULTS AND DISCUSSION

A WSD project has a large basket of results that vary greatly in form, magnitude, and quality. The data are, therefore, segregated in sets. This helps in reprocessing, collating and analysing within a particular set as well as in association with other sets that lead to appropriate process analysis. Examples from some of the generally accepted sets will be presented below.

Risk from Rainfall Variations

Rainfall and its distribution within a year and over the years profoundly affects the performance of a plant or crop. For micro-level works, which are the common features in WSD, a record of 10 to 25 years that immediately precedes is more relevant than a long record extending to the distant past (Das, 1967). In a 10-year period, record drought visited the six representative locations of the Siwaliks of Punjab State (Fig. 1) between 62.5 and 100% of the years. In the critical crop month of July, rainfall at Pathankot varied from a high of 620 mm to a low of 59.3 mm. In August, rainfall ranged from 406.2 to 38 mm. Variations or departures from the normal need to be studied. Similarly, for January and February, the winter crop rainfall ranged from 125.6 to 4.6 mm and 101 mm to 0.0 mm respectively. Trends for seasonal rainfall for wet as well as winter crop seasons exhibited similar wide variations (CES, 2002a).

Risk from Changing Land Use

Distribution of the geographical area under different uses is a significant pointer to watershed degradation. The area under perennial vegetation or green area vis-à-vis areas under crop and other uses reveals land use alienation, or extent of area taken out of natural land use of green cover. The details given in Table 1 from a study in the Siwaliks area of Punjab

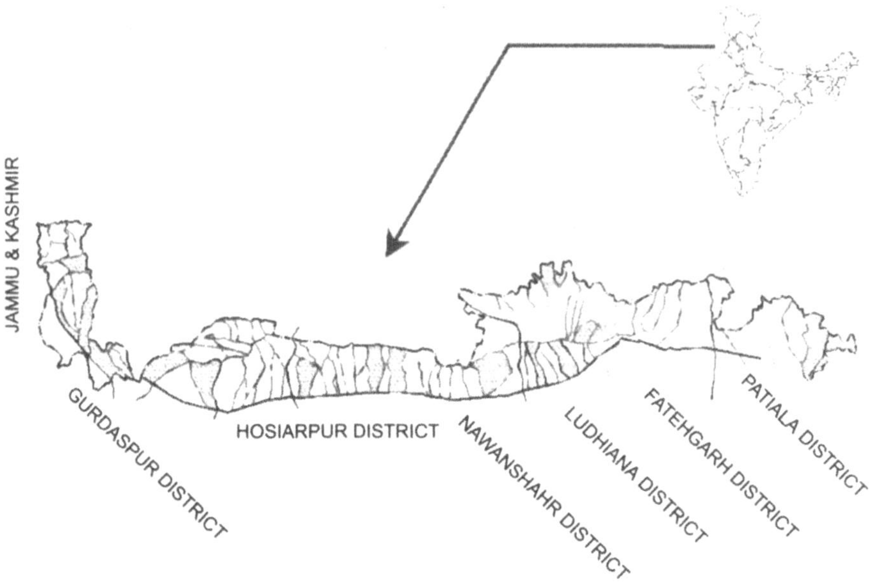

Fig. 1 Sub-watershed area in Siwaliks of Punjab under IWDP (Hills-II).

(Note: The above is a rough sketch map of India and does not aim to show political boundaries)

Table 1 Distribution of micro-watersheds under different adequacy rating classes of green area (percentage of total number of micro-watersheds).

District	*< 20% green area (Poor)*	*20 to 40% green area (Fair)*	*40 to 60% of green area (Good)*	*> 60% green area (Very Good)*
Gurdaspur	30	20	30	20
Hoshiarpur	50	12.5	12.5	25
Nawan Shahr	-	-	50	50
Ropar	-	-	33	67
Patiala	100	-	-	-

show that, in micro-watersheds (MWS) of district Patiala, land is increasingly being alienated. Alienation is least in MWS of district Ropar. Therefore, without any neutralizing intervention the risk of degradation is also very high in Patiala notwithstanding its generally flat topography (CES, 2002a).

Risk from Soil Changes

Data on soil characteristics provide information on the land's productivity, proneness to erosion, and moisture status. Variations in

Table 2 Soil organic carbon in the soils of MWS in three districts of Punjab (range in percentage).

MWS of District	*Agriculture*	*Forest or common land*
Patiala	0.480-0.900	0.330-0.030
Ropar	0.750-0.180	0.570-0.120
Nawan Shahr	0.150-0.270	0.120-0.030

values of soil organic carbon (SOC) at different sites and under different land use are given in Table 2. These values showed that soils of the project area could not have had reasonable productivity (CES, 2002a). For good crop productivity, SOC should be greater than 1% and for favourable hydrologic response of MWS, SOC in forest areas should be around 2%. Risk due to soil was very high to moderate in agricultural lands, while for forest soils it was very high.

Level and Intensity of Treatment Packages

Soil and water conservation (SWC) treatments were given to areas of three watersheds of eastern India that could produce biomass. They included land treatments, cultivation methods, and afforestation. The picture of three watersheds would reveal the relative utility of the packages (Table 3) (LUCI, 1999).

The high-intensity package for Karkara watershed came from three large check dams with a total of 84.4 ha-m surface storage that could hold 4.82% of incident rainfall. Small earthfill dams and gully plugs were put up as part of a medium-intensity package across the flow direction in series and in a networking formation in Karma watershed and created 71.5 ha-m of surface storage. The irrigation potential thus created in Karkara watershed is 450 ha in wet season and 175 ha in winter season, while the irrigation potential created in Karma is 360 ha for wet season and 120 ha in winter season (LUCI, 1999). This has reduced the risk of erosion and degradation.

Table 3 Conservation and protection provided vs. three intensity levels of treatment in three watersheds.

Watershed	*Pre-project*	*Post- project*	
	% of safe area	*% of safe area*	*Level of treatment*
Mahesha*	0	0	None
Karma	0	60	Medium
Karkara	0	72	High

*Control watershed

Silt Deposited due to Drainage Line Structures

Drainage line treatments on upper reaches generally comprise brushwood check dams (Fig. 2) and dry stone masonry structures (Fig. 3), while crate wire stone check dams are dominant on lower reaches. At foothills and plains near the banks of rivulets, crate wire stone spurs, live hedge spurs and wire crated training wall (Fig. 4) are the obvious choice.

Dry stone masonry structures and brushwood check dams hold back eroded materials and also reduce the velocity of flow, enhancing percolation of rain water into the soil profile and substrata below over an extensive area and for a prolonged duration. The extent of stream cross-sections behind these structures are filled up by the eroded materials carried by the runoff. The aggregate volumes of these deposits represent the potential sediments prevented from flowing down the drainage system and out of the watershed. These volumes are computed from the filled-up depth behind each structure and the distance of the fills upstream of the structure. The volume of deposits obtained through a field survey-cum-computation for two MWS in each of five river catchments in the Siwaliks of Himachal Pradesh are given in Table 4 (CES, 2002b).

The drainage line structures could reduce the potential sediment production from these catchments at rates ranging from 0.13 to 8.60 t/ha.

Fig. 2 Brushwood check dam across a gully reduces the velocity of flow and soil erosion in Amraiyon Bakarla MWS of Himachal Pradesh.

Fig. 3 Dry stone masonry structure plugs the gully head in Aduwal MWS, Himachal Pradesh.

Fig. 4 Crate wire stone training wall in the stream at Amraiyon Bakarla MWS, Himachal Pradesh, protected 2 ha of arable land.

Table 4 Sediments held back by drainage line structures in five Siwalik catchments of Himachal Pradesh.

Catchments	*Area (ha)*	*Dry stone structures and brushwood check dams (No.)*	*Deposited silt (cum)*	*Weight of silt (tons)*	*Silt held back (t/ha)*
Markanda	1,440	521	1,620	2,430	1.68
Ghaggar	1,037	175	92	138	0.13
Sirsa	1,078	568	2,495	3,743	3.47
Swan	2,086	592	8,493	12,739	6.11
Chakki	1,283	924	7,358	11,037	8.60

Table 5 Sediments held back and the density of structures in MWS of five Siwalik catchments of Himachal Pradesh (CES 2003).

Catchments	*Potential sediments held back (t/ha)*	*Density of drainage line structures (No./ha) or average area served by one unit of structure (ha/unit of structure)*
Markanda	1.68	2.43
Ghaggar	0.13	5.93
Sirsa	3.47	1.90
Swan	6.11	3.52
Chakki	8.60	1.39

The variations can be explained from the differences in the density of structures or the number of structures put up on each hectare, as shown in Table 5. A larger number of structures constructed on one hectare or a smaller area served by one unit means higher density and thus greater capacity to hold back the eroded materials.

Improved Groundwater Status

The increase in recharge of groundwater and consequent increase in well-irrigated area were substantial (Table 6).

Table 6 Improved groundwater availability and increase in irrigated area in the watersheds of Damodar Valley Project.

Watershed	*No. of respondents*	*No. of wells*	*Pre-project (ha)*		*Post-project (ha)*	
			Irrigated area	*Total crop land*	*Irrigated area*	*Total crop land*
Karkara	26	21	14.02	62.43	31.14	62.43
Karma	27	29	22.52	62.19	46.67	62.19

A community-based natural resources management project supported by the European Union and implemented by the NGO N.M. Sadguru Water and Development Foundation (NMSWDF) in four districts of Gujarat State recorded that well water levels rose by 2-10 m because of project interventions (Singh and Gupta, 1997).

Enhanced Irrigation Potential

In two years, project intervention in the Siwaliks of Himachal Pradesh has helped increase area under irrigation by 752 ha from 98 water harvesting structures of various types with a total storage capacity of 60 ha-m. The diversion structures or lifting systems that were put in use are illustrated in Figs. 5 and 6.

Improved Channel Flow

Owing to project activities carried out in the first two years, there have been perceptible changes in flow duration as well as volume in the streams of some of the MWS in the Siwaliks of Himachal Pradesh (Table 7, CES, 2003).

Increased Biomass Productivity

Project interventions in some districts in Gujarat helped enhance water availability that resulted in increase in acreage under high-yielding cereals, and better remunerative pulses and vegetables. Average yield increased by 3-8 times under irrigation and 2-4 times in rainfed areas. Consequently, cropping intensity rose from 50% to 100% (NMSWDF, 1999).

Another significant impact has been in the form of trees owned by the households of 60 out of 300 programme villages. From about 1,000 trees raised by individuals, each household is getting biomass worth $45 per year, while remote sensing data revealed that these villages attained

Table 7 Improved flow duration and flow volume due to watershed development activities (obtained by participatory village transect survey).

Micro-watershed	*Drainage channel*	*Flow duration months*	*Flow duration months*	*Increase in flow volume (%)*
		1999	*2001*	
Amrayion	Amraiyon Khad	12	12	10
	Jamli Khad	12	12	20
Bankala	Rukhri Khad	9	10	10
Tikri Bhojpur	Tikri khad	12	12	10
Koti Kalyanpur	Mohan nala	12	12	20

additional vegetation cover to the tune of 35%. Besides, this has improved the availability of fuelwood, fodder and small timber for house construction and thus greatly reduced the dependency of the inhabitants on forest (NMSWDF, 1999).

Employment Opportunity

There was substantial gain in employment due to project implementation in both Karkara and Karma watersheds. But the increases varied with the intensity of the package of treatments. In Karkara watershed with a higher level of treatments the increase was to the tune of 258% as against 75% in Karma watershed with medium level of treatments. The large increase in Karkara watershed was due to intensification in cultivation and other diversified avenues. In the cultivation sector, the variations in the growth in employment from pre-project to post-project were as given in Table 8 (LUCI, 1999).

Because of lower treatment intensity there was little diversification in employment opportunities in Karma watershed and thus the villagers continued to work in cultivation. Total increase was far below the desired level. Total employment on each hectare of agricultural land in Karkara watershed was substantial, even though each person still had to work on at least 3 ha to have full employment of 180 person-days. In Karma watershed, a person had to work on at least on 12 ha to be fully employed (LUCI, 1999).

Institution Building

The higher-level project inputs in Karkara watershed included a variety of community-based organizations, including 22 women's self-help groups, one watershed development committee, four village agricultural development committees, one inter-village horticultural development committee, an inter-village improved seed and artificial insemination committee, and user groups for irrigation and fishery. These organizations created their own funds through subscriptions, project

Table 8 Increase in employment opportunity (in person-days) due to implementation of project activities.

Period	*Karkara watershed*		*Karma watershed*	
	Total	*Per ha agricultural area (515 ha)*	*Total*	*Per ha agricultural area (1.455 ha)*
Pre-project	16.410	31.86	11.130	7.65
Post-project	30.990	60.17	21.360	14.68
Percent increase		88.85		91.93

contribution for labour provided, sale proceeds, service costs, and other means. Many successfully obtained credit for their members from banks through their accounts at a lower rate of interest. They also recovered the loan amounts from their members and repaid the dues to the banks in time (LUCI, 1999).

WSD activities in four districts of Gujarat not only strengthen the grassroots level institutions, but also created better skill among the beneficiaries as well as a strong force of trained volunteers who have empowered the villages to take new initiatives in maintaining and operating assets created through the project (NMSWDF, 1999). An example is the lift irrigation committees formed and functioning in the project districts.

Economic Status

The post-project family income in three watersheds of Damodar Valley area is given in the Table 9.

In spite of the high level of project interventions in Karkara watershed, the average family income is the lowest and equal to that of Karma watershed. This showed that the project interventions and benefits were not equitably distributed. In fact, only 4 villages out of 13 received irrigation water created by three relatively large check dams and the lift irrigation systems in Karkar. There was no comparable package for other villages where irrigation potential either did not exist or could not be created.

The distribution of families under different annual income classes revealed that about 50% of the families were still below the poverty line in Karkara watershed, which had received high-intensity treatments. This showed that the level of watershed development was still inadequate and the desired objective of distributed equity not achieved, as 8 out of 13 villages did not receive the benefit of created irrigation potential or comparable alternatives (LUCI, 1999).

Table 9 Variations in family income due to intensity of treatments.

Watershed	*Income (INR/family/yr)*		
	Lowest	*Highest*	*Average*
Karkara	2,000	132,000	32,000
Karma	2,000	100,000	44,000
Mahesha	5,500	62,000	32,100

Note: $1 = INR45.

For the project in four districts of Gujarat the benefit-cost ratio was 2.9 inclusive of indirect benefits (Nyborg et al., 1993), while Mitchell (1994) found a ratio of 1.85 excluding indirect benefits.

Fig. 5 A distribution chamber at village Malpur in Aduwal MWS of Himachal Pradesh, which lifts water from water harvesting structures for irrigation at higher elevation fields.

Fig. 6 Diversion structure in Pargana MWS of Himachal Pradesh drawing water from Jabbar Khad and irrigating about 40 ha land.

Environmental Improvement

Natural regeneration of forest areas close to habitations was studied to assess quality of stand (Das, 1999). Due to the proximity of the settlements and high demands for fuelwood, the forests were getting depleted and their quality had been deteriorating. The village agricultural committees had taken steps to protect the forest from grazing and over-exploitation. The vegetation was surveyed to determine the extent of natural regeneration induced through these efforts. The sites identified were Chorah and Sargaon in Karkara watershed and Dhori Chhagaria in Karma watershed. Nested quadrates and line transects were used. Sargaon site had the maximum of 20 species, Dhori Chhagaria 13, and Chorah 10. Only five herbaceous species were seen and the most common one was *Cynodon dactylon*. *Shorea robusta,* the dominant species of the region, was the highest (65%) at Chorah and the least (10.5%) at Sargaon. At Dhori Chhagaria it was 30%. *Madhuca indica,* the most revered tree of the region, was found only at Sargaon and constituted only 5.4% of the total population. Other important species found were *Carissa opeca* (a shrub), *Diospyros melanoxylon, Terminalia tomentosa,* and *Butea monosperma*. To determine the differences in quality of composition of forests, the species diversity index (SDI) and similarity index (SI) values were computed (LUCI, 1999). These are shown in Table 10.

The site at Sargaon had a higher SDI because it was further from the villages. At the other two sites, SDI values were low because of greater anthropogenic interference due to proximity of the villages. Protection alone cannot raise the SDI significantly within a short period. It would be desirable to accelerate the process of natural regeneration by planting with the dominant species of the region, such as *Shorea robusta, Syzygium cumini, Carissa opeca, Anogeissus latifolia, Acacia catechu,* and *Emblica officinalis,* besides *Madhuca indica,* for which a special drive should be mounted (LUCI, 1999).

Enlarged Green Area

The extent of land area under any perennial vegetation is defined as green area. It includes areas under forest, grassland or silvipasture, horticultural

Table 10 Species diversity index and similarity index of three forest sites near three villages in Damodar Valley.

Site	*SDI, %*	*SI*
Sargaon	29.20	0.47 with Chorah
		0.30 with Dhori
Chorah	3.44	0.52 with Dhori
Dhori Chhagaria	2.35	

and commercial plantation of trees or shrubs, fuelwood or energy plantations. For gross green area, the simple area is taken without any consideration of ecological attributes of the standing stock. Effective/ equivalent gross area takes into account the ecological quality of the green stand.

The perception of watershed degradation rests on generation, exhaustion or degradation, and regeneration of abiotic attributes by interaction of life forms living in the ecosystem. As sun rays, the primary energy source, effect these interactions the balance in the energy flux of a watershed ensures its stability against degradation. Thus, a ratio of the total production of energy (P) to total utilization of energy or respiration could be taken as an indicator to measure watershed degradation. This has been simplified as a ratio of O_2/CO_2 or O_2 released by plants to CO_2 absorbed by plants (Odum, 1975). But it was not easy to apply this indicator at field level because measurement of O_2 and CO_2 even at micro-level was a problem (Ambasht, 1990). Thus, the ratio has been transformed into an area rationale or area under perennial cover to those under other uses (Das, 2003), and increase in green area due to project activities is taken as an indicator for environmental regeneration of the watershed.

Increase in green area after project implementation was detected in a number of MWS of the Siwaliks of Himachal Pradesh. These increases were due to land use shifts to trees, horticultural plants, or pastures. From the ecological point of view, the enlarged green area is a favourable change. Actual increase in 10 MWS can be seen from Table 11 (CES, 2003).

Table 11 Total increase in area under green or perennial cover during mid-term evaluation as compared to that during baseline survey.

Micro-watershed	*Total increase in area under green cover (ha)*	*% increase*
Amriyon-Bakarla	10.70	1.71
Bankala	36.96	8.32
Tikri Bhojpur	22.89	4.71
Koti Kalyanpur	23.41	4.15
Dhang Phalsi	20.00	2.32
Aduwal	41.00	18.80
Jadla-1	48.80	6.49
Bhadsali	43.34	6.09
Thora Baloon	45.77	5.76
Bari Khad	44.15	9.80
Average	33.70	6.82

Gross Green Area and Effective Green Area

The physical extent of area under perennial covers such as forests on government, community and private lands, grazing lands, and land under horticultural plants constitutes the gross green area (GGA). But the ground realities indicate that much of the gross green area has suffered depletion in terms of trees, shrubs and grasses and also has become deficient in ecological attributes owing to various types of degradation. The equivalent or effective green area (EGA) gives due weight to the number, composition, and growth parameters such as girth and height. The EGA may be much smaller than the GGA. This would be evident from the values computed for GGA and EGA of some micro-watersheds both for the pre- and mid-project period (Table 12).

In the pre-project period, EGA varied from 16.34% of GGA for Jadla-1 to 44.89% of GGA for Thora Baloon MWS. In the mid-project period, EGA values registered a significant increase, ranging from 24.25% of the GGA for Jadla-1 to 82.95% for Aduwal. The improvements were due to enhanced planting through the project.

Land Use Shifts

Project interventions induce changes in land use. In 10 sample villages of Uttaranchal significant shifts have taken place through integrated watershed management, as is evident from Table 13 (CES, 2003).

- Land under perennial green cover increased by 43.84 ha or 4.1%.
- There was a beneficial decrease of poor quality grazing or community land by 300.90 ha or 46.05%.
- There was a beneficial decrease of unproductive fallow or cultivable waste by 33.96 ha or 20.24%.

Table 12 Gross green area vs effective green area (ha) during pre-project and mid-project periods for some MWS of Himachal Pradesh.

Micro-watershed	*Pre-project*		*Mid-project*	
	GGA	*EGA*	*GGA*	*EGA*
Amraiyan	535.37	168.78	546.44	239.13
Tikri	270.10	83.35	292.99	188.50
Aduwal	88.00	43.30	129.00	107.00
Jadla-1	418.00	68.31	466.80	113.21
Thora Baloon	356.42	160.07	394.42	241.07

Table 13 Land use shifts in 10 villages in Uttaranchal due to watershed development activities (areas in ha).

Type of land use	*Pre-project*	*Post-project*
Forests/afforested lands	400 (forests)	537.50 (forest + afforested)
Silvipasture and fodder grass	-	92.90
Fuelwood plantation	-	73.50
Grazing/community land (poor return)	653.43	352.53
Orchards	9.14	42.80
Trees in homestead (ha)	12.75	19.43
Fallow and cultivable waste	167.75	133.79
Total	**1,242.57**	**1,252.45**

Reduced Poor Quality of Bovines and Enhanced Fodder Availability

Ten sampled villages in Uttaranchal under an integrated watershed development programme registered reduction in poor quality of the local bovine population due to various project interventions, as shown in Table 14 (CES, 2003).

The population of poor quality bovines reduced by 1,544 or 49.74% of the pre-project population. Fodder demand of the 1,544 heads, at 7 kg per head per day, comes to 3,945 tons per year. In other words, for the reduced population of 1,560 heads, the increased availability is 2.53 t/yr or 6.93 kg/day. Thus, the fodder requirement of these cattle could be met almost entirely from existing sources without serious risk to the environment.

Rural Connectivity

Connectivity of the villages in the rural areas of hills is of paramount importance for the well-being of local people. Although connectivity exists in the remote villages in the hills, it still takes considerable time to reach markets, health centres, schools, and livestock in nearby villages

Table 14 Number of bovines of poor quality in pre- and post-project period in 10 sampled villages of Uttaranchal.

Bovine type	*Number of heads*	
	Pre-project	*Post project*
Local cow	1,441	810
Local buffalo	294	284
Local bull/ox/male buffalo	1,112	247
Stray bull	257	219
Total	**3,104**	**1,560**

and towns. People have accorded top priority to the strengthening of existing roads, footpaths, culverts, and footbridges (Fig. 7). Connectivity in many villages of the project area has improved, and people are selling their agricultural produce at better prices and earning a better livelihood. Thousands of people and animals benefited.

Cost Sharing of Beneficiaries

The packages of interventions for the Gujarat project were decided through PRA. They included land levelling, terracing, earthen bunds, graded contour bunds, dry stone bunds or walls, continuous contour and staggered trenches, dry stone gully plugs, grade stabilizers, gabion wire crated structures, and afforestation activities on community land, wasteland, and field boundaries. The implementation was mainly through the farmers, who shared the cost of interventions. They contributed 40% of the total cost in the form of labour. Out of their contribution, 25% went to the village community fund, which was used to advance loans to the needy.

Since the whole model is evolved and built around people's needs, people have developed the capacity to manage various activities,

Fig. 7 A footbridge that connects villages in remote area, thus improving the connectivity of villagers, particularly when bringing agricultural produce to market.

Fig. 8 Harvesting of rainwater in the series of terraces in Village Thuthi Kankasia, Jhalod Taluka of Gujarat.

including the cost of maintenance, repair, and operation, efficiently and effectively.

CONCLUSIONS

The details from actual field work, as discussed in preceding paragraphs, have established the necessity for a participatory model in assessing multiple impacts of various project interventions as well as of the structured interactive schedules for collection of data at village, household, and specific site and beneficiary levels. The results and analysis of data for the selected indicators of important parametric areas discussed above showed that useful outcomes or durable effects could be inferred with a reasonable degree of reliability.

Data schedules designed along with tools and techniques chosen call for hard and intensive homework before their application in the field. It was also clear that participatory modules, involving beneficiaries and functionaries at grassroots level, can provide a fair assessment of on-site and immediate benefits and only on a few easily perceived indicators.

In order to assess both on-site and off-site impact and provide insight into process details over a longer time-frame, participatory modules should involve on-site beneficiaries as well as experts from outside. In addition to the beneficiary responses and collective assignment of a qualitative description or assignment of an agreed quantified value to a specific impact, determination of the degree of impact would call for application of some scientific principles as well as tools and techniques. This would stand close scrutiny by a third party due to back-up data and process analysis, which are needed for investment decisions and for giving confidence in replicating the project evaluated.

However, people's participation will be the key to monitoring and evaluation even in such mixed modules. In watershed development projects, a combination of scientific techniques and participatory techniques will continue to be a practical necessity.

References

Ambasht, R.S. 1990. Plant Ecology. Students' Friends & Co., Lanka, Varanasi, India.

Consulting Engineering Services (CES), India. 2002a. Final Base Line Survey and Current Status Report on IWDP (Hills-II), Punjab, New Delhi.

Consulting Engineering Services (CES). 2002b. Final Base Line Survey Report on IWDP (Hills-II), Himachal Pradesh, New Delhi.

Consulting Engineering Services (CES), India. 2003. Final First Quarterly Impact Evaluation (for 50 completed villages) Report. IWDP (Hills-II), Uttaranchal, New Delhi.

Das, D.C. 1999. M&E of afforestation and eco-development projects – Ecological, social, economic and managerial indicators: current efforts and future options. Agricultural Finance Corporation, Ltd., New Delhi. RC-63: 16-37.

Das, D.C. 2003. Land Use Alienation, Green Area and Eco-Index in Watershed Development. International Conference on Sustainable Management of Natural Reources. Banaras Hindu University, Varanasi, February 2003.

Das, D.C., B. Raghunath and G. Poornachandran, G. 1966. Rainfall and climatic associates in relation to soil and water conservation at Ootacamund (Nilgiris) Part I: Acceptability of data, amount and its distribution. J. Indian Soc. Agr. Eng (ISAE), III(2): 1-14.

Land Use Consultants International (LUCI). 1999. Impact and sustainability of watershed management programme in Bihar, Prepared for RODECO, Germany. Tech. paper No. IGBP/WSM/120/99.

Mitchell, R. 1994. A benefit-cost analysis: managing the economics of the environment. Paper presented at the National Workshop on "Economics and Environment", 27-29 January 1994, New Delhi. 12 pp.

N.M. Sadguru Water and Development Foundation (NMSWDF). 1999. Annual Report, NMSWDF, Dahod, Gujarat.

Nyborg, I., R. Mitchell, J. Studsrod and S. Phadake. 1993. Mid-term evaluation of Sadguru Water and Development Foundation's Five Year Plan, funded by European Union. Unpublished, 61 pp.

Odum, E.P. 1975. The Link between the Natural Sciences and Social Sciences. Oxford & IBH Publ. Co., New Delhi.

Singh, K. and K.K. Gupta. 1997. The Sadguru Model of Community Based Natural Resources Management. Occasional Paper Series Publication No. 14, Institute of Rural Management, Anand, 50 pp.

World Bank. 1999. Regional Environmental Assessment for the proposed integrated watershed development projects (Hill-II). Prepared by Lea Associates South Asia Pvt. Ltd, New Delhi.

9

Monitoring and Evaluation of Soil Conservation and Watershed Development Projects in Sri Lanka

A.N. Jayakody[1], W.A.D.P. Wanigasundara[2] and E.R.N. Gunawardene[3]

[1]Department of Soil Science, Faculty of Agriculture, University of Peradeniya, Peradeniya, Sri Lanka. E-mail: anaja@pdn.ac.lk

[2]Department of Agricultural Extension, University of Peradeniya, Peradeniya, Sri Lanka. E-mail: dpwanige@pdn.ac.lk

[3]Department of Agricultural Engineering, University of Peradeniya, Peradeniya, Sri Lanka. E-mail: nimalgun@pdn.ac.lk

ABSTRACT

This chapter is a compilation of monitoring and evaluation (M&E) activities on major soil conservation and watershed management/development projects executed and being executed in Sri Lanka. It is enriched with relevant background information as well as procedures.

The clearing of the hilly slopes of the Central Highlands of Sri Lanka for planting coffee and tea during the 19th century led to the reduction of water retaining capacity, humidity, and coolness of the environment, which made conservation measures necessary thereafter. Soil erosion was one of the most damaging impacts. Natural resources legislation in Sri Lanka thus dates back to 1840, along with the introduction of the Land Encroachment Ordinance, followed by another 11 Parliamentary Acts to date.

Soil conservation and watershed management and development activities in Sri Lanka are generally launched through foreign-funded, large-scale projects or programmes implemented by mandated ministries or agencies as well as other interested agencies such as research institutes and NGOs. Since 1975, there have been 18 foreign-funded projects. The main project successfully completed in 2006 was the Upper Watershed Management Project covering the Mahaweli upper catchment area, the M&E procedure of which is especially elaborated here. In addition, the M&E procedures of some selected mandated agencies are included for the purpose of comparison.

Finally, the attempts of the State in institutionalizing M&E pertaining to relevant fields are also highlighted and linked with some conclusions and recommendations.

INTRODUCTION

The geomorphic regions of Sri Lanka are basically characterized by considering the elevation above mean sea level (asl) and the nature of the terrain. The regions are Coastal fringe, Central Highlands, Southwest country, East and Southeast, and North-Central Lowland. This tropical island has an area of 65,510 km^2 and a population of around 18 million. Monthly mean maximum and minimum temperature ranges are 18.7°C to 33.1°C and 7.7°C to 23.8°C respectively. The rainfall ranges of the major climatic zones, namely dry, intermediate and wet zones, are 875-1,875 mm, 1,875-2,500 mm, and 2,500-5,000 mm respectively.

The Central Highlands region, covering around one-third of Sri Lanka and comprising mid-country (600-1,200 m asl) and up-country (>1,200 m asl), has become its most important watershed, and soil conservation is a major concern there.

The headwaters of all major Sri Lankan rivers originate from the Central Highlands, creating a radial network comparable to a tent with a central mast. Most of the lands in the highland region were covered by natural vegetation many centuries ago. But these have been considerably disturbed and eroded by illegal encroachment of lands for cultivation and settlements, establishment of tea plantations, increased cultivation of vegetables, urban expansion, building, and road construction. Most of these operations have neglected the need to protect soils or conserve water for power generation and agriculture.

The mountain crests, valleys and slopes of the highlands were cleared for planting coffee and tea during the 19th century, reducing their water retaining capacity and changing the humidity and coolness of the environment. In many locations, rain water is rapidly drained off as a result of the reduced absorption capacity of the soils, thus increasing the degree of soil erosion.

The extent of annual erosion generally ranges from 0.5 t ha^{-1} under Kandyan forest gardens to around 70 t ha^{-1} under extreme situations such as tobacco cultivation without soil conservation measures (Krishnarajah,

1984; Stocking, 1986). These events obviously decrease water conservation in the midst of a reported decrease in rainfall as compared to previous centuries (Atukorala, 1995). On the other hand, high siltation rates are evident in many areas of the low country and also in man-made reservoirs in the up- and mid-country. Land degradation due to soil erosion is considered one of the five major environmental issues in Sri Lanka (SOE, 2000). It is specifically noted as the main degradation process in the Central Highlands, accompanied by reduction of water storage (Nayakakorale, 1998). Thus, soil conservation and watershed management activities have become essential.

In Article 28 of the Constitution of Sri Lanka, it is stated that "It is the duty of every person in Sri Lanka to protect nature and conserve its riches." There is also a long history of natural resources legislation in Sri Lanka in the Land Encroachment Ordinance of 1840, Irrigation Ordinance of 1856, Forest Ordinance of 1907, Land Development Ordinance of 1935, State Lands Ordinance of 1947, Crown Land Ordinance of 1947, Soil Conservation Act of 1951 (amended in 1996), Water Resources Act of 1964, Land Grants Act of 1979, Agrarian Services Act of 1979, Mahaweli Authority of Sri Lanka Act of 1979, and the National Environment Act of 1980.

To achieve the objectives of this legislation, soil conservation and watershed development projects have been launched in Sri Lanka by a number of governmental, non-governmental, and private institutions. Various M&E procedures are attached to these projects to help meet responsibilities towards the people as well as towards funding organizations.

PRINCIPLES AND CONCEPTS OF M&E

At all stages of soil conservation and watershed development, people, materials and money are needed, especially for planning, implementation, monitoring, and evaluation. The operation of the project needs to be monitored over the entire project cycle to enable the implementers to understand how well they are reaching the set targets, objectives and goals. Hence, M&E usually appears in the work plan involving all stakeholders. It would generally answer the questions "Where are we?" and "What have we achieved?" It could also be understood as situation analyses upon which adjustments could be made as needed.

However, for M&E to be meaningful there should be appropriate techniques. The conventional techniques adopted in Sri Lanka are reviews of available documents, surveys, interviews and discussions, observations, participatory documentation, transect walks, and fine analysis of the problem.

One important dimension of M&E is to answer the question "What do we want to achieve?" which derives from goal setting. In fact, the set objectives become the foundations to understand the requirements for M&E in judging the progress of the tasks underway. Hence, it is obvious that there should be a set of monitoring indicators that are explicit, pertinent, and objectively verifiable. The four types of indicators considered in the Sri Lankan context are *input, output, outcome* and *impact*; for example, amount of money spent, number of check dams constructed, total land extent developed, and alleviation of poverty of the stakeholder community, respectively.

Most significant concerns for M&E are to be able to complete the project in time, to do the right things in the right order, to identify responsibilities, and to sense the possible alterations and modifications needed. Frequently, monitoring and periodical evaluations are conducted during the initial stages of implementation, when most of the planned activities are initiated and when it is not too late to change the implementation procedure if required.

As there are positive and negative forces for M&E, the project committees discuss the strengths and weaknesses of their approaches to bringing out opportunities and threats so that negative forces can be overcome. It is necessary that the type of these people as well as their training and encouragement should be at the highest standard for optimum M&E. Gathering information is an integral component of M&E and this can be done in written or in verbal form. In participatory monitoring at village level, there is a dilemma in reporting as some villagers cannot read and write to the level required to complete questionnaires. Hence, projects may sometimes be compelled to choose visual and verbal procedures for gathering information. Though verbal reporting can be low cost and quick, there are constraints in storage, replication, and consistency of information. Hence, Sri Lankan monitors mostly prefer the survey procedure of information collecting where a trained person interviews the stakeholder using a questionnaire and notes down the facts on a prescribed form.

The evaluation segment of M&E generally involves a value judgement. Evaluations are conducted during and after implementations of projects, enabling the project to review the strategies according to changing environments progressively.

TYPES OF SOIL CONSERVATION AND WATERSHED DEVELOPMENT ACTIVITIES

Soil conservation and watershed development projects in Sri Lanka fall into three main categories:

1. Large-scale projects at national level assisted by foreign agencies,
2. Programs implemented by mandated agencies such as *Hadabima* Authority, Government Departments of Irrigation, Agriculture and Agrarian Services,
3. Small-scale projects and programmes of other agencies such as Tea Research Institute, Tea Small Holding Development Authority, and NGOs.

These agencies perform diverse procedures for M&E purposes. Projects of the first category usually have their own specific techniques and procedures meeting the expectations of the funding organizations, while the government also has direct intervention through specific mechanisms earmarked for M&E activity. At times, the national, provincial, and divisional government officers vested with the responsibility of approving and releasing funds for constructions and services undertake M&E on contract basis.

Mostly the permanently recruited officers perform M&E of the second category of activities. They are usually well trained to inform stakeholders on implementation of the tasks and subsequently to assess and release subsidies. Almost all are field officers who must visit the fields at regular intervals. The subsidies are generally released in installments based on completion of conservation related tasks.

The third category of projects are operated at micro level, where M&E is undertaken according to specific objectives related to soil and water conservation. In the subsidy programmes of the tea sector, attention is given to new planting and replanting of tea and a serious emphasis is laid on soil conservation. These organizations use their extension staff for the monitoring of activities where they are required to provide detailed periodical reports.

PRESENT SITUATION OF SOIL CONSERVATION AND WATERSHED DEVELOPMENT ACTIVITIES

In relation to soil conservation, the measures adopted by the tea plantation sector are considered satisfactory due to intensive monitoring by plantation managers and lesser financial constraints.

However, the small allotments of traditional smallholders are not well looked after, as most of the owners are poor or do not have clear title to the land and are less aware of soil conservation. Therefore, most of the soil conservation and watershed development activities of such lands are handled at the national level. A significant feature of most national-level projects is central funding through the government with or without foreign assistance.

There have been approximately 18 foreign-funded projects to promote conservation of natural resources since 1975. The principal sponsors have been the governments of Germany, Japan, the Netherlands, Norway, Denmark, Sweden, the United States of America, and the United Kingdom as well as the Asian Development Bank and the World Bank. Examples of some main projects launched since 1975 are Watershed Management Project, Reforestation and Watershed Management Project, Upper *Mahaweli* Watershed Project, Swedish Cooperative Center Environment Project, Shared Control Natural Resource Project, Mahaweli Environment Rehabilitation Project, Forest and Land Use Mapping Project, Soil Conservation Projects in Ratnapura and Nuwara Eliya, Reforestation Project of the Integrated Rural Development Project, Environmental Action 1 Project, Sustainable Natural Resources Management Project, and Upper Watershed Management Project.

These projects have been launched in the form of mixed large- or small-scale activities under the supervision of about 40 institutions or departments belonging to several line ministries related to agriculture, forests, lands, livestock, water, and environment.

Apart from these projects, the Hadabima Authority of the Ministry of Agriculture has been involved in soil conservation and rehabilitation activities for around 20 years. However, the Hadabima Authority concentrated on the conservation of lands belonging to settlements. There are also national policies and plans formulated to secure soil conservation and watershed development over the long term, such as the National Conservation Strategy of 1988, National Forestry Policy of 1995, National Land Use Policy of 1996, National Water Resources Policy of 2000, National Environment Action Plan of 1992, and Forestry Sector Master Plan of 1995.

As public money and resources have been invested for such activities, there are different mechanisms, linked to project objectives, to monitor and evaluate the processes as well as short- and long-term impacts. A failure of a large-scale project would have repercussions for every citizen of the country, dropping the general living standard. In fact, implementation of an uncertain project without proper M&E would be disadvantageous for everyone. Therefore, physical, financial, process, and impact M&E have been blended into all projects in Sri Lanka to minimize the risks of project failures so that the stakeholders, the donors and the State are satisfied. Satisfaction is usually achieved through the participation of stakeholders themselves, internal supervisory staff, external consultants, and evaluators adopting their own and donor procedural guidelines. However, it should be noted that each project has been considered unique and hence has its own M&E procedures. Some examples are included in the latter part of this chapter.

Most national projects are designed as a top-down process involving material and financial assistance to stakeholders, mainly farmers. In some projects, simple implements are provided free to voluntary labour, whereas in others, payments are made for the work executed, such as construction of stone terraces, cutting of drains, or planting of double hedgerows, vetiver and grass strips. Payments are always made in installments subsequent to the monitoring of performance. Hence, it is not easy to evade monitoring,though there are sometimes considerable shortcomings.

In the process of monitoring, among the most important and commonly adopted tools in Sri Lanka are awareness meetings at which stakeholders will be educated on responsibility for self-monitoring based on a participatory approach. The importance of soil conservation and watershed development is passed on to stakeholders through study-circle activities, school programmes, films, and videos. These include attempts taken towards community empowerment that would later become *in situ* monitoring. Such procedures provide a foundation for *in situ* monitoring networks covering large regions as well.

Apart from agricultural projects, disturbances to soils, lands, and water bodies by construction activities are monitored, assessed, and evaluated through legislation stipulated by the Central Environment Authority. During the course of such actions, institutions such as the Natural Resources and Management Center of the Department of Agriculture, universities, research institutes, and other qualified agencies are given the responsibility to strictly monitor on-going activities at the field level, aiming at elimination of failures.

M&E OF MAJOR PROJECTS

The first watershed management project in Sri Lanka started in 1975, was FAO funded, and ran for nearly eight years. The Land and Water Use Division of the Department of Agriculture was the implementing agency. Two large catchments were instrumented to measure runoff and soil loss. Several small plot studies were also conducted to find the effect of different crops on soil erosion. As this project was mainly of a pilot nature, M&E was limited to periodic progress reports with a summary of results.

The Reforestation and Watershed Management Project commenced in 1980 and was implemented by the Forest Department with the financial and technical assistance of USAID. It focused on institutional development of the Forest Department, establishment of forest plantations for fuel wood in the dry zone, and introduction of vegetative cover on erodible slopes of the Central Highlands (Bandaratilake, 1989).

As the Forest Department has implemented it using only its internal labour force, the progress monitoring was mainly confined to existing internal mechanisms.

Monitoring of watershed development projects became somewhat more complicated as it advanced with the introduction of participatory management to programmes where people were considered as stakeholders, implementers, and beneficiaries simultaneously.

The Upper Mahaweli Watershed Management Project (GTZ, 1990), Swedish Cooperative Center's Environment Project (Gibbon et al., 1998), and Shared Control Natural Resource Project (Jinapala et al., 1998; Makadawara, 1998), are considered as major projects, which have adopted different M&E techniques. The Upper Mahaweli Watershed Project used maps to indicate areas of interventions and provide tabulated information such as number of training programmes conducted and number of cuttings and planting materials distributed. The Swedish Cooperative Center's Project obtained the services of external consultants on a regular basis to prepare M&E reports, whereas the Shared Control Natural Resources Project used questionnaire surveys for M&E.

M&E OF UPPER MAHAWELI WATERSHED MANAGEMENT PROJECT (UWMP)

The most sophisticated and comprehensive M&E ever carried out for watershed development projects in Sri Lanka has been employed for the UWMP, implemented by the Ministry of Environment and Natural Resources and funded by the ADB. This project was launched in 1998 and successfully completed in 2006. The detailed monitoring programme adopted for the soil conservation component of the project was reported by Gunawardene and Esser (2000). However, the procedure in a nutshell will be presented here as an example of appropriate M&E. The institutions for implementation and procedures to be followed were planned beforehand. Existing administrative set-up for reaching the village level was drawn in to ensure the credibility of implementation, monitoring and evaluation.

Intervention of on-farm and off-farm soil conservation activities in four "critical sub-watersheds" was identified as one of the activities of the UWMP. "Critical sub-watersheds" are considered to be "watersheds with significant areas exposed to active soil erosion" and have been tentatively identified on the basis of digital maps showing land use and slope categories.

Implementation of soil conservation measures on farmers' fields comes under on-farm soil conservation. These measures include identification of lands for intervention, preparation of farm plans, marking contour lines, adoption of recommended soil conservation measures, provision of grass and vetiver cuttings, and payments to farmers for soil conservation work completed in their fields. The recommended on-farm soil conservation measures for the project area are construction of stone walls, soil bunds, different types of terraces and drains (lock and spill, diversion and leader paved with grass and stones at times), and establishment of double hedgerows, vetiver and grass strips.

The Provincial Departments of Agriculture are the primary institutions responsible for the implementation of on-farm soil conservation work in each province. The Provincial Director of the department is the officer in charge of this work. The Director undertakes this work through the Provincial Deputy Director of Agriculture, Assistant Directors of Agriculture of each Divisional Secretariat, and Agricultural Instructors (AI) of each Agrarian Services Division. The Project staff facilitate the work at various stages of implementation.

Given below is the procedure followed for the implementation, monitoring and evaluation of the on-farm soil conservation work in the project area.

1. The interventions are based on *Grama Niladhari* (GN) division, the area assigned to a permanent government officer or *Niladhari* linking the information at village level to the Divisional Secretariat. The annual task for each province is decided and agreed upon at a provincial workshop followed by another workshop conducted by the Provincial Director of Agriculture with his staff. The detailed plan for each Agriculture Instructor (AI) range is to be worked out at this second workshop.
2. A Participatory Rural Appraisal is conducted for each GN division. A field manager of the project with five or six Social Mobilizers (SMs) participate in this endeavour with representatives from the implementation agencies and an AI. This becomes the official indication for the farmers in the GN division that the soil conservation work has started.
3. Usually the soil conservation work starts from the upper reaches of GN divisions and moves to the lower reaches. This facilitates the proper designing of drainage canals. The most critical areas are identified after meeting local officers such as the AI, Divisional Officers of the Agrarian Services, and *Govi Niyamakas* (Agricultural Research and Production Assistants) of the GN division. The information from the Participatory Rural Appraisal was also used for verifications.

4. Once the critical areas were identified, SMs met the farmers and got their help to prepare the farm plans and fill up the data sheets required by the project consultants. The AI marks the proposed conservation measures on the farm plan and also on the ground along the contours after agreeing with the farmer implementing the work.
5. The SMs hand over the farm plan with the data sheet to the project office to be entered into the soil conservation database. Once data entry is completed, the form is returned to the SMs.
6. The SMs were requested to maintain regular contact with the farmers and monitor progress periodically. They also help farmers obtaining necessary inputs such as vetiver or grass cuttings and act as liaison officers to the AIs.
7. Once a farmer completes 50% of the planned implantation work, the AI measures and evaluates the quality of work done and the information is entered in the data sheet maintained by the SM. Once again, the data sheet is brought back to the project to be included in the database.
8. The payments due to the farmer are prepared by the AI and given to the Provincial Agricultural Office through the Assistant or Deputy Director of Agriculture. The Provincial Director of Agriculture in each province then arranges to pay the farmers through the Deputy Director of the Provincial Department of Agriculture and Assistant Director of Agriculture in respective Divisional Secretariats.
9. The final payments are made in the same way once the farmer completes the remaining 50% of soil conservation work.

In the M&E process, the SMs were trained to use clipboards, farm data forms, grid sheets, volume calculation forms, compasses, Global Positioning System, and school mathematical sets for measuring slopes and drawing farm plans. The field managers use topographic maps (1:50,000), scientific calculators, and basic cartographic equipment. The officers of the Provincial Department of Agriculture use road tracers to mark the contours in farmers' fields.

QUALITY ASSURANCE IN UWMP

Consultants usually make brief visits to some of the areas in which soil conservation works have been completed. Up to now the work has been satisfactory except in some places where contours were not marked along the contour lines, stone walls were not constructed according to the specifications, and walls were raised only to prevent animals from coming to the farms.

A biannual monitoring programme by the Natural Resources Management Center (NRMC, the organization mandated for soil conservation work in Sri Lanka) of the Department of Agriculture is expected to guarantee the quality of work executed. The NRMC is requested to visit sample sites and submit an evaluation report to the project authority. It also conducted field days to ascertain the shortcomings and regular training sessions for stakeholders. In this endeavour, the qualitative as well as quantitative aspects of UWMP were monitored and evaluated by categorizing the overall performance of farmers as good (all advice followed as recommended), satisfactory (some recommendations were not considered but measurable or acceptable), and unsatisfactory (considerably short of recommendations), expressed in percentages. The evaluation team comprises a soil scientist, an agronomist, a soil surveyor, a soil conservation specialist, and a land evaluator.

DATABASE FOR M&E

A database for storing intervention data, calculating payments to farmers, and monitoring the implementation progress was designed to match the forms used for data collection by the SMs. Provision was also made for updating payment rates without influencing records of previous payments. The data on the following parameters were entered at the planning, halfway, and final stages of implementation:

- File number with SM code
- Name of farmer
- Address
- GN division
- Divisional Secretariat division
- Province
- Date
- Point location
- Formal farm area in ha
- Actual farm area in ha
- Land tenure
- Irrigation facility
- Cropping seasons
- Slope in %
- Intervention area in ha
- Stone walls in m^3
- Single tree terrace
- Soil terrace in m^3
- Soil bund in m^3
- Diversion drain in m^3
- Lock and spill drain in m^3
- Double hedgerows in m
- Vetiver strips in m
- Grass strips in m
- Stone paved leader drain in m
- Grass paved leader drain in m
- Specific remarks

The database was used to receive the following comprehensive information as outputs in printed form:

- Amounts to be paid to individual farmers for half-completed structures including expenditure for each GN division
- Amounts to be paid to individual farmers for completed structures including expenditure for each GN division
- Amounts to be paid to individual farmers for additional completed structures
- Amounts to be paid to individual farmers for all completed structures
- Lists of land tenure, intervention areas, and completed structures

The database also provided summary reports as and when required and facilitated the export of files for point locations for each farm with ID number, intervention area, and date of completion as well as all raw data. Point locations transferred data to the Mahaweli Authority of Sri Lanka for the preparation of maps showing the areas exposed to erosion and locations of interventions. Raw data may be of interest to a wider range of users such as ministries and educational and research institutes.

The summary reports include the following:

- Conservation structures sorted and summarized for the entire project area, districts, Divisional Secretariat divisions and GN divisions
- Conservation structures sorted and summarized for watersheds, sub-watersheds and GN divisions
- Payments made to farmers for conservation structures sorted and summarized for the entire project area, districts, Divisional Secretariat divisions and GN divisions

AN EXAMPLE FOR IMPACT ASSESSMENT

One way of finding the impact of soil erosion control measures was to find the gradual decrease of sediment load carried by the stream leaving the watershed. Sediment discharges from micro- (< 1 ha), mini- (1-5 ha), and sub-watersheds (about 25,000 ha) were monitored by the UWMP through a contract with the Mahaweli Authority as it is well equipped for M&E activities. This organization has the necessary equipment and expertise for monitoring activity. Data for the monitoring site of the sub-watershed was available for a period of five years before the implementation of the project. This data provided the necessary background information on which to base post-intervention observations.

The mini- and micro-watersheds provided data that were directly linked to the intervention measures. At each mini- and micro-watershed, the instrumentation consisted of an H-flume, a flow recorder, a silting basin at the entrance to the H-flume to collect the bed load, and a mixing tank below the outlet of the H-flume to ensure proper mixing of suspended material.

This stream gauging and sediment monitoring programme had run for six years, and the time series data provided an indicator of the project's impact on soil conservation. Any reduction of sediment load carried by the river will extend the life span of the multipurpose Rantembe reservoir constructed at the lower reaches of the river. The investment in the UWMP was justified partly by the reduction of silting rate so that benefits accrued from power generation would help to set off the cost of the project.

M&E BY HADABIMA AUTHORITY

The major activity of the *Hadabima* Authority of the Ministry of Agriculture is to improve land productivity through soil and water conservation of marginal slope lands subject to severe erosion in the past. The M&E procedures of the Authority have been altered from time to time in accordance with the suggestions of periodic evaluation reports. From 1978 to 1980, the Authority had a special monitoring unit working in collaboration with its internal audit unit at Provincial and Divisional Secretariat levels.

Thereafter, monitoring was performed through the monthly meetings of the *Hadabima* staff. The need for a monitoring unit was felt again when the *Hadabima* Authority was vested with responsibility for launching a project linked to the World Food Programme ran until 1994, major aspects of which were soil conservation and fertility improvement. External monitors were also drawn in during that period. At present, monitoring is again being done through the monthly meetings.

The field officers named at *Hadabima* as Unit Managers bring the information based on their field visits in the form of written reports to monthly meetings. Most of the time, external evaluators are drawn in for preparation of evaluation reports for which the evaluators visit intervention locations selected at random and make observations and measurements blended with verbal information of the stakeholders. For example, in 1999, a group of evaluators of the University of Peradeniya evaluated the soil and water conservation programme. The evaluation team followed a procedure focused on input/output relationships. Five percent of the stakeholders chosen at random were evaluated using questionnaires with field measurements. The views of the Unit Managers were also given due consideration. It was evident that a satisfactory message had gone to the stakeholders via Unit Managers pertaining to the importance of soil conservation (Jayakody and Wanigasundara, 2000). However, as a non-uniformity of monitoring activities was observed among different regions, it was suggested that the knowledge of Unit Managers be upgraded through a special educational programme, which was completed at the Post-graduate Institute of Agriculture of the University of Peradeniya.

NATIONAL INSTITUTIONALIZATION OF M&E

In August 2002, the Ministry of Policy Development and Implementation established several Sector Monitoring Committees that were linked to the National Operation Room at the Central Bank of Sri Lanka, where the National Database is kept. The Ministry of Plan and Plan Implementation is mandated to monitor, evaluate and review the development agenda of the government. Sector committees were established to monitor and evaluate the performance of foreign-funded projects and other related activities. The soil conservation and watershed development projects were allocated to the Agriculture Sector and Environment, Land and Natural Resources Monitoring Committees, respectively. The task of the monitoring committees was to look into project performances and draw the attention of the Ministry for required adjustments. The Monitoring Committees were expected to identify the problems, issues and constraints that inhibit the effectiveness of projects and recommend appropriate measures to the policy makers.

The M&E mechanism of the Agriculture Sector is divided into ministry and sub-national levels. The activities at the sub-national level are further divided into provincial, district and divisional levels. These are mostly looked at on a monthly basis through the sittings of the relevant groups and committees. The important outcomes will be passed on to the Ministry to take necessary action and be considered in policy formulations.

CONCLUSIONS AND RECOMMENDATIONS

It is apparent that there should be diverse modes of M&E for soil conservation and watershed development projects depending mainly on the magnitude and nature of each project as well as its procedure of implementation. Establishment of databases as done for UWMP in Sri Lanka has to be considered a very important component of the entire exercise, as these databases will provide immediate access to information when and where required.

A national-level compilation of critical data and progress information through a central M&E body such as the National Operation Room in Sri Lanka provided a sense of responsibility to the people and institutions involved in execution of activities at all levels. This has to be considered also as a procedure for incorporating transparency and verification of effectiveness of the projects to be used by policy makers at different stages. Management information needs and data for future planning could also be extracted from such central databases. This type of central pooling of information would be more beneficial for a broad spectrum of users than having information in individual project offices satisfying only the donor agencies.

The M&E of projects should be increasingly geared towards the level of beneficiary responses through participatory approaches. The periodic meetings held with stakeholders in Sri Lanka to educate them and to receive their feedback on successes and failures may be advantageous for many others as well. In addition, stakeholders would gain a sense of responsibility and the importance of active participation towards national development.

References

Atukorala, G.K. 1995. Country paper on Sri Lanka. pp. 228-240. *In:* Soil Conservation and Watershed Protection in Asia and the Pacific. Asian Productivity Organization, Tokyo.

Bandaratilake, H.M. 1989. Problems and status of watershed management in Sri Lanka. Paper prepared for the Government Consultation Meeting, Forest Department, Colombo, 25 pp.

Gibbon, D., A.A. Kodituwakku, A. Lecamwasam and S.G. Girahagama. 1998. Evaluation of the Swedish Cooperative Center's Environment Project in Sri Lanka. Department of Rural Development Studies, Swedish University of Agricultural Sciences, Uppsala, Sweden, 18 pp.

GTZ. 1990. Upper Mahaweli Watershed Management in Sri Lanka. Report on the project progress review. Upper Mahaweli Environment and Forest Conservation Division, Polgolla, Kandy, 32 pp.

Gunawardena, N. and K. Esser. 2000. Soil conservation and hydrological monitoring. Report submitted to the Upper Watershed Management Project. Ministry of Environment and Natural Resources, 39 pp.

Jayakody, A.N. and W.A.D.P. Wanigasundara. 2000. Evaluation Report of the Development Programme - 1999, Hadabima Authority of Sri Lanka, Faculty of Agriculture, University of Peradeniya, Peradeniya, Sri Lanka.

Jinapala, K.D., J. Merrey and P.G. Somaratna. 1998. Institutions for shared management of land and water on watersheds. pp. 67-88. *In:* M. Samad, N.T.S. Wijesekere and S. Birch. [eds.] Proceedings of the National Water Conference on Status and Future Directions of Water Research in Sri Lanka. BMICH, Colombo, Sri Lanka, November 4-6, 1998. International Water Management Institute, Colombo, Sri Lanka.

Krishnarajah, P. 1984. Erosion and the degradation of the environment, Annual Sessions of the Soil Science Society of Sri Lanka, 1-10.

Makadawara, S.M.B. 1998. *Uma Oya* Catchment Conservation. Report submitted to the Upper Mahaweli Environment and Forest Conservation Division, Polgolla, Kandy, 32 pp.

Nayakakorale, H.B. 1998. Human induced soil degradation status in Sri Lanka. J. Soil Sci. Soc. Sri Lanka 10: 1-35.

SOE. 2000. Status of the Environment Report Sri Lanka, First draft, Ministry of Forestry and Environment.

Stocking, M.A. 1986. Soil conservation in land use planning in Sri Lanka, Consultant's Report No. 3, Colombo, Sri Lanka, UNDP/FAO Project, SRL 84/032.

10

Regional Monitoring and Evaluation of Soil and Water Conservation — A Case Study on Loess Plateau, China

Rui Li, Qinke Yang, Xiaoping Zhang, Zhongmin Wen and Fei Wang
Institute of Soil and Water Conservation (ISWC) CAS and MWR, Northwest Sci-Tech University of Agriculture and Forestry, Yangling 712100, Shaanxi, China. E-mail: lirui@ms.iswc.ac.cn

ABSTRACT

Loess Plateau in China is well known in the world for its unique landscape and severe soil erosion. To meet the needs of decision-makers, the Soil and Water Conservation Monitoring Information System (SWCMIS) was developed, which is oriented to the current situation of soil erosion and assists decision-making on soil conservation for this region. The system consists of two parts: database of soil erosion and conservation including data collection and processing and a knowledge library. The Multi-level Remote Sensing Monitoring Information System in the Loess Plateau consists of three levels. Because soil erosion is a complex process and is affected by many factors, such as climate, landform, soil, vegetation and human activities, it is necessary to integrate remotely sensed data and other data, such as DEM (digital elevation model), observed data, the fluvial sediments and runoff plot data, as well as social economy.

Based on the database and results of related research, a model of soil erosion at a regional scale was developed. It is an exponential function correlating erosive ability with rainfall in the rainy season, gully density, ratio of slope land, content of soil particles > 0.25 mm, and vegetation cover. The main procedures of modeling include making base-maps by overlaying maps of the main factors to extract parameters from remote sensing data and thematic maps or observations, establishing a statistical regression model, and evaluating and predicting the regional situation of soil and water conservation on the Loess Plateau. The results are surprisingly statistically significant.

INTRODUCTION

China suffers from severe soil erosion. From the second remote sensing investigation, the area affected by soil erosion was 3.56 million km^2 at the end of the 1990s, of which 1.65 million km^2 was affected by water erosion and 1.91 million km^2 by wind erosion. A large number of experiments were done by Chinese researchers during the long struggle against soil erosion (Xianmo Zhu et al., 1992; Qinke Yang, 1994). The comprehensive management of smaller watersheds has resulted in 2.0 million ha of cropland converted to forest and grass, 100,000 km^2 of natural forest and rangeland have been protected, and 40 million people have been relieved of poverty. These efforts provide a sound foundation for the comprehensive control of soil erosion in China (Wenzhi Yang and Congzu Yu, 1992; Guiqing Song and Zhijie Quan, 1996). In recent years, the State Council authorized the National Eco-Environmental Construction Programme and the National Eco-Environmental Protection Plan. This means that with the development of the economy in China, soil and water conservation has moved into a new stage. Regional monitoring and evaluation of soil and water conservation has been considered a key technology. Loess Plateau is taken as a case study of a monitoring system and evaluation model (Poesen et al., 1996; Rui Li et al., 1998).

STUDY AREA

The Loess Plateau region is located in the middle reaches of the Yellow River. The landform is loess hills, sand-loess hills and loess tableland, with a gully density of 4-6 km km^{-2} (Figs. 1 and 2). The inter-gully area takes about 40-60% of the total. The climate is typical continental climate, with mean annual temperature of 6.6-14.3°C and mean annual rainfall of 250-700 mm. The precipitation is concentrated in summer, which accounts for 50-70% of the annual precipitation. The population density is 40-270 people km^{-2}. The Loess Plateau in China is well known for its unique landscape and severe soil erosion. This region has been cultivated for nearly 6,000 years. During the last few centuries, especially in the last hundred years, natural vegetation has been destroyed because of

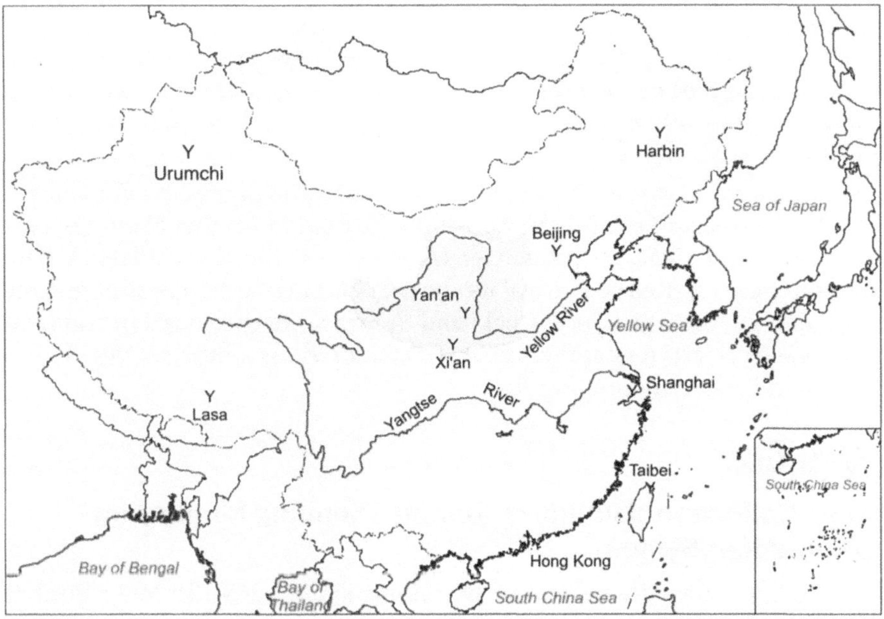

Fig. 1 Location of Loess Plateau (in yellow), China.

Fig. 2 Landscape of Loess Plateau in 1999 (Photo: ISWC).

increasing population (Guiqing et al., 1991). From the investigation of the Yellow River basin in 2000, the area of soil erosion by water was 574×10^3 km^2 and the area of soil erosion by wind was 130×10^3 km^2. Tables 1 and 2 show the intensity of erosion by water and by wind, respectively, of the Yellow River basin.

In the development strategy of western China, the Loess Plateau has held an increasingly important position. The biggest national energy base of heavy and chemical industries is located on the Loess Plateau. This region grows apples because of its special climate. It is rich in sunlight and heat energy, the soil layer is thick, and there are vast areas of land suitable for forestry, fruit trees and grass. It is considered one of the key regions for ecological rehabilitation in China.

METHODS

Data Collection: Multi-level Remote Sensing Monitoring Information System

In order to meet the needs of decision-makers, a Multi-level Remote Sensing Monitoring Information System was developed to supply

Table 1 Intensity of soil erosion by water of the Yellow River basin.

Grade of erosion	*Amount of soil loss, tons km^{-2} yr^{-1}*	*Area, 10^3 km^2*	*Proportion of total erosion area, %*
Very slight	< 1,000	259	45.1
Slight	1,000-2,500	104	18.2
Moderate	2,500-5,000	91	15.8
Severe	5,000-8,000	72	12.6
Very severe	8,000-15,000	35	6.1
Extreme	> 15,000	13	2.2

Source: Bureau of the Upper and Middle Reaches of Yellow River Committee (2000).

Table 2 Intensity of soil erosion by wind of the Yellow River basin.

Grade of erosion	*Amount of soil loss, tons km^{-2} yr^{-1}*	*Area, 10^3 km^2*	*Proportion of total erosion area, %*
Very slight	< 200	18	14.2
Slight	200-2,500	31	24.0
Moderate	2,500-5,000	33	25.5
Severe	5,000-8,000	19	14.8
Very severe	8,000-15,000	11	8.9
Extreme	> 15,000	16	12.6

Source: Bureau of the Upper and Middle Reaches of Yellow River Committee (2000).

information on the most current situation of soil and water erosion of this region. Generally speaking the system consists of three levels. The top level is based on environmental satellite imagery, such as FENGYONG (China), NOAA (AVHRR) and MODIS. It is focused on the general tendency of the whole plateau region to change, especially in terms of shifting forest boundaries in the southern part and desertification in the northern part. The middle level uses LANDSAT (TM) or SPOT to monitor the key regions, such as key catchments for soil conservation, mines, and main areas of exploitation. The dynamic changes of land use, land degradation and key engineering works are monitored and analysed at this level. The lowest level is focused on small catchments and observation stations distributed in different regions. The main data are regular aerial photographs of large scale (1:10,000). Land use, soil erosion, conservation practices and their effects are monitored in detail. From the basic monitoring of small catchments to the middle regions, then to the whole plateau area, a monitoring network has been formed. Remotely sensed data give more accurate and updated information (Feng Jiao et al., 1998).

Ground observation stations have been established in different regions. The stations collect local ground data including change of land use, soil conservation practices, sedimentation of small watersheds, and other data related to soil erosion.

Data Processing: GIS-based Integrated Information System

Regional soil erosion is a complicated and synthetic process affected by many factors including geomorphological conditions, soils, climate, hydrology, land use, vegetation, and soil conservation measures. In this study, the following data were collected and used:

- Precipitation data: All the rainfall data from 178 stations (1955-1986) in Loess Plateau were collected. Using this data, we calculated the mean annual rainfall in the rainy season (July to October) and measured the latitude and longitude of each station.
- Soils data: Information was based on the Soil Map of Loess Plateau (produced by the Chinese Academy of Sciences in 1991). The content of soil organisms was extracted from several soil monographs.
- Gully density data were extracted from the soil erosion map of Loess Plateau.
- Land use data: The ratios of cropland and forest/grass were extracted from land use map (1:250,000) in ARC/Info coverage format.

Precipitation, sediment and gully intensities are in point format; they are processed in ARC/Info and SUFFER software to generate a contour map. All data have been transformed into common projection, using Albers projection.

In this study we divided the study area into 3,380 map units, according to the TM imagery. All the parameters were pre-processed into maps and then integrated into each of the basic polygons. In the processing of integration, position data and topology relationship were put into a standard basic information unit map; the boundaries of the basic unit do not change, that is the number of units is fixed. Attributes can be added again and again. The basic information units have key words to relate many kinds of attribute data layers.

Regional Erosion Evaluating Model

Based on the systematic analysis of genesis, evolution, and related factors of soil erosion at the multi-scales, the general form of the models for regional erosion should be as follows:

$$A = f\,(Q, S, g, v, c)$$

where Q = hydrology factor, S = soil factor, g = geomorphology factor, v = vegetation factor, and c = soil conservation measure factor.

It is an exponential function correlating erosive ability, rainfall in the rainy season, gully density, ratio of slope land, and content of soil particles > 0.25 mm, and it is an exponential correlation with vegetation cover. This model can be regressed to give the following mathematical formula:

$$L = 0.4735P^{0.9282}.S^{-0.08855}.G^{2.2666}.M^{0.07254}.c^{-0.00047}$$

where L = erosive intensity (t km^{-2}.a), P = rainfall in rainy season (mm), S = content of soil particle > 0.25 mm (g kg^{-1}), G = gully density (km km^{-2}), M = percent of slope land, C = vegetation cover (%).

Multi-correlation statistics are R = 0.9369, F = 2984.64 >> $F_{0.05}$ = 2.21. The statistical correlation significance is remarkable. From the regression result, we can see that the erosive intensity shows positive correlation with rainfall in rainy season, gully density, and ratio of slope land, while it presents negative correlation with the content of soil particles and vegetation cover (Burough, 1998; Kirkby et al., 1996).

RESULTS AND DISCUSSION

General Changes of Regional Erosion Status

From the remote sensing imagery analysis, during the last 10-15 years the erosion area has been reduced and the intensity is diminished. Compared with the investigation results before 1990, the total eroded area reduced 52×10^3 km^2 (the area of very severe soil erosion reduced 17×10^3 km^2 and the area of extreme soil erosion reduced 20×10^3 km^2). The main reason is that soil and water conservation practices have become increasingly important. Some eroded areas have been controlled by various soil and water conservation measures, including terraced fields (13 million ha), reforested areas (43 million ha), fruit tree cover (5 million ha), and grass cover (4.3 million ha). In addition to this, 1,403 dams and reservoirs and 100,000 check dams have been built. It is estimated that these measures can potentially collect 1.5 billion tons of eroded soil and 300 million tons of sediment are prevented from being transported into the Yellow River.

Selection and Integration of Data

The current satellite observation system can provide abundant data for dynamic monitoring of the environment. But a satisfying analysis cannot be obtained from remotely sensed data alone because soil erosion is a complex process affected by climate, relief, soil, vegetation and various kinds of human activities. Therefore, it is necessary to integrate remotely sensed data and other data, such as DEM, observed data, the fluvial sediments and runoff plot data, as well as social economy. In addition, the appropriate data should be chosen for each study objective.

Image Interpretation Procedures

Although manual interpretation and mapping is still a basic method, computer-assisted classification represents the direction of future development. It is difficult for common image processing software to identify the same objects with different spectra or different objects with the same spectrum. Especially in the Loess Plateau, the topography is so complex that the spectral features depend on landforms to some degree. So it is necessary to develop specific image processing technology for this case study by:

- Integrating remotely sensed data with others such as DEM, field station observation data, thematic maps.
- Pre-processing before classification. The PCA, Edge-extraction, LOG/EXP, and CURVATURE have been used for different regions.
- Interface-classification of image procedures.
- Post-processing to improve the result of classification.

Procedures for Soil Erosion Evaluation on a Regional Scale

The following procedures are recommended for a similar study.

- Make a base map (or map-unit map): A map-unit map is the base map for this study. Each map-unit has uniform landscape conditions and erosion. It is the basic unit of data collection and analysis. The map was made through overlaying maps of the main factors mentioned above.
- Extract the parameters from remote sensing data and thematic maps or observing data, and integrate all the parameters into the basic unit map derived from the first step. In this way a parameter database will be built.
- Establish a statistical regression model between sediment and the factors with correlative analysis and geo-statistics method.
- Use the model to evaluate and predict the regional situation of soil and water conservation on the Loess Plateau.

References

Bureau of the Upper and Middle Reaches of Yellow River Committee. 2000. Unpublished Report. Bureau of the Upper and Middle Reaches of Yellow River Committee, China.

Burough, P.A. 1998. Dynamic modeling and geo-computation. *In:* D. Karssenberg and P.A. Burrough, Faculty of Geographical Sciences. [eds.] Environmental Modeling in GIS. Utrecht University, The Netherlands.

Feng Jiao, Xiaoping Zhang and Rui Li. 1998. Application of GPS in soil and water conservation. Bull. Soil Water Conservat. 18(5): 32-34.

Guiqing Song, Li Rui and Jiang Zhongshan. 1991. Atlas of Experimental and Demonstrative Areas for Integrative Controlling in the Loess Plateau. Press of Survey and Mapping, Beijing.

Guiqing Song and Zhijie Quan. 1996. Theory and Practice on the Land Resources of the Loess Plateau. Hydropower Press, Beijing, China.

Kirkby, M.J., A.G. Imeson, G. Bergkamp and L.H. Cammeraat. 1996. Scaling up processes and models from the field plot to the watershed and regional areas. J. Soil Water Conservat. 51(5): 391-96.

Poesen, J.W., J. Boardman, B. Wilcox and C. Valentin. 1996. Soil erosion monitoring and experimentation for global change studies. J. Soil Water Conservat. 51(5): 386-390.

Qinke Yang. 1994. The classes and regions of soil erosion in China. *In:* Soil Science Study in Modern Time. China's Agriculture Science and Technology Publishing House, Beijing

Rui Li, Qinke Yang and Yong'an Zhao. 1998. Application of spatial information technology in soil and water conservation of China. Bull. Soil Water Conservat. 18(5): 1-5.

Wenzhi Yang and Congzu Yu. 1992. The Regional Reclamation and Appraisal of the Loess Plateau. The Science Press, Beijing.

Xianmo Zhu, Daizhong Cheng and QinkeYang. 1999. 1:15,000,000 soil erosion map of China. p. 200. *In:* Atlas of Physical Geography of PRC (2nd ed.). China's Cartographic Publishing House, Beijing.

11

Impact Monitoring as the Key for Testing the Long-term Watershed Development Project Hypothesis and Its Success – as Demonstrated in the Changar Project in India

Rajan Kotru
Indo-German Changar Eco-Development Project. Palampur, H.P., India.
E-mail: rkotru@gmx.net, rkotru@gtzindia.com

ABSTRACT

Over the last decade India has made huge efforts to address the issues of alleviating poverty, redeeming ecological balance, and creating economic avenues through watershed development programmes. Monitoring and evaluation is an integral part of project management that provides a continuous flow of decision-oriented information for planning, implementation and overall steering of such programmes. Within the project-specific monitoring and evaluation, the purpose of impact monitoring is to *measure the trend or the benefits achieved on the basis of development objectives set.* Since impact can be change of attitude, behaviour, skills of target groups (e.g. organizational capacity of a community), and change in their economic, socio-cultural, and socio-ecological environment, its monitoring is of prime importance to reflect the success or failure of project interventions.

The Changar project area consists of 37 ecologically degraded micro-watersheds, which have numerous sub-catchments. The concept of eco-development[1] designed by the project is being implemented on the principles of watershed development. Village-based planning became a focus for precise and verifiable indicators of the project goal, purpose and results. Participatory impact monitoring was introduced (for instance by involving the community in spring-head water flow or recharge observations by giving minor incentives). Similarly, meteorological observatories were monitored by school-going children against an annual incentive (in kind) to the school. The aim was not only to collect technical data but also to generate interest and motivation among the communities towards the project work and its subsequent monitoring. The data collected was used for re-planning, reorientation and impact measurement.

INTRODUCTION

Substantial international funds support poverty alleviation and environmental rehabilitation programmes in developing nations. In India, these are supplemented by several national programmes allocating funds for the overall development of rural communities. Funds are spent under various developmental programmes (e.g. rural development, forestry). Over the last decade India has made huge efforts to address the issues of alleviating poverty, restoring ecological balance and creating economic avenues through watershed development programmes. Such projects are being proposed and implemented in most of the geographical regions of the country. Till recently, however, the focus was mostly on target achievement and critics have insisted that performance of such projects has not always generated the desired impact.[2] Hence, in terms of project management, a distinct shift from the concern for project performance towards concern for objective achievements is also visible (Ojha, 1998). To measure the impact of project interventions the incorporation of monitoring and evaluation (M&E) in the project management is now widespread.

The introduction of AURA-Format in the German Agency for Technical Cooperation (GTZ) clearly underlines the concern of bilateral technical agencies also as we have started shifting from a project approach to a programme approach, aiming to achieve milestones and impact during the active implementation phase of a programme. Hence, data and

[1]It has four components: (1) community/community organization's mobilization for sustainable natural resource management, (2) natural resource management programme development, (3) capacity building of major stakeholders, (4) establishment of Himachal Pradesh Eco-Development Society.

[2]Impact is understood as a lasting and significant effect at the goal level, which is only identifiable some significant time after project completion (Gohl et al., 1993).

information – based on experiences and observations – are not only collected but also evaluated so that planning is efficient and corrective measures are taken in time. Moreover, among the donors there is a genuine concern – after the implementation is over – that the project completion reports include evaluation of the impacts that actually occurred and the effectiveness of mitigation measures (World Bank, 1991). The European Commission (CEC, 1993) stresses the analysis of results and impact of the project during and after implementation with a view to taking remedial actions or giving recommendations for the guidance of similar projects in future. This is especially the case for projects in which a number of implementation phases are involved. The current efforts of coordination among various international donors in India are part of the attempts made to create convergence in development philosophy and synergy in implementation so that a greater and quicker impact is achieved.

Monitoring and Evaluation and Project Management

Monitoring and evaluation is an integral part of project management. It provides a continuous flow of decision-oriented information for planning, implementation and overall steering of the project. Within the project-specific M&E, the purpose of impact monitoring (IM) is to *measure the trend or the benefits achieved on the basis of development objectives set.* This facilitates timely decision-making (e.g. re-planning, corrective actions) and the assessment whether the project strategy and its implementation are in line with the objectives set. Thus, IM sets the pathway to measure whether project interventions are *preventing, minimizing, mitigating, or compensating for adverse impacts* (Narayan, 1993), in other words, to assess which positive impacts resulted or which objectives were met.

Impact Monitoring in Practice

Since impact can be change of attitude, behaviour, skills of target groups (e.g. organizational capacity of a community), and change in their economic, socio-cultural, and socio-ecological environment, its monitoring is of prime importance to reflect the success or failure of project interventions. In spite of donor stipulations and attempts to incorporate IM in the overall framework of M&E, it has seldom been sincerely implemented owing to many practical difficulties:

- Partner institutions are not interested in IM.
- The project planning matrix or logical framework as the basis for IM remains a donor document.
- The focus is usually on physical targets and annual reports.
- The impact of a project is seen as a long-term effect and not considered during implementation.

- In inter-sectoral projects it is often difficult to decide which activity or sector can be of multiple use concerning impact-related data.
- Impact indicators are often not fully specified.
- There is no consensus on whether external or internal evaluators should measure impact.
- In the beginning of a project, financial monitoring, process and output dominate.
- The Indian auditing system rarely uses field checks to assess impact or quantitative monitoring.
- Impact monitoring based on community participation suffers because community awareness and mobilization is often slow and uneven in terms of commitment.
- The trust of line departments in communities to monitor is often lacking.

Often, partners associate IM with fault-finding rather than fact-finding. Most of the concepts are theoretical and unrealistic, so implementation remains ineffective and impractical. A simple, applicable and easily understandable methodology has remained evasive. Furthermore, it has to be pointed out that the project approach or method for IM is greatly influenced by the project background (e.g. goal, purpose specifications) and project design (e.g. target group). Hence, standardized techniques for IM can act only as a guideline. The IM concept, therefore, has to follow a balanced approach linked to the project planning document and requirements of both the donor and the target group (e.g. implementing counterpart or agency). In this chapter, the background of the GTZ-supported IGCEDP (Indo-German Changar Eco-Development Project) is presented and the IM concept followed is elaborated.

CHANGAR PROJECT BACKGROUND

Implemented through the Himachal Pradesh Eco-Development Society®, Palampur (HPEDS), IGCEDP has operated since 1994, when the GTZ entered the project, and its Phase II ends in 2006 (Phase II, 1999-2006). The project area is located in the state of Himachal Pradesh, India. It consists of 37 ecologically degraded micro-watersheds, which have numerous sub-catchments called mini-microwatersheds (MMWS) for the purpose of this project. The concept of eco-development designed[3] by the project is being implemented on the principles of watershed development (WSD), to

[3]It has four components: 1) Community/Community Organization's mobilization for sustainable natural resource management 2) Natural Resource Management Programme Development 3) Capacity Building of major stakeholders 4) Establishment of Himachal Pradesh Eco-Development Society

improve the management of existing land uses of micro-watersheds. In this context, WSD is a dynamic process, which integrates local people's problems and solutions with the help of experts. The project aims at the rehabilitation of degraded MMWS (average area per MMWS = 500 ha, average 5 villages per MMWS) by involving local village communities or user groups, mobilized through elaborated extension strategy, in the planning and management of their resources. Village Development Committees (VDCs) or Panchayats[4] as active institutions act as mobilizing, planning and implementation bodies for sustainable natural resource development and management. Hence, the watershed approach is followed village by village to cover the overall area of a particular MMWS. The advantage here is that even if all the villages of a MMWS do not equally get sensitized simultaneously, the watershed approach is applied for each village nevertheless.

The broad data of the project region are given in Table 1. The altitude of the project area varies from 500 m to 1,300 m above sea level. Geologically it belongs to the Shiwaliks or Outer Himalayas. The Changar region (*changar* is the local term for remoteness, rugged topography and water scarcity) falls in the warm sub-humid to humid agro-ecological zone, with annual rainfall of 1,500 mm to 2,000 mm, almost 75% of which falls between June and September. Natural erosive processes are accelerated by human impact. The main technical components of IGCEDP are afforestation with need-based trees and eco-income generation, soil conservation, water resource development, livestock management, and grazing land improvement. These components have the strongest bearing on ecological rehabilitation and subsequent eco-development.

Project Planning Matrix (PPM) and Impact Monitoring

In the context of GTZ projects, IM is defined as the measurement of goal and purpose achievement (reflected in PPM). Thus, it is related to changed

Table 1 Features of the Changar Project Area (2000).

Project area	439 km^2
Number of villages	593
Total population	130,000
Population growth	0.8% per annum
Average land holding	0.74 ha
Livestock population	80,000
Household size	4.8 persons

[4]Panchayat is the formal grassroots organization responsible for village administration.

conditions for the target groups by project interventions (GTZ, Summary Status Report, 1996). Natural resource development and capacity-building of the population in the target area provide basic monitoring parameters. The first project-phase (1994-1999) PPM was drafted in 1994 and was streamlined in 1997 in view of experiences and field realities. Modifications were necessitated in Phase II - as village-based planning became the focus - for overall consistency of the logic of the PPM and for more precise and verifiable indicators of the project goal, purpose and results. The development hypothesis is evident from PPM (Phase II) in Table 2 (only up to purpose level).

The long-term goal is to improve living conditions without destroying the resource base further. At the purpose level, the community's capacity-building comes to the fore as it manages its natural resources. The indicators further qualify the PPM as they set the stage for IM. The indicators at different levels were used to tailor the project- and situation-specific IM.

M&E IN THE CHANGAR PROJECT

In order to be able to compare the planned and actual facts and impacts of the activities carried out, M&E is conducted on a decentralized basis by the project M&E Unit responsible for the collection and processing of data. The system is divided into three parts:

1. Activity monitoring
2. Financial monitoring
3. Impact monitoring (eco- and socioeconomic)

Activity and Financial Monitoring

Activity and financial monitoring are closely related to the annual budget and physical targets. They are implemented according to a three-year village-based Integrated Resource Management Plan (IRMP), on the basis of participatory bottom-up annual planning, which in turn finds use in reviewing achievements (results, activities) and planning future activities and strategy. The relevant M&E documents are:

1. Project Planning Matrix (PPM, derived from goal-oriented project planning)
2. Plan of operation (derived from PPM)
3. Annual plan of operation (derived from village-based planning and plan of operation)
4. Fortnightly planning (project review meetings at headquarters as well as in field offices)

Table 2

Goal: Reduced ecological degradation in Changar area.		
Development goal: Improved natural resource management has led to better living conditions in the project area.	Drinking water is available during the whole year in 30% of the small catchments treated by 03/2004.	
Purpose: Village groups or organizations manage their natural resources sustainably with support of HPEDS and other institutions.	1. Women and men organized in village groups (active women members at least 40%) are increasingly taking over responsibility for implementing and updating their Integrated Resource Management Plans (IRMP). Number of village groups increased from 25 in 03/2000 to 40 in 03/2002 and to 100 in 03/2004. 2. At least 100 villages groups are jointly managing their community plantations with the Forest Department under JFM policy and as per IRMP by 03/2004. 3. Government and non-governmental organizations adopted the HPEDS approach while working in Changar. At least two such organizations are supporting the implementation of IRMP in at least four villages per contact office by 3/2004 4. At least 50% of the households in the project area benefit during 10 months of the year from an improved availability of drinking water for human beings and animals by 03/2004. 5. The grass production from six community areas increases on average by 75% up to 12/2002 in every contact office.	1. The H.P. government continues to support the concept of HPEDS and makes available the necessary personnel and finances. 2. The Indian government continues the decentralization process (Panchayati Raj Act) and it is implemented at village level. 3. The Forest Department continues to implement the JFM policy. 4. Ali stakeholders including external organizations support the village land-use planning approach.

Financial achievements are regularly monitored by sector on a monthly basis and by activity on a quarterly basis. The procedures and details for activity and financial monitoring are not elaborated further. However, their role in supplementing the IM findings is in no way underestimated.

The IM Concept

The IM concept was established to survey not only the long-term but also the short-term impact of project interventions. According to the PPM, the IM in IGCEDP was designed up to the goal level. In addition to the annual performance monitoring, impact assessment relied on specialized studies conducted through external experts. The baseline information pertaining to various hierarchical levels of PPM was established in the initial stages up to 1997. IM is developed as an ongoing process within the project for which information is established for longitudinal monitoring (following hierarchical levels of PPM) supplemented by studies by evaluators (internal or external) and participatory reflections of the beneficiaries. Indicators are developed for monitoring over periodic intervals to ascertain impact over time. The set of indicators have the characteristics of indicators as defined by Bollom (1998), which measure changes in regard to relevant objectives.

Impact monitoring has been sub-divided into eco-impact monitoring and farm budgeting (Table 3).

In IGCEDP the space for multifarious eco-impact-oriented data collection was wide (Table 3). Thus information regarding the pre-project grass yield on degraded community land, water recharge of spring-heads, and characteristics of degraded upper soil had to be collected on the basis of baseline in-depth studies. Participatory impact monitoring was introduced (for instance by involving the community in spring-head water flow or recharge observations by giving minor incentives). Similarly, meteorological observatories were monitored by school teachers and children against an annual incentive (in kind) to the school. The aim was not only to collect data but also to generate interest and motivation among the communities towards the project work and its subsequent monitoring. As communities' perceptions of change must also gradually flow into IM and as project staff are the major counterpart of these people, it was deemed necessary to use this combination rather than taking external evaluators. The focus fields of the IM including procedures are given in Table 3. Moreover, the survey procedures and remarks elaborate on the utility of each IM type.

Table 3 Features of IM types in IGCEDP.

IM types	*Survey method /Data source*	*Remarks*
Eco-impact:		
Plantation monitoring	Selected plantations, fixed sample plots, annual data collection Responsibility: project and community (with incentive)	The fixed sample plots on community and private land have progressively been developed as data source for multiple aspects concerning social, institutional, ecological and economical impact, over the short as well as long term.
Soil monitoring	Fixed sample plots of plantation Responsibility: project Not yet repeated	Soil samples are taken from upper and lower profiles and analysed through project soil-test kit with random samples analysed through other institutions. It is to be used for indicating eco-rehabilitation and soil improvement.
Grass production	Fixed sample plots of plantation Responsibility: project and people (no incentives) Few annually repeated	It is useful in depicting eco-rehabilitation, yield increase, economic benefit, quality of fodder, community-based institutional arrangements and plantation management interventions required. Good for short-term impact.
Water recharge Water flow	Selected spring-heads in treated catchments (preferably in plantation monitoring area) Responsibility: people (minor incentive) and partly project Annual data during monsoon	It reflects the improvement in water availability by soil and water conservation measures (including afforestation) and thus also eco-rehabilitation.
Surface runoff monitoring	In one selected catchment comparing degraded and treated site (H-flumes/V notches) Responsibility: project and people (minor incentives) Annual data in rainy season	It directly reflects the changes in surface runoff, silt load and preferably watershed-water balance if other parameters are available (e.g. climatic data). Good for short-term impact caused by afforestation and soil/water conservation measures.
Climatic appraisal	In three meteorological observatories installed in schools Responsibility: people and school (incentive, not cash)	This data, even if it does not meet the full requirements of the official Indian Meteorological Stations, has provided invaluable baseline data abou[t]

(Table 3 Contd.)

(Table 3 Contd.)

	Daily data collection Baseline character Local microclimate	the heterogeneous climate of Changar. The inter-sectoral data is useful for innovative activities and can be used in water balance calculations as impact assessment. Its maintenance and management reflect the community's advancement in self-management.
Photomonitoring	Plantation monitoring sites and soil conservation engineering measures Responsibility: project	The objects here are the plantation sites and soil and water conservation engineering measures in these plantations and on other specific locations (roadside). Visual display of eco-rehabilitation and management of assets by the community.
Farm budgeting: Household survey	Snapshot survey based on various household types Responsibility: project	It was partly repeated in 2003.
Animal census	Snapshot survey Responsibility: project and community (minor incentives)	Locally trained girls were responsible for data collection and this data acts as a baseline on livestock population, composition and productivity dynamics. It was repeated in 2001.
Lactation monitoring	Selected household/animal types for one full year Responsibility: project and community (minor incentive)	Locally trained girls were responsible for data collection and this data acts as a baseline on milk yield of various animal types, fodder requirements and seasonality aspects.
Fuel wood survey	Selected resource farmers Responsibility: resource farmers (no incentive)/project Not yet repeated (baseline)	In selected MMWS resource farmers were motivated to observe how much fuel wood they require in summer and winter with observation period lasting seven continuous days in each season.
Leaf fodder survey	Selected resource farmers Responsibility: resource farmers (no incentive)/project Not yet repeated (baseline)	In selected MMWS resource farmers were motivated to observe how much fodder they require in summer and winter with observation period lasting seven continuous days.

In addition to the IM types described in Table 3, early warning systems had to be set for assessing the progress of local communities towards self-management in natural resource management. Hence, a preliminary institutional impact assessment was made (VDC evaluation). The survey was designed by an external consultant and the data was collected by the M&E cell. This was supplemented by a target group impact assessment (resource-poor and other beneficiaries) by an external evaluator.

To achieve an early impact of multi-sectoral activities within the smaller watershed, four MMWS were developed as models for replication, for the financial phase, within which IM was focused. The IM approach and guidelines as understood by the project are summed up in Table 4. These reflect the IGCEDP's understanding of IM and various experiences and findings made since 1994.

A SNAPSHOT OF METHODOLOGY

The photo sequence in Fig. 1 shows the process and result from the improvement of the project area from 1995 to 2003. The effectiveness of photo-monitoring as one of the tools to assess visible or physical change through watershed treatment is demonstrated. The sensitization of local communities and their involvement in planning and implementation in turn leads to greater participation in monitoring and management.

In the context of IGCEDP, progressive IM is seen as a gradual elaboration of monitoring on the basis of one major sectoral activity (e.g. community forestry), which can reflect on various aspects (e.g. production, biodiversity, social changes) linked to other inter-sectoral activities subsequently. It combines quasi-experimental data with other modes of information (e.g. consultant reports, community's view) to measure the impact. Similar procedures can be followed in other sectors. Progressive IM is purposely demonstrated below on the basis of forestry interventions (e.g. afforestation), since in the Changar project it allowed inferences to be drawn on social, institutional, economic and ecological impact (in terms of results achieved). It is true that a direct social and institutional impact assessment can reflect on the impact of any particular technical input such as forestry interventions. But experience showed that community mobilization is a slow process and cannot be accelerated through top-down procedures. On the other hand, the technical inputs in the field are generally given at faster rate and can be easily measured or analysed (e.g. supplemented by quarterly or annual activity monitoring data). Hence, IM prioritized the forestry sector. Moreover, along with the soil and water conservation measures (engineering and bio engineering) it constituted over 70% of the annual budget spent.

Table 4 IM approach and guidelines.

IM guideline	*Approaches and status*
• IM is a major tool for short-term decision-making with a clear long-term focus on assessment of project goal and purpose. • IM is an ongoing process in the project during implementation. Rapid appraisals by outsiders are often biased in understanding the project fundamentals and information collection. • IM consists of two parts: eco-impact and farm budgeting monitoring with an overall reflection on self-development of beneficiaries, village institutions and economic betterment of target groups. • IM data collection is systematic and based on applied science with gradual incorporation of target groups according to the project stage. People's observations, studies and reports do the balancing act. • OVIs should be individual indicators and can be used for drawing inferences on social, institutional, economic and ecological issues. • Village integrated resource management plans rather than PPM can be the basis for IM.	• Baselines for both ecological and economic IM are established for temporal monitoring of the identified indicators. Information is generated through periodic surveys and observations. • Quasi-experimental designs are used for eco-impact monitoring. Data is collected by villagers and analysed by the project IM cell. • The data generated through participatory rural appraisal has led to a comprehensive village integrated resource management planning that sheds light on indicators at the goal and purpose levels. • The IM feedback is mostly based on eco-impact monitoring and farm budgeting appraisals, most of which get reflected in regular monitoring of project management staff. • A gradual shift towards community involvement and individual beneficiary in measuring or assessing impact. • Monthly meetings of project management staff, fortnightly planning, intensive field trips, interaction with people, in-depth studies involving beneficiaries and selective use of consultants are other approaches to IM in use. • Experience indicates the need to select a primary activity with maximum impact on goal and purpose level and focus on it. • The usefulness of IM is recognized only in the later stages of the project. • Initial and minor incentives (no cash) towards people- or individual-based IM acts as a catalyst for participatory IM.

Forestry-related indicators are evident in PPM at all the major levels (i.e. goal, purpose and result). Selected plantations of particular MMWS (including four test MMWS, i.e. where intensive work was done to develop replicable models in Phase I) are being monitored on the basis of

Fig. 1 Planning and implementation by local community leads to responsibility in monitoring (top). Effect of engineering measures in revegetation near Sakri Village, in 1995 (middle left) and 2003 (middle right). Rehabilitation of river bed at Jalag II, from 1995 (bottom left) to 2003 (bottom right).

systematic network and slope-wise equal number of sample plots (upper, middle and lower). The sampling was done on fixed sample plots (marked) with number of trees kept constant at N = 6 or N = 10. In a few cases the constant sample plot area approach (radius = 5 m) was also applied. The data is collected annually (i.e. after culmination of annual growth). The primary aim was to monitor the long-term growth and survival of various tree species. Moreover, the improvement of the ecological situation could be monitored gradually as fixed sample plots were used to deliver multifarious data: changes in species composition, upper soil thickness, and increase in grass yield (e.g. savings made in grass purchases). Thus, information is available on the production of biomass and rehabilitation of degraded land. Moreover, if plantation enclosures are respected by communities (with no browsing damage on trees, no broken fences), social development in terms of the community's contribution to protection and management of plantations is obvious. Similarly, it reflects whether the VDC has been able to act as a responsive institution.

The procedure of plantation monitoring was initiated in selected MMWS treated intensively (technically), which also meant intensive extension. Most of the village-based plantations within such MMWS are monitored (the data from the plantation journal prepared by the field staff is considered as an additional monitoring input and good for cross-checks also). Other monitoring aspects are gradually linked to plantation monitoring (e.g. spring-heads that fall in the treated catchment or plantation are monitored). In combination, the data from all villages give a comprehensive basis for IM for the whole MMWS. The plantation monitoring procedure is as follows:

- Selection of plantation (in a treated site of MMWS, also baseline data collection).
- Fixed number of sample plots (marked in treated area).
- Data collection on multiple aspects (survival percentage of trees, growth in height and diameter, natural regeneration, damage on trees, grass yield, biodiversity, nutrient content, damage to fence).

The idea of focusing on only one major activity with various impacts, as was the case in forestry, has proved a very successful tool of IM. The initial high input of the project IM cell subsides gradually as community and individuals at village level can easily verify through observations whether grass yield on the fixed sample plots has increased or decreased or whether a weed has infested the sample plot area over the course of time. Moreover, the data from fixed sample plots can be easily calibrated by documented or verbal information and impact inputs such as regular individual or community-based group reflections (e.g. VDC monthly meeting, project review meetings), field staff observations (through field work, visits, plantation journal), and external reports and experiences (e.g. impact assessment).

CONCLUSIONS

The following conclusions are drawn on the basis of experience so far:

1. Definitions at goal, purpose and result levels should be clear and simple if efficient and individual indicators are to be carved out for effective IM.
2. The activity that has a dominant role in achieving the goal or purpose[5] can be used to spell out indicators so that the base for testing development hypothesis is set early on.
3. Short- and long-term IM assessment can be combined through a set of individual indicators.
4. PPM based on goal-oriented project planning needs to be updated during the early years on the basis of baseline consolidation in the field and with various target groups.
5. IM is not only a tool to show the positive or negative impact of a particular activity but also a major means for short-term decision-making for corrective measures (e.g. failure of a tree species or non-acceptance of a fenced plantation by community, i.e. cattle-grazing despite ban).
6. The monitoring of plantation development on the basis of certain indicators ultimately provides reflection on the process of community skill-building for the management of a plantation. It can be asserted that if the short-term indications of the right trend are evident, then the long-term goal is likely to be achievable.
7. External evaluators can perceive impact only to a limited extent as they do not have data or information about the evolutionary processes of a change unless documented. Moreover, rapid appraisals can easily lead to a biased opinion.
8. A gradual shift towards larger involvement of community and individual farmers in M&E and IM is justified, as community mobilization is a slow process. But it is necessary to validate or verify the people's observations and information.
9. An integrated resource management plan as designed by the project is ideal to form the basis for a community-based M&E system and impact evaluation subsequently.
10. Intensive field visits and interaction with the communities and user groups on project work are an effective source of impact-related information.

---

[5]In IGCEDP, even livestock management could be used to measure the trend in goal achievement. Hence, if milk yield increases that would signify a successful breeding programme, better production and quality of fodder, and even better livestock management.

11. The partner, if involved from the start, realizes the importance of IM. In this case, the regular supply of appraisal notes on different indicators justifies the effectiveness of IM to the partner.
12. The minor incentive given as cash or kind has helped to generate community interest in the project and its purpose and hence in the impact evaluation.
13. Intensive follow-up in the early years is very productive in showing the sharing and caring sense, which can be supplemented by regular training and other inputs so that IM can be institutionalized at the grassroots level.

RECOMMENDATIONS

Several recommendations can be made on the basis of this study:

- In a self-help or participatory project the task of monitoring and evaluation, especially various aspects of IM (eco-rehabilitation, social benefits, economical improvement), must gradually be shifted to the community.
- Projects should design their IM concept and calibrate it with observations and reports of evaluators after some time. Similarly, project internal mechanisms should be used fully (i.e. monthly or weekly planning meetings at headquarters, field-office planning meetings and mid-term reviews).
- Projects should not defer IM until post-project evaluation.
- During implementation, the project must continuously test and validate the indicators defined in the PPM and lay the fundamentals of IM accordingly. It is important that the baseline data be consolidated within first few years of the project.
- Training and know-how in IM, not necessarily highly technical, must be provided to the community and the community must know how to assess certain developments (e.g. if they perceive that enclosure of a degraded community land has increased production of grass, then why and by how much or which new plant species have invaded the treated sites).
- Quasi-experimental IM methods should be tried but while collecting such data people's observations must be incorporated.
- Simple, realistic and understandable indicators, which are cost-efficient, must be formulated (preferably giving short- and long-term impact trends). If available, local knowledge should be incorporated in indicators.

- The data of participatory rural appraisal, which leads to comprehensive village integrated resource management planning at village or watershed level, must be used for periodic review at goal and purpose level.
- Selected IM types can be conducted through consultants or practical trainees, as proposed by the project.
- IM should avoid agenda-orientation or donor-driven thrust, as that proliferates a target-centred approach.
- Participatory IM needs gradual consolidation and should be consolidated with the maturity of communities for self-management and management of natural resources.
- IM must be flexible and situation-oriented, i.e. it has to take the type and stage of the project into consideration.
- The urge for quantitative data must not unjustifiably replace qualitative indicators (e.g. survival rate of a plantation is not enough, growth and quality also must be taken into account).
- Data analysis and findings must also be interpreted and disseminated into the communities.
- Establishment of fixed sample plots on which various kinds of data can be measured (or sample households or resource farmers) is practical and must be further explored.
- The activity best suited to give an early track of positive or negative changes must be prioritized.
- IM should be treated separately within M&E and must have an expert staff.
- Participatory IM and evaluation is relevant only if a project focuses on village self-management groups or users.
- IM findings should be available, easily understandable and acceptable to donors as well.
- The IM even if quasi-experimental must be simple, as must instruments used to collect data (e.g. automatic rain gauges are difficult for the community to operate).

PROGRESSIVE IM APPLICATION FOR WSD

A MMWS consists of several villages or communities living in it. Hence, even if WSD is meant for the whole watershed, the inter-sectoral interventions start in each village. The overall impact of various treatments is only visible when the whole MMWS is treated. By having a village-based watershed approach, the IGCEDP experience clearly shows that through various and simultaneous inter-sectoral activities a culmination of impact is possible at MMWS level - if all villages or communities adopt it - within a short span of time (approximately five years). This is especially the case if forestry and other soil and water conservation measures (with community participation) are combined. Smaller units (MMWS of 500 ha) can prove ideal to focus on the development hypothesis set out in the PPM. The progressive IM therefore relies on one major activity or sectoral intervention and gradually interlinks it to other sectors. For example, forestry interventions have proved useful to reflect on social and institutional impacts. Similarly, livestock management can throw light on forestry impact. In this regard, a quasi-experimental method of data collection can be a relevant option. Test micro-watersheds in IGCEDP are ideal, among other things, for demonstrating the impact on water balance or water resource development. The involvement of communities is a must and just, but it can only happen with the maturity of a particular community towards WSD (e.g. involvement in planning, implementation, and ultimately M&E). With this the long-term impact trends can be assessed in a watershed and corrective measures, if needed, can be applied in time.

References

Anonymous. 1996. Regional Working Group in Participatory Monitoring and Evaluation. Summary Status Report, Kathmandu.

Bollom, Michael W. 1998. Impact indicators. An alternative tool for the evaluation of watershed management. IGBP "Watershed" GTZ, Delhi.

Commission of European Communities. 1993. Manual Project Cycle Management. Integrated Approach and Logical Framework. Commission of European Communities. Evaluation Unit Methods and Instruments for Project Cycle Management, Brussels.

Gohl, E., D. Germann, J. Prey and U. Schmidt. 1993. A short guide to participatory impact monitoring. FAKT-Stuttgart (Draft).

Indo-German Changar Eco-Development Project. 1997. Global Challenges - Local Visions. Experiences in people's-centred eco-development in the Himalayan foothills. Indo-German Changar Eco-Development Project, Palampur, H.P., India.

Indo-German Changar Eco-Development Project. 1997. Project Planning Matrix and Plan of Operations of the first phase (revised). Indo-German Changar Eco-Development Project. Palampur, H.P., India

Kotru, R. and H. Wienold. 1995. Concept for economical and ecological impact monitoring and evaluation. Indo-German Changar Eco-Development Project. Palampur, H.P., India.

Narayan, D. 1993. Participatory Evaluation: Tools for managing change in water and sanitation. World Bank Technical Paper No. 203. World Bank, Washington D.C.

Ojha, D.P. 1998. Impact Monitoring Approaches and Indicators. Experience of GTZ supported multi-sectoral rural development project in Asia (Draft Report). Impact Monitoring Unit, German Technical Cooperation (GTZ), Kathmandu, Nepal.

World Bank. 1991. Environmental Assessment Sourcebook. Volume 1: Policies, Procedures, and Cross-Sectoral Issues. Environment Department, World Bank.

12

Linking Monitoring to Evaluation in a Soil and Water Conservation Project in Southern Mali

Ferko Bodnar[1], Jan de Graaff[1], Sean Healy[2] and Bertus Wennink[3]

[1]Erosion and Soil & Water Conservation Group, Wageningen University and Reaserch Centre, The Netherlands. E-mail: luciefer.ko@hetnet.nl, jan.degraaff@wur.nl

[2]French Ministry of Agriculture/CERDI, Clermont-Ferrand, France. E-mail: s.healy@u-clermont1.fr

[3]Royal Tropical Institute, Amsterdam, The Netherlands. E-mail: b.wennink@kit.nl

ABSTRACT

In recent discussions it is argued that a more timely evaluation of impact is needed to steer projects to more effective and efficient approaches. This requires more from the monitoring system. Following the monitoring and evaluation (M&E) system and the logical framework of a soil and water conservation (SWC) project in southern Mali, we found that baseline information and monitoring were insufficient at the goal level and long-term targets were missing at the purpose level. This hindered the evaluation of impact on land degradation – one of the project goals. Complementary monitoring by the SWC project and a project-external M&E Unit made it possible to evaluate impact on crop yield – another project goal. Efficiency was evaluated by comparing the costs for the SWC project with the benefits (additional cotton yield) for farmers. SWC projects should document baseline data and assure monitoring at all levels in the logframe, and set long-term targets at the purpose level. Monitoring will be more efficient when the SWC project collaborates with a project-external

M&E Unit, especially for general indicators at the project goal level (e.g., crop yield). Specific indicators for the SWC project (e.g., land degradation) could be monitored by the SWC project in the fields of the M&E Unit sample.

THE NEED FOR MORE TIMELY EVALUATIONS

Evaluations and the Logical Framework

Project evaluations serve three main purposes: to give timely feedback for improvements, account for the expenses incurred, and draw lessons for future projects. Evaluations refer to a project strategy and project plan, which are best visualized in a logical framework, or logframe. The logframe was first developed in the late 1960s for USAID to facilitate monitoring and evaluation but was later much used and adapted by other organizations as a project planning tool (USAID, 1980; GTZ, 1988; Crawford and Bryce, 2003).

As an example, we describe a typical logframe for a soil and water conservation (SWC) project with 4 columns x 5 rows. Vertically from the bottom row upwards, the logframe presents the project strategy of how inputs (money) contribute to project activities (training sessions), to project outputs (farmers capable of protecting fields), to project purposes (fields protected from erosion), which will then contribute to the wider goal (reduced degradation, improved production). Other logframe examples may combine inputs and activities or present inputs separately in the project document. Going from the left column to the right, each narrative summary in the first column is accompanied by objectively verifiable indicators in the second column, which should include baseline data describing the situation at the start of the project and long-term targets describing the desired situation. However, targets at the goal level, which are affected also by other factors, are often not quantified. The third column describes the means of verification: how and by whom is progress monitored and reported. The last column gives the assumptions and risks: conditions outside project control that need to be fulfilled for the project to succeed.

In addition to the long-term targets in the logframe, annual targets may be specified in annual work plans, giving details at the input, activity and output level. Annual targets are often based on the capacity of the project (budget and personnel) and on the participation of beneficiaries.

Figures 1-4 show soil erosion and some conservation measures in Mali.

Fig. 1 Tree roots exposed by soil erosion.

Fig. 2 Gully erosion in a cotton field.

Fig. 3 Live fence as border for SWC.

Fig. 4 Stone barrier for SWC.

Monitoring and Evaluation

We will use the following definitions to distinguish monitoring and evaluation:

- Monitoring is the on-going data collection and comparison of achievements with targets within each horizontal level in the logframe.
- Evaluation is the analysis of the contribution of project achievements at one level in the logframe to a higher level.

In this chapter, we consider the aggregation of annual achievements to cumulative achievements that can be compared with long-term targets as part of monitoring. Concerning evaluation, we are especially interested in the contribution of project outputs to project purposes (outcome) and the contribution of project outcome to project goal (impact).

Five Aspects of Evaluation

Complete project evaluations, after project closure, typically look at the following five aspects (OECD, 2002):

- Effectiveness in the fulfilment of project objectives.
- Impact, whereby intended and unintended, positive and negative effects of the project are considered.
- Efficiency, comparing the project achievements with their costs.
- Relevance, comparing project achievements with the needs and priorities of the beneficiaries, with the country policy, and with the donor policy.
- Sustainability, assessing whether benefits will continue after project closure.

Changes in M&E Approach

Traditionally, monitoring and evaluation were distinct activities. Monitoring was an ongoing internal comparison of achievements with plans, for example on an annual basis. One step further in the analysis is to aggregate annual achievements to cumulative achievements over a longer period that can be compared with long-term targets. Evaluation was often an external, more in-depth analysis of how project outputs contributed to project outcome (achievement purposes) and project impact (achievement goal), validating the vertical logic in the project strategy. Typically, an evaluation was done at project appraisal, halfway through the project, and after project closure (Casley and Kumar, 1987; van de Putte, 1995).

In current discussions, however, the need is felt for more timely internal feedback during project implementation. This feedback should emphasize less the comparison of achievements with plans at the activity and project output level than the comparison of project outcome and impact with project purpose and goal. This means that the project strategy itself, including the assumptions, should be regularly validated (Cameron, 1993). Early impact assessment is not only valuable to improve the project strategy but also facilitates capacity building through learning-by-doing (Hauge and Mackay, 2004) and serves as a convincing extension tool (Nill, 1999). Different terms are used for this internal impact assessment: *ongoing evaluation, impact monitoring* (Vahlhaus and Kuby, 2001), *impact monitoring and assessment* (Herweg and Steiner, 2002), and *performance measurement* (Binnendijk, 2001). The need for a more timely and regular assessment of project outcome and, as far as possible, project impact, requires on the one hand more in-depth analyses and on the other hand relatively simple methods that can be done on a more regular basis during project implementation.

The Problem: The Attribution Gap

Vahlhaus and Kuby (2001) and Herweg and Steiner (2002) argued that one should not expect from individual projects to evaluate the impact they have on the achievements of the higher-level development goal. They suggest, in what they call the GTZ model, two evaluation steps: one from project activity upward, the other from higher development goal downward.

On the one hand, project evaluations should be limited to the highest level of outcomes that can unambiguously be attributed to the project: the *direct benefits* experienced by beneficiaries who use the project outputs. These direct benefits are also called farm-level effects (van de Fliert and Braun, 2002). In the example of a SWC project, the project would thus evaluate, for example in case studies, whether adopters of SWC experience a reduction in erosion or an improvement of crop yields, without assessing land degradation or agricultural production at the district or province level. Douthwaite et al. (2003) argue that the evaluation of direct benefits requires more detailed cause-effect steps than usually presented in the logframe and propose an *impact pathway evaluation*.

On the other hand, impact at the higher goal level should be assessed in larger project-independent evaluations (e.g. regional surveys on agricultural production) without attributing impact to individual projects. The gap between these bottom-up and top-down evaluations, with respect to the cause-effect chain, is called the "attribution gap" (Vahlhaus and Kuby, 2001). Douthwaite et al. (2003) suggest that this attribution gap may be bridged by the evaluator identifying "plausible links" between project outputs and higher-level development changes. Although they acknowledge the importance of bridging the attribution gap during project implementation, they foresee this second step as an ex-post impact assessment.

The Challenge: Adjusting Monitoring for Timely Evaluations

The objective of this chapter is to suggest improvements in the monitoring system for a more timely evaluation, during project implementation. From the five aspects of a complete evaluation, our focus will be on the effectiveness, intended impact, and efficiency. This means that somehow we need to "fill" the attribution gap between project outcome and impact.

Although some authors have defined effectiveness as the project ability to achieve what was planned, comparing targets and achievements within one horizontal level in the logframe (Binnendijk, 2001; Guijt and Woodhill, 2002), we prefer the definition of effectiveness as the contribution of project activities and outputs in achieving project purposes (Leeuwen et al., 2000; MDF, 2003).

The intended impact is the project contribution to the achievement of the goal. In the remainder of this chapter we will not look at unintended impact.

Evaluating the efficiency includes a comparison of costs incurred with output, outcome or impact achieved (Graaff, 1996; Guijt and Woodhill, 2002; OECD, 2002). A distinction is made between financial efficiency from the viewpoint of a participant and economic efficiency from the viewpoint of a society as a whole (Casley and Kumar, 1987).

Taking a SWC project in southern Mali as a case study, we will follow three specific questions: (1) How are indicators monitored at different levels of the logframe, especially at purpose and goal level? (2) How was internal monitoring (from project activity upwards) complemented by external monitoring (from goal downwards)? (3) To what extent was it possible to evaluate effectiveness, impact, and efficiency of the SWC project?

METHODS

First, we describe the SWC project in the cotton-producing region of southern Mali, its logframe, and the different structures involved in M&E.

For this case study, we presented a simplified project logframe and focused on the output, purpose, and goal level. We chose a few indicators that we had ample information on. At the purpose level, we focused on the adoption of two erosion control measures: stone rows and live fences. At the output level, we focused on the number of villages with a trained SWC village team. For a simple evaluation of efficiency, we compared the costs of the SWC project with the benefits for farmers in terms of improved cotton yield.

To assess the relevance of the monitored indicators, we verified the availability of reference values: baseline, long-term targets and short-term targets, and the possibility of aggregating annual achievements to cumulative achievements that can be compared with long-term targets. For evaluations, we considered possibilities to attribute outcome and impact to project interventions.

THE SWC PROJECT IN SOUTHERN MALI

Southern Mali, where total cotton and cereal production have shown a spectacular increase over the last decades, has a number of land degradation problems. Expanding agriculture, increasing livestock numbers, and firewood demand have led to localized overexploitation of the agro-sylvo-pastoral resources. Although slopes of agricultural fields are gentle, 1-2% on average, the long slope length, low soil cover, and poor soil structure result in substantial water runoff and sheet erosion (Hijkoop et al., 1991). Furthermore, in the prevailing cropping systems nutrient balances for the cultivated area are generally negative, even for crop rotations with fertilized cash crops (van der Pol, 1992).

In the framework of a Dutch bilateral aid programme, a SWC project was initiated in 1986, known under the name "Projet Lutte Anti-Erosive" (PLAE). It was decided to incorporate this project in the CMDT (Compagnie Malienne pour le Développement des Textiles) that was in charge of the management of the cotton production and marketing chain. CMDT was considered the only organization capable of reaching the majority of the rural population in southern Mali. In 1992, the project became an institutionalized SWC unit of the CMDT (IOV, 1994). In 1998, most donor funding to the SWC programme had stopped (IOB, 2000). Although in 1996 the project approach was replaced by a programme approach, and the SWC programme continued even after donor funding had stopped, we will refer to the whole period as the SWC project.

The CMDT in southern Mali had a two-fold mission: on the one hand a commercial mission to organize activities related to cotton (inputs on credit, extension, processing, export) and on the other hand a public mission of rural development (e.g. alphabetization, water supply, management of inland valleys), consigned by the Mali State. This public mission was withdrawn from the CMDT in 2002 when a reform process of the cotton sector in Mali started.

Although the SWC project in southern Mali did not use the same planning format during its evolution over the years, the concept of the logical framework will be used to facilitate the discussion of the M&E system applied. The logical framework of the SWC project can be summarized as follows (PLAE, 1986, 1989; Mourik et al, 1993; CMDT, 1995):

- The goal was to reduce the degradation of the ecosystem, to increase agricultural production (crop yields), and to intensify agriculture (increasing inputs and outputs per hectare without increasing the cultivated area per person).
- Four purposes contributing to the goal were defined: capacity building of the CMDT extension service, awareness raising of the rural population, farmer adoption of improved practices, and documentation and dissemination of experiences. Our focus is on the farmer adoption of stone rows and live fences.
- The project outputs that contribute to farmer adoption concerned an enhanced capacity of farmers for adopting SWC measures. The main output was the presence of a trained SWC village team.
- Several project activities were involved, including awareness meetings, training sessions, field visits, and supply of planting materials. For the remainder of this chapter, we will not look at this detailed activity level.

The SWC project used a specifically developed SWC village approach. One of the main project activities that was part of the approach was the installation and training of a "SWC village team". We will refer to villages with a trained SWC village team as "SWC villages" and villages without a trained SWC village team as "non-SWC villages". One of the first activities of the SWC village team was the elaboration of an "erosion inventory and assessment" indicating which areas of the village suffered from erosion and a "land management scheme" indicating where erosion control measures should be installed. During each annual review and planning meeting, the evaluation of the previous season's activities was linked to the planning of the next season's activities.

For the monitoring, three CMDT structures were involved: (1) The Statistics Unit collected agronomic data, with a strong focus on cotton, through the general extension agents. Data included area under different crops, input use, and production and were aggregated to district, regional, and national level. (2) The SWC Unit monitored SWC training and farmer adoption of SWC measures at the village level, through SWC specialists helped by general extension workers and the farmers in the SWC village team. In annual activity reports, annual work plans were being compared with annual achievements. From 1986 to 1995, only targeted SWC villages were monitored. From 1996 onwards, SWC activities in all villages were monitored, distinguishing targeted (SWC villages) and un-targeted villages (non-SWC villages). (3) A separate M&E Unit monitored detailed agronomic data in a sample of 54 villages, through specialized enumerators that were not involved in extension. Data were collected in a systematic way and stored in an un-aggregated database so that more complex analysis and evaluations could be made upon demand. The M&E Unit monitored also the actual adoption of SWC practices at the farm level (which is the cumulative result of activities in previous years). In general, the results of the M&E Unit monitoring were used for policy making and strategic planning at CMDT.

MONITORING OF THE SWC PROJECT

The simplified monitoring framework is given in Fig. 5. We discuss three levels of monitoring: the output level (M1), the purpose level (M2), and the goal level (M3). Note that we have left out the input level - the money spent on each of the project activities - and that we will not discuss the monitoring of project activities in this chapter.

Monitoring involves the comparison of achievements with plans and, preferably, also with baseline and long-term targets. The SWC annual work plans gave details for the output and purpose level. Based on the annual work plan, progress can be monitored following simple indicators.

Monitoring Project Outputs: SWC Village Teams Installed and Trained (M1)

The output indicator, the percentage of villages with SWC village teams, has clear reference values: the baseline at the start of the SWC project was 0% and the long-term target was 100%. The SWC annual work plan can be quite certain about the project outputs, which are, by definition, under project control. Annual targets were set according to personnel and budget available.

Annual achievements are reported in the SWC annual activity reports. Taking the annual activity report of 1998/1999 as an example, 763 SWC teams were installed and trained. Cumulative achievements can be calculated by adding up annual achievements. A total of 3,217 village teams were trained from 1986 to 2000 (PLAE, 1987, 1988, 1992, 1995; Schrader and Wennink, 1996; CMDT, 1997, 1998a, 1999a, 2001). If we

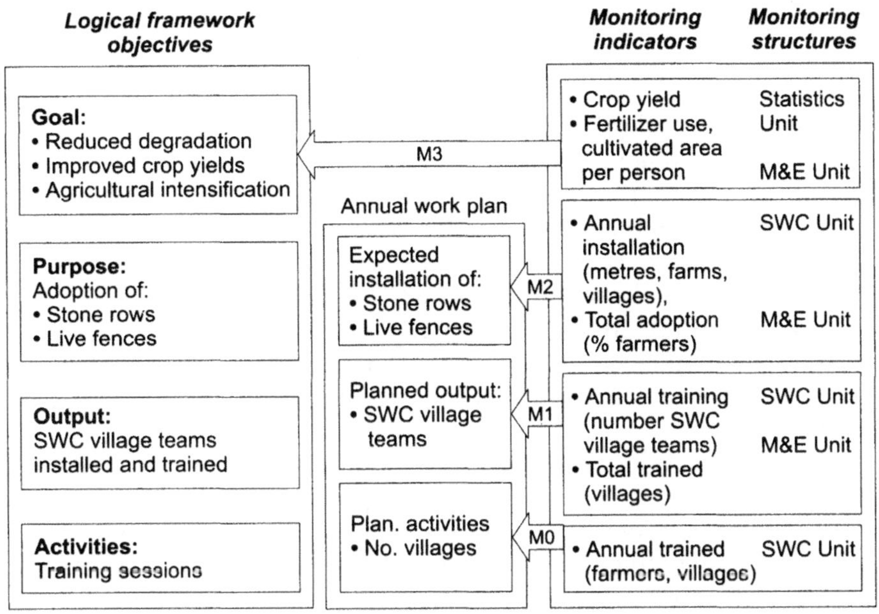

Fig. 5 Monitoring in relation to the logical framework of the SWC project in southern Mali: monitoring indicators at the levels of activity (M0), output (M1), purpose (M2), and goal (M3).

assume one team per village, 64% of the total of 5,054 villages in the CMDT area would have a SWC village team in 2000. However, there are villages with more than one village team. From 1999/2000 onwards, the annual activity report also indicated the cumulative number of targeted villages. In 2000, 2,562 villages, or 51%, effectively had one or more SWC village teams. In the M&E Unit sample of 54 villages, 60% of the villages had a SWC village team in 1999/2000 (CMDT, 2000a).

Monitoring Indicators Related to the Project Purpose: Farmer Adoption of SWC Measures (M2)

The baseline adoption of erosion control measures, at the start of the project, was negligible.

Setting long-term targets for the adoption of erosion control measures, both in terms of percentage of farmers and in terms of quantity of erosion control measures, would need an inventory of erosion and erosion risk. This was done at the village level, where, after an erosion inventory was drawn up, a land management scheme was developed with long-term targets for erosion control measures to be installed. It is at the village level that the relation between area to protect from soil erosion and quantity of erosion control measures is clear. However, the village land management schemes were not documented, nor were they aggregated to long-term targets for the whole SWC project.

The SWC annual work plan cannot be very specific about the project outcome (by definition not under project control). Farmers adopt SWC measures on a voluntary basis and for different reasons. Nevertheless, the annual work plan does include the expected quantitative results, based on the participatory village planning of SWC activities.

As an example, the annual activity report 1998/1999 is presented, where annual achievements are compared with annual plans (Table 1). A distinction is made between achievements in SWC villages and non-SWC villages.

Cumulative physical achievements can be calculated by adding up figures of the different annual activity reports. About 23,000 km of live fence and 7,500 km of stone rows were installed between 1986 and 2000. In fact, the total length of live fences actually present in 2000 was less than 23,000 km because part of the live fences had died or had been replaced and was thus counted double.

No cumulative adoption in terms of number of farmers can be derived from the annual activity reports from the SWC Unit. We may know the number of farmers who installed live fences in a certain year, but we don't know whether they were installing for the first time that year or continuing the installation they had started in previous years. Similarly, no cumulative adoption can be calculated for the number of villages.

Cumulative adoption is presented in the annual reports by the M&E Unit. Their sample is relatively small and they regularly change part of

Table 1 Comparing annual achievements with annual targets for the installation of live fences, stone rows and cattle pens. Annual Activity Report SWC Unit 1998/1999.

		Planned	*Realized*			*% realized*
			SWC villages	*Non-SWC villages*	*Total*	
	Villages		2,400	2,654	5,054	
Live fences	Metres	2,940,957	2,345,029	1,004,807	3,349,836	114%
	Farms		11,455	4,448	15,903	
	Villages		1,379	1,025	2,404	
Stone rows	Metres	770,234	484,596	32,552	517,148	67%
	Farms		4,028	544	4,572	
	Villages		814	97	911	

Source: SWC Unit (CMDT 1999a).

their sample villages and their sample fields. Therefore, an average over several years, 1997-2000, gives a more reliable estimation of the adoption. According to their reports, 15% of the farmers have installed live fences and 10% have fields protected by stone rows (CMDT, 1998b, 1999b, 2000a).

However, without long-term targets, the cumulative achievements are difficult to interpret.

Monitoring Indicators Related to Project Goal (M3)

Three project goals are distinguished: (1) reduced land degradation, (2) increased crop yields, and (3) agricultural intensification. For each goal, we discuss the availability of baseline data and monitoring results. Generally, no targets were set for the goal-level indicators, which are affected also by other, non-project factors.

For land degradation, no baseline data were available, except from a few specific or localized studies (Jansen and Diarra, 1992; van der Pol, 1992). The erosion inventory done in SWC villages could have served as a "first measurement at starting point" baseline survey but this information was not documented. Land degradation was not monitored at all.

For crop yields, no specific baseline information was collected in targeted SWC villages. The Statistics Unit monitored total cultivated area, production and crop yields. As an example, in 1999/2000, there were 442,469 ha under cotton cultivation, producing a total of 429,990 tons of cotton, with an average yield of 972 kg cotton ha^{-1}. For the CMDT, the main reference values are the production and crop yield of the previous season and the production targeted for the current season (CMDT, 2000b). Yield trends show a decline in cotton and maize yields between 1993 and 2002, which can be attributed partly to a decline in rainfall (Bodnar et al., 2005).

For agricultural intensification, no specific baseline information was collected in targeted SWC villages. In 1996, indicators related to agricultural intensification were included as performance indicators of the donor-funded SWC programme, which were accompanied by baseline data and targets for 1997 and 1998, aggregated per region. Chemical fertilizer doses on cotton should increase and the cultivated area per person (distinguishing total, cotton and cereal area) should not increase any further (Schrader, 1997). The M&E Unit monitored input use and cultivated area per person. On average, 125 kg complex fertilizer and 51 kg urea were used per ha cotton. On average, 0.62 ha was cultivated per person (CMDT, 2000a). Average fertilizer doses in the CMDT zone have increased between 1994 and 1997 due to the higher percentage of land area under fertilized cash crops (cotton and maize) and due to higher doses applied to these crops (Doucouré and Healy, 1999). However, the cultivated area per person has increased as well, following the increased level of mechanization and encouraged by the devaluation of the currency in 1994, which resulted in a more profitable ratio of cotton prices versus fertilizer and labour prices (Giraudy and Niang, 1996). This shows how project impact is affected by non-project factors (Bodnar et al., 2005).

Summarized, most "gaps" in the monitoring system occur at the goal level, especially for land degradation (Table 2).

Evaluation of the SWC Project

We distinguish three evaluation steps, as diagrammed in Fig. 6: effectiveness or the contribution of project output to project purposes (E1); impact or the contribution of project purposes to the project goal (E2); and efficiency or comparison of project costs with benefits of improved production (E3).

Evaluating Effectiveness: Effect of Farmer Training on Adoption of SWC Measures (E1)

Up to 1995, no comparison was possible between SWC villages and non-SWC villages, which were not monitored by the SWC Unit.

Table 2 Availability of information on planning and monitoring in relation to the logframe of the SWC project of southern Mali

Logical framework levels	*Planning*			*Monitoring*	
	Baseline data	*Long-term targets*	*Annual targets*	*Annual achievements*	*Cumulative achievements*
Goal:		*	*		
• Reduced degradation	v	(–)	(–)	–	–
• Improved crop yields	–	(–)	(×)	+	o
• Agricultural intensification	–	(1996)	(1996)	+	+
Purpose:					
• Adoption of erosion control measures	+	v	+	+	o
Output:					
• Villages with SWC village team	+	+	+	+	+
Activity:					
• Training sessions	+	+	+	+	+

* A SWC project is not expected to quantify targets at the goal level.
v: Only at the village level, not documented and not aggregated to regional level;
+: Sufficient; -: Insufficient; o: Monitored but not compared with targets;
x: Annual targets for production were set by the CMDT, but not in relation to SWC.
1996: Targets set in 1996 for 1997 and 1998.

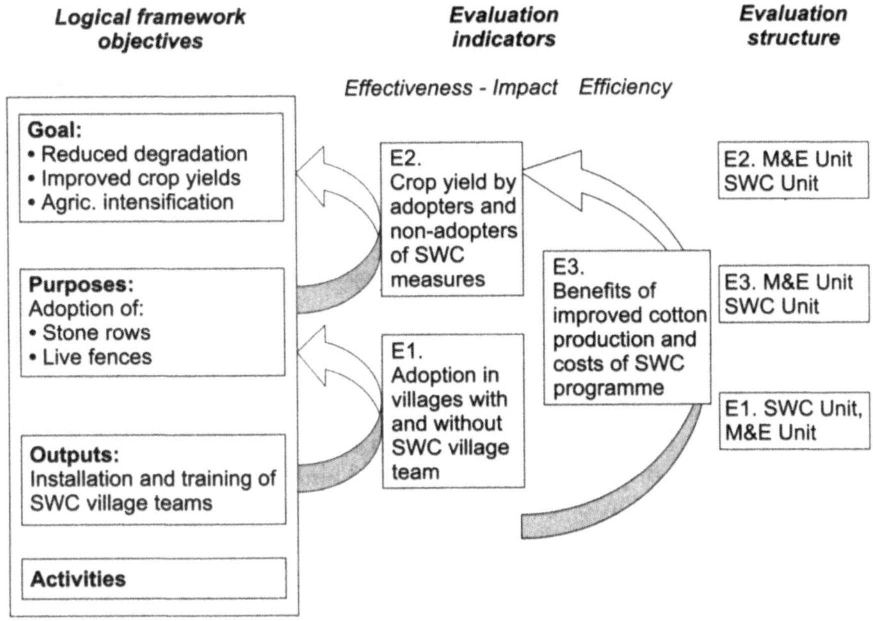

Fig. 6 Evaluation in relation to the logical framework of the SWC project in southern Mali: evaluating effectiveness (E1), impact (E2), and efficiency (E3).

In 1994, the SWC Unit and the M&E Unit undertook a joint study whereby performances in villages with and without a SWC village team were compared (Giraudy et al., 1996). Adoption in villages with SWC village team was systematically higher than in villages without SWC village team.

From 1996 onwards, the SWC Unit distinguished achievements in villages with and without SWC team, which allowed evaluation of the effect of the SWC village team on the adoption of SWC measures. However, part of the adoption in untargeted villages can be attributed to the SWC project as well, especially because adoption of erosion control measures was negligible before the project started.

Using the M&E Unit data, averaged over three years, a positive effect is found of the specific SWC village approach and the presence of a trained SWC village team on the cumulative adoption of SWC measures (Table 3). Adoption of erosion control measures is more than double in villages with SWC village team.

Attribution to the project requires a comparison of with and without or before and after project intervention. In the initial phases of adoption, we compared adoption in SWC villages with adoption in non-SWC

Table 3 The effect of a SWC village team on the adoption of some erosion control and compost-making measures, in percentage of farmers, averaged for 1997-1999.

	Non-SWC villages	*SWC villages*
Number of villages in sample	*22*	*32*
Live fences	5%	24%
Stone rows	7%	11%
Any erosion control measure	16%	40%

Source: Additional analysis by M&E Unit (CMDT 1998b, 1999b, 2000a).

villages and attributed the difference in adoption to the project. However, adoption in untargeted villages picked up under the influence of neighbouring targeted villages and informal information networks between farmers and trained general extension workers. In 2001 and 2002, the difference in adoption between targeted and untargeted villages had disappeared; adoption of erosion control measures reached about 54% (Bodnar et al., 2006). Because adoption before the SWC project was negligible, we may attribute adoption of erosion control measures in all villages to the SWC project.

Evaluating Project Impact: Effect of Adoption of SWC on Cotton Production (E2)

Comparing achievements with targets within one level of the logframe without attributing change to project activities says little about project impact. When, in 1996, the Dutch donor and the CMDT agreed on a number of performance indicators also at the goal level, they did not foresee analyses to identify factors, including SWC activities, that explain change. For example, fertilizer use and cultivated area per person increased, but we do not know whether and how this was affected by SWC activities: the attribution gap remained.

From the three goals – to reduce degradation, to increase agricultural production, and to intensify agriculture – we focus here on cotton production. What is the impact of the adoption of erosion control measures on cotton production?

The most simplistic evaluation of impact is to compare cotton yields obtained by farmers in villages with and without SWC village team. This comparison, however, showed no significant difference.

A comparison of the cotton yield obtained by farmers with and without erosion control measures, to whichever village they belonged, showed a systematic positive difference. Between 1997 and 1999, farmers with erosion control measures obtained on average 12% higher cotton yields (CMDT, 2002). However, we have to be careful in drawing conclusions. First of all, the degree of adoption of erosion control measures was known only at the farm level, not at the field level, so the monitored cotton fields may not have been protected by erosion control measures. Second, the improved crop yields may be due to factors other than erosion control measures: farmers who adopt SWC measures often produce more compost as well and they may generally be better equipped or have more resources. The CMDT used an indicator to classify farms in four *farm types*, according to the possession of ploughing equipment and cattle. Farm type can be used as covariate when analysing the effect of erosion control measures on crop yield.

After discussions between the SWC Unit and the M&E Unit in 1999, the M&E Unit adapted the monitoring format and monitored the presence of erosion control measures at the field level. This enabled a comparison of crop yields from fields with and without erosion control measures, whereby other factors (rainfall, fertilizer and compost use, farm type) were taken into account. In 2000, 2001 and 2002, a significantly higher cotton yield was obtained in fields with erosion control measures (Table 4).

Table 4 Comparison of cotton yield in fields with and without erosion control (EC) measures, taking into account rainfall, fertilization and farm type as covariates.

	2000		*2001*		*2002*		*Average*	
	N	*kg ha^{-1}*	*N*	*kg ha^{-1}*	*N*	*kg ha^{-1}*	*N*	*kg ha^{-1}*
Without EC	251	984	541	1,067	561	934	1,353	986
With EC	75	1,068	147	1,198	164	1,009	386	1093
% fields with EC	23%		21%		23%		22%	
% yield difference		9%		12%		8%		11%
Significance (P)		0.120		0.000		0.017		0.000

N: number of fields in M&E sample.

Evaluating Efficiency: Comparing the Benefits of Improved Cotton Production with the Costs of the SWC Project (E3)

No regular evaluations have been undertaken to assess the efficiency of the SWC project. After having analysed effectiveness and impact in the previous section, we can extrapolate the impact to southern Mali and try to assess the efficiency of the SWC project, comparing the value of the additional cotton production with the costs of the project. The basis for this evaluation is weak, and the results described below serve only as an illustration of how the efficiency of such a SWC project could be evaluated.

If we assume that all adoption of erosion control measures, in both villages with and without SWC village team, is due to the SWC project, the additional cotton production due to the SWC project can be extrapolated to about 2.4% of the total production in southern Mali. For the year 1999/2000, with 453,000 ha under cotton, this would mean an additional cotton production of about 10,000 tons. This figure excludes the additional yield due to increased compost use, which is also an effect of the SWC project.

The annual operational costs of the SWC project are estimated for two different periods: the project period with extensive external assistance in the beginning and the programme period with limited external assistance after donor funding to the CMDT had stopped.

- During the first two project phases, with several expatriates assisting the development of the SWC project, €3,484,000 was spent from 1986 to 1992, or an average of €580,667 per year (IOV, 1994).
- In 2000/2001, 46 full-time Malian staff were involved in the SWC programme and about 630 extension staff spent an estimated 10% of their time on SWC. The total salaries and operating costs were budgeted at €762,195, but not all of this was actually spent (Derlon, pers. comm.). This does not include the German-funded technical assistance, estimated at €55,000.

On the basis of these rough data, a simple comparison can be made between the annual costs of the SWC project and the annual financial benefits for farmers (Table 5).

From this simplified efficiency analysis, we would conclude that the benefits to farmers exceeded the costs of the project, for either the donor or the CMDT, even without taking into account the effect of increased compost use.

A number of important aspects were not considered: the costs farmers incur for SWC (mainly labour investments), the benefits for the CMDT

Table 5 Estimated annual benefits (1999/2000) for farmers and annual costs of the SWC project (1986-1992 and 2000/2001).

Estimated annual benefits in cotton production and farmer income	
• Additional cotton production	9,945 t yr^{-1}
• Average cotton price for farmers (1999-2001)	€263 t^{-1}
• Additional cotton value for farmers	€2,628,000 yr^{-1}
Estimated annual cost of the SWC project	
• Average cost of SWC project 1986-1992 (Donor)	€580,667 yr^{-1}
• Cost of SWC programme 2000/2001 (CMDT)	€762,195 yr^{-1}

Source: M&E Unit and SWC Unit (IOV 1994, CMDT 2002).

(profit) and for the Malian government (taxes) through increased cotton export, and the long-term benefits from reduced land degradation. This simplified evaluation served only as an illustration of how the efficiency of a SWC project could be estimated.

CONCLUSIONS AND RECOMMENDATIONS

Conclusions from the SWC project in southern Mali and recommendations for the M&E system of SWC projects in general are presented for: (1) project monitoring at the goal level, (2) setting long-term targets at the purpose level, (3) documentation and aggregation of village-level data, (4) collaboration between project monitoring and external monitoring, and (5) evaluations attributing change to project interventions.

Project Monitoring at Goal Level

To evaluate impact and efficiency, baseline information and monitoring are needed for the indicators at all levels in the logframe, including the goal level. The goal of the SWC project in southern Mali was to reduce land degradation, to increase agricultural production, and to intensify agriculture. Indicators at the goal level were not accompanied by baseline data nor were they monitored by the SWC project.

It is understandable that SWC projects do not want to be trapped in ambitious targets at the goal level, because the goal is affected also by other influences out of project control. Nevertheless, it is necessary to document a baseline situation and monitor change to validate assumptions of how project purpose would contribute to project goal. Because erosion control measures are taken especially in fields with erosion problems, often with lower crop yields, specific baseline information is needed for a good comparison of trends in fields with and without erosion control measures.

In the SWC villages in southern Mali, erosion inventories were made. These inventories indicated the areas with erosion symptoms (mainly gullies) and areas prone to erosion (on steep slopes, under bare plateaus). If these inventories had been documented before the project started, they could have served as baseline studies. Because gullies were already used as indicators in the erosion inventory, they could also have been monitored regularly and related to the presence of erosion control measures.

Setting Long-term Targets at the Purpose Level

Indicators at the purpose level - adoption of erosion control measures - were monitored but were not accompanied by long-term targets.

It is difficult for the SWC project to set long-term targets for farmer adoption of SWC measures because farmers install erosion control measures voluntarily. Besides short-term targets, which were set in a participatory planning process, long-term targets are needed to compare the cumulative achievements with.

The results of the land management schemes in the SWC villages could have been documented and aggregated as long-term targets. A note of caution: these long-term targets should not be used as fixed targets. The goal is to reduce land degradation. Therefore, evaluating impact on land degradation is more important than comparing achieved adoption with long-term targets of adoption.

Documentation and Aggregation of Village-level Data

In the previous two sections we argued that the results of the erosion inventories and land management schemes in the SWC villages in southern Mali should have been documented to serve as baseline situation of land degradation and as long-term targets for erosion control adoption. Aggregation of these detailed results may be difficult.

We propose to document the detailed information per village and to aggregate information over all villages in a simplified form. Four indicators could be attributed to each village: (1) the baseline severity of erosion at the start of project activities, in classes from *no erosion* to *very severe erosion*; (2) the year erosion control started; (3) the degree of actual erosion control, in classes from *not under control* to *fully under control*; and (4) the actual erosion severity.

Villages could then be classified according to the erosion severity at the start of SWC activities. Within each baseline erosion class we could present in annual monitoring reports the total number of villages, the average degree of erosion control, and the average actual erosion severity. More complex evaluations are needed to analyse relations between the severity of erosion at the start of SWC activities, the number of years villages have been active in SWC, the cumulative adoption of erosion control measures, and the impact on actual severity of erosion.

Complementary Monitoring of SWC Unit and M&E Unit

The SWC Unit compared annual targets with annual achievements and focused on project outputs and purposes. Their monitoring is well complemented with the monitoring by the (project-external) M&E Unit, which focused on some of the indicators related to project goal and on cumulative project achievements at the purpose and output level. The great advantage of involving the M&E Unit in SWC project monitoring was the large number of monitored fields and the limited additional effort (costs) needed to monitor at the goal level. This monitoring by the M&E Unit worked well for the agronomic data (area, production, input use: indicators covering two of the three project goals) and reasonably well for simple adoption indicators (presence or absence of erosion control measures).

However, information on land degradation (the third project goal) or more detailed information on the erosion control measures would have been too complicated or too much extra work for the enumerators of the M&E Unit and is better done by SWC project personnel.

Evaluation: Attributing Change to Project Activities

Monitoring of indicators at the goal level does not automatically allow the attribution of change to project interventions.

In southern Mali, this became possible only after the M&E Unit had adjusted its monitoring format and considered the SWC output (village with trained SWC village team) and SWC purpose (fields with erosion control measures). This matching of monitoring formats made it possible to fill the attribution gap and to evaluate the SWC project effectiveness and impact. Efficiency, here simplified to comparing project costs and benefits to farmers from additional cotton revenue, could be evaluated after effectiveness and impact had been evaluated. Because land degradation was not monitored by the SWC Unit or by the M&E Unit, the project impact on land degradation could not be evaluated.

A further collaboration between a SWC project and a project-external M&E Unit could ensure timely evaluation of impact and steer the project to more effective and efficient approaches. Specific SWC impact indicators (e.g. land degradation) should then be monitored by SWC personnel in villages and fields that are in the sample of the M&E Unit.

References

Binnendijk, A. 2001. Results based management in the development co-operation agencies: A review of experience, Background report. OECD DAC. Paris.

Bodnar, F., F. van der Pol and D. Babin. 2005. Chapter 4 *In:* F. Bodnar, F. Monitoring for impact: Evaluating 20 years of soil and water conservation in southern Mali. PhD thesis. Wageningen University, Wageningen, 217 pp.

Bodnar, F., T. Schrader and W. van Campen. 2006. Land Degradation and Development, 17: 479-494.

Cameron, J. 1993. The challenges for monitoring and evaluation in the 1990s. Project Appraisal, 8: 91-96.

Casley, J. and K. Kumar. 1987. Project Monitoring and Evaluation in Agriculture. A joint study by The World Bank, International Fund for Agricultural Development and Food and Agriculture Organisation. John Hopkins University Press, Baltimore.

CMDT. 1995. Programmes détaillés. Proposition d'un programme d'appui des bailleurs de fonds auprès de la CMDT (1996-2000). Actions vocation centrale. CMDT, DTDR. CMDT. Bamako.

CMDT. 1997. Réalisations Maintien du Potentiel Productif 1996/1997. DDRS, CMDT. Koutiala.

CMDT. 1998a. Réalisations Maintien du Potentiel Productif 1997/1998. DDRS, CMDT. Koutiala.

CMDT. 1998b. Annuaire statistique 97/98. Résultats de l'enquête agricole permanente. SE, DPCG, CMDT. Bamako.

CMDT. 1999a. Réalisations Maintien du Potentiel Productif 1998/1999. DDRS, CMDT. Koutiala.

CMDT. 1999b. Annuaire statistique 98/99. Résultats de l'enquête agricole permanente. SE, DPCG, CMDT. Bamako.

CMDT. 2000a. Annuaire statistique 99/00. Résultats de l'enquête agricole permanente. Suivi Evaluation, DPCG, CMDT. Bamako.

CMDT. 2000b. Bilan de commercialisation, activités du développement rural et industrielles 1999/2000. Suivi Opérationnel, CMDT. Bamako.

CMDT. 2001. Réalisations Maintien du Potentiel Productif. Résumé 1999/2000. DDRS, CMDT. Koutiala.

CMDT. 2002. Impact du programme maintien du potentiel productif sur la durabilité du systèmes de production au Mali Sud. DDRS et SE, CMDT. Bamako.

Crawford, P. and P. Bryce. 2003. Project monitoring and evaluation: a method for enhancing the efficiency and effectiveness of aid project implementation. Int. J. Project Manag. 21: 363-373.

Doucouré, C.O. and S. Healy. 1999. Evolution des systèmes de production de 94/95 à 97/98. Impact sur les revenus paysans. Suivi Evaluation, DPCG, CMDT. Bamako.

Douthwaite, B., T. Kuby, E. van de Fliert and S. Schulz. 2003. Impact pathway evaluation: an approach for achieving and attributing impact in complex systems. Agr. Syst. 78: 243-265.

Giraudy, F. and M. Niang. 1996. Impact de la dévaluation sur les systèmes de production et les revenus paysans en zone Mali-Sud. CMDT. Bamako.

Giraudy, F., T. Schrader, A. Maiga and A. Niang. 1996. Enquête sur les techniques de maintien du potentiel productif. Synthèse. DDRS et SE, CMDT. Bamako.

Graaff, J. de. 1996. The Price of Soil Erosion. An Economic Evaluation of Soil Conservation and Watershed Development. PhD thesis, Wageningen Agricultural University. Mansholt Studies No 3. Backhuys Publishers, Leiden, 300 pp.

GTZ. 1988. ZOPP. An introduction to the method. GTZ, Eschborn.

Guijt, I. and J. Woodhill. 2002. A guide for project M&E. IFAD, Office of Evaluation and Studies, Rome.

Hauge, A.O. and K. Mackay. 2004. Monitoring and evaluation for results: Lessons from Uganda. Poverty reduction and economic development — Findings (World Bank

Institute) 242, September 2004. pp. 1-4. www.worldbank.org/afr/findings/english/find242.pdf.

Herweg, K. and K. Steiner. 2002. Impact Monitoring and Assessment. Instruments for Use in Rural Development Projects with a focus on sustainable land management. Volume 1: Procedure. CDE & GTZ, Bern.

Hijkoop, J., P. van der Poel and B. Kaya. 1991. Une lutte de longue haleine. IER / KIT, Sikasso / Amsterdam.

IOB. 2000. Révue de la coopération entre le Mali et les Pays Bas 1994-2000. Direction évaluation de la politique et de opérations. Den Haag.

IOV. 1994. Mali: Evaluatie van de Nederlandse hulp aan Mali, 1975-1992. Ministerie van Buitenlandse Zaken. Den Haag.

Jansen, L. and S. Diarra. 1992. Mali-Sud, étude diachronique des surfaces agricoles. KIT. Amsterdam.

Leeuwen, B.v., H. Gilhuis, M. Kleinenberg, C. Mann, M. Mwaura, A. Roberts, G. Saha and A. Timlin. 2000. Building Bridges in PME. ICCO, Zeist.

MDF. 2003. Course on project cycle management: Monitoring and evaluation of project portfolios. MDF Training and Consultancy BV, Ede.

Mourik, D.v., M. Niang and A. Mohammedoune. 1993. Système de suivi évaluation CMDT pour la LAE et GRN. Préparé dans le cadre d'une mission d'appui à la CMDT pour le PLAE. DDRS, CMDT. Bamako.

Nill, D. 1999. Suivi agro-écologique des mesures de gestion des ressources naturelles. Diguettes, digues filtrantes et foyers améliorés au Projet Aménagement des Ouadis dans le Ouaddaï-Biltine (PAO) - Tchad. *In:* GTZ. [ed.] Mesurer les effects des projets: Suivi d'impact et calcul de renstabilité économique. GTZ.

OECD. 2002. Glossary of key terms in evaluation and results based management. DAC/OECD, Paris.

PLAE. 1986. Plan d'opération du PLAE dans la zone Mali-Sud, 1986-1989. CMDT - IRRT. Sikasso.

PLAE. 1987. Rapport bilan première campagne mai 1986 - septembre 1987. CMDT - IRRT. Koutiala.

PLAE. 1988. Rapport bilan campagne octobre 1987 - septembre 1988. CMDT-IRRT. Koutiala.

PLAE. 1989. Plan d'opération du PLAE dans la zone Mali-Sud, deuxième phase 1989-1993. DG, CMDT. Bamako.

PLAE. 1992. Rapport bilan campagnes octobre 1988 - septembre 1989 et octobre 1989 - septembre 1990. CMDT - KIT. Koutiala.

PLAE. 1995. Rapport bilan du programme lutte anti érosive. Camapgne octobre 1993 - septembre 1994. PLAE - KIT. Koutiala.

Schrader, T. 1997. Le système de suivi de la CMDT et les indicateurs de performance des programmes appuyés par la Coopération néerlandaise. DTDR, CMDT. Bamako.

Schrader, T.H. and B.H. Wennink. 1996. La lutte anti érosive en zone CMDT. Rapport final du PLAE. CMDT - KIT. Bamako, Amsterdam.

USAID. 1980. Design and evaluation of aid-assisted projects. USAID, Training and Development Division, Office of Personnel Management, Washington, D.C.

Vahlhaus, M. and T. Kuby. 2001. Guidelines for impact monitoring in economic and employment promotion projects with special reference to poverty reduction impacts. Part 1: Why do impact monitoring? GTZ, Eschborn.

van de Fliert, E. and A. Braun. 2002. Conceptualizing integrative, farmer participatory research for sustainable agriculture: From opportunities to impact. Agr. Human Values 19: 25-28.

van der Pol, F. 1992. Soil mining: an unseen contributor to farm income in southern Mali. Royal Tropical Institute, Amsterdam.

van de Putte, R.A. 1995. The design of monitoring and evaluation. pp. 76-85. *In:* P. van Tilburg and J. de Haan. [eds.] Controlling Development. Systems of Monitoring & Evaluation and Management Information for Project Planning in Developing Countries. Tilburg University Press, Tilburg.

Part 3

Physical Parameters in M&E

13

An Alternative Way to Assess Water Erosion of Cultivated Land

Robert Evans

Research Fellow, Mellish Clark Building, Room MEL 105, Anglia Ruskin University, East Road, Cambridge CB1 1 PT, UK. E-mail: r.evans@anglia.ac.uk

ABSTRACT

In the early 1980s, the Ministry of Agriculture in England and Wales took the decision to assess if water erosion was a problem, deciding to answer the question through field-based assessment rather than by using plot experiments. Although giving valuable information on the rates, frequency and extent of erosion, as well as delivery of sediment out of catchments, the results do not allow rates to be related to individual parameters, so that rates of erosion can be predicted. The results from the monitoring scheme explain why farmers think erosion is of little importance. The rates of erosion in England and Wales are compared with those measured in the field in other countries. Although, within any particular environment, mean and maximum rates cover a wide range of values, mean values generally relate well to climate and soils. Maximum values reflect rainfall intensities and amounts falling in rare storms. A field-based approach such as that described here provides a rapid and realistic way to assess erosion and the results can be validly compared across a wide range of environments. Such a technique can be used to monitor and evaluate soil conservation and development projects or programmes.

INTRODUCTION

This chapter is a shortened and somewhat revised version of a paper first published in 2002 (Evans, 2002). It gives the reasons why a new way of assessing erosion is needed, outlines a field-based method, and gives a short account of the results of a field-based monitoring scheme carried out in England and Wales. The results from that project are compared with other field-based measurements made elsewhere in the world and some conclusions are drawn. Such field-based assessments of erosion could be useful for monitoring and evaluating soil conservation and development projects or programmes.

THE CASE FOR FIELD ASSESSMENT

For many years, and in many parts of the world, soil erosion by water has been assessed from data gathered from small plots extrapolated to the wider landscape. The method mostly used has been based on the Universal Soil Loss Equation (USLE) or the "Revised" version of it (RUSLE). However, this approach has recently been questioned (Evans, 1993, 1995; Boardman, 1996; Herweg, 1996; Stocking and Murnaghan, 2001). Lal (1997: 1010) has noted, "... there is a need to develop and standardize methodology for assessment of soil erosion rate.... The scale at which such measurements are made is important. The scale should preferably be the 'watershed' of [the] 'landscape unit'."

At the 1999 conference of the International Soil Conservation Organization (ISCO) at Purdue University, West Lafayette, Indiana, USA, the scarcity of good quality data on soil erosion was considered by some participants to be the primary reason why soil degradation by erosion is not being adequately addressed. Thus, there is little information good enough to persuade governments and multi-national organizations to take action to conserve soils (Dumanski, 1999; El-Swaify, 1999a,b; Scherr, 1999). However, there was little appreciation that, with regard to soil erosion, there may be a link with the (poor) prediction of erosion rates based on the RUSLE or other similar experimental plot-based models, and farmers' and government's perceptions that erosion is not a serious problem that needs to be tackled. For example, if farmers are told by researchers that plot experiments and models predict that water erosion is important in stripping soil from slopes and that agricultural productivity will be affected, but they see in the field neither water-cut channels nor sandy depositional features, nor any short-term impacts on crop yields, farmers will consider that erosion is not much of a problem.

While the plot-based approach can aid the understanding of the processes and factors that govern water erosion, and can be "good" science that gives statistically valid results, it is of little help in predicting rates of water erosion in cultivated fields or the landscape as a whole or in predicting the extent and frequency of occurrence of accelerated erosion. The reasons are set out in more detail elsewhere (Evans, 1993, 1995) and outlined below. As Boardman (1998: 46) has recently commented, "Plot experiments are a poor basis for regional generalization: monitoring schemes are preferred but are uncommon."

A major drawback of plot experiments is that the runoff and soil carried in a plot are collected by directing the flow of water over the lower edge of the plot and so via a rapid fall in height into containers, or they are discarded. Such a rapid increase in gradient provides a potent "driver" to the erosion system that would not usually be there in the field unless a ditch or stream was adjacent to the foot of the slope. Compared to measured rates of erosion from plots, the USLE overpredicts at low rates, but underpredicts at high rates (Risse et al., 1993; Boardman, 1996; Reyes et al., 1999).

To account for variations in slope-form and slope angle within a field or a landscape a whole series of RUSLE predictions is needed to estimate the different rates of erosion that will occur in those segments. This is no easy task but may become more feasible using digital elevation models incorporated into a Geographical Information System. However, the time and cost of producing such models at the appropriate (large) scale may be prohibitive.

Furthermore, it may be that RUSLE is adequate only for short slope segments and not for fields or landscapes with their more complex slope patterns. Plot-based predictions have not been compared with rates measured in a field or fields within a landscape so it is not known if the erosion processes (splash erosion, sheet or interrill erosion, and smaller rill erosion) that drive RUSLE-based predictions are adequate to predict what is actually happening in farmers' fields where larger rills, ephemeral gullies, and gullies may also be occurring.

The RUSLE can only be used for predicting erosion rates on slopes and is not able to predict for those situations where channels (ephemeral gullies) cut into valley floors and depressions. Where erosion is largely confined to valley floors and depressions, therefore, in clayey landscapes in England, for example (Evans, 2002), the RUSLE is of little relevance. In the vast majority of instances ephemeral gullies reach to the edges of fields and their sediment loads are transported directly into ditches and streams. It is likely that such erosion is a major contributor of sediment to streams.

FIELD-BASED ASSESSMENT OF WATER EROSION

Field assessment of water erosion is based on two major, not unreasonable assumptions.

1. Over the short-term splash and sheet wash are of minor importance in redistributing soil within a field other than over a distance of a few metres (Evans, 1990a).
2. Following on from the first, it is rills and gullies that redistribute soil within a field or a landscape (Evans, 1990a).

In the USA, as in Europe, it is becoming more accepted that "Interrill erosion plays a very limited part in directly affecting topography or in affecting field operations" (Laflen and Roose, 1998: 39) and that "Nearly all land degradation caused by water erosion is due to channels" (Laflen and Roose, 1998: 42).

There may be exceptions where sheet wash can entrain high sediment loads. In England and Wales this process is found mainly on the shallow silty soils over chalk in southeastern England. But even then, much of the soil is often deposited within the field and these deposits are easily found. Colborne (1987, personal communication), in his study of the relative importance of splash, sheet and rill erosion on erodible silty soils in Somerset, England, found that splash and sheet erosion, as a proportion of the total amount of soil eroded, was inversely and exponentially related to volumes of soil transported downslope and trapped in Gerlach troughs.

The locating, mapping and measuring or estimating in the field of channel erosion (rills and gullies) and downslope deposition is not difficult and such techniques have now been used worldwide (Evans, 2002). Herweg (1996) and Stocking and Murnaghan (2001) have proposed field-based techniques for mapping channel erosion in their manuals aimed particularly at less developed countries. Similar techniques have been in use in England and Wales for more than two decades and many localities have been monitored (Evans, 2002). Traverses across landscapes to locate eroded fields can be made by vehicles or on foot (Boardman, 1990; Evans and McLaren, 1994) or air photos can be interpreted and then checked by fieldwork. Alternatively, fields can be selected by random sampling techniques and then surveyed for erosional and depositional features. Both kinds of survey have been used in England.

Channel lengths can easily be measured or estimated in the field and their cross-sections measured at appropriate intervals along their profiles to arrive at a volumetric assessment of erosion with a reasonable level of accuracy (Evans and Boardman, 1994). Estimates can also usefully be made from ground photos (Watson and Evans, 1991). The volume eroded can easily be converted to mass if the bulk density of the topsoil is known.

Commonly in the UK topsoil bulk density is about 1,300 kg m^{-3}. Volumes of sediment deposited are often easily estimated where fans have formed, for the area can be rapidly estimated and its mean depth arrived at by sampling the depth of sediment at a number of points. However, sometimes depositional features can be complex and difficult to map or cannot be found because the sediments have been transported out of the field.

SOME RESULTS FROM A FIELD-BASED SURVEY OF EROSION IN ENGLAND AND WALES

The Survey

In the early 1980s a decision was taken on how water erosion should be monitored and estimated in England and Wales. It was decided that a field survey of erosion was needed to assess the extent, rates and frequency of erosion, as data from plot-based experiments would not satisfactorily give this information. At that time it was considered that plot experiments gave a good understanding of the factors that controlled erosion but could not answer the question - is erosion a problem? The work described here was carried out between 1982 and 1986. The 17 surveyed traverses covered a wide variety of soil landscapes, representative of much of arable England and Wales, and topsoil textures ranged from clay to sand. For further details of the scheme and its results see Evans (1990a, 1996, 2002).

Air photos were obtained of each locality in late spring or early summer, when much of the erosion that occurred in autumn- and spring-sown crops would be recorded. The photos were interpreted by one person, and field checking was carried out in late summer when the cereal crops had been harvested. About 700 km^2 of land was surveyed each year and a total of 1,700 eroded fields identified. It did not take long to interpret the air photos and carry out the fieldwork each year. On average, each year it took 5.1 (range 4-7) working days to interpret the air photos and 23.3 (16-32.25) days to do the fieldwork.

Some Results

The locality most affected by erosion was the sandland of Nottinghamshire, central England, where, on average, 14% of the arable landscape eroded each year, with a range of 1.5-24.0% (Fig. 1). Eight other localities where erosion was widespread (5-10% of fields) also had erodible topsoils containing high sand or silt contents and often grew a wide range of crops that were vulnerable to erosion in both winter and spring storms. Erosion occurred most widely in soils drilled to

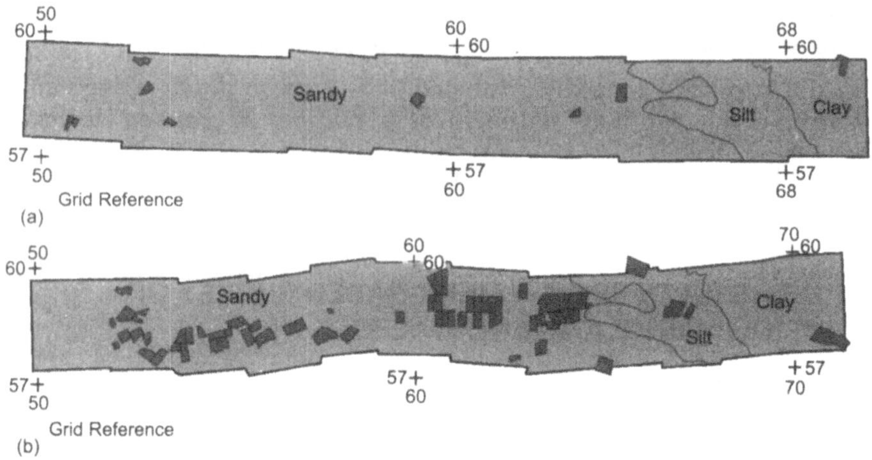

Fig. 1 Extent of erosion in Nottinghamshire sand-land in (a) drier and (b) wetter years.

Table 1 Water erosion and crop type, 1982-1986.

Crop	*% occurrence erosion (1)*	*% national crop area (2)*	*Ratio 1:2*
Winter cereals	42.8	61.5	0.69
Spring cereals	11.5	13.9	0.83
Oilseed rape	1.5	5.4	0.29
Ley grass (grass < 5 yrs old)	0.2	4.7	0.05
Sugar beet	18.4	4.5	4.05
Potatoes	10.6	3.2	3.28
Market garden	6.3	3.2	2.00
Peas	1.0	1.3	0.76
Bare soil/fallow	1.5	1.1	1.36
Field beans	0.4	0.9	0.39
Kale	0.7	0.6	1.23
Maize	1.6	0.4	4.02
Hops	0.5	0.1	4.17
Other	3.0	1.1	2.79

autumn-sown cereals (Table 1) but was disproportionately frequent under hops, sugar beet, maize and potatoes and rarely in late summer-sown oilseed rape and, especially, reseeded grasses.

Rates of erosion were highest (4-5 m^3 ha^{-1}) in fields in localities with topsoils containing high silt or fine sand contents and were about half those rates on sandy soils. In many other localities rates of erosion were

Table 2

Crop	Mean volume eroded (m^3 ha^{-1})	Median volume eroded (m^3 ha^{-1})
Market garden	5.08	1.47
Maize	4.48	1.00
Ley grass	4.09	1.14
Hops	3.92	1.01
Sugar beet	3.04	0.92
Other	2.83	1.09
Potatoes	2.53	1.01
Kale	2.10	1.41
Oilseed rape	1.92	0.30
Winter cereals	1.85	0.68
Spring cereals	1.75	0.71
Bare soil/fallow	1.61	0.27
Peas	1.21	0.91
Field beans	0.47	0.22

about 1.0 m^3 ha^{-1}. Highest rates of erosion by crop type were found under market garden crops, maize, ley grasses, hops and sugar beet (Table 2). Median values were often much lower than the mean values of erosion, indicating a positively skewed statistical distribution of the data.

Mapping where erosion took place in the field and estimating rates of erosion for the areas directly affected showed that, compared to rates averaged out for the whole, transect rates were very high in the small parts of the field affected by channelling. Thus, in a wet year, ratios of mean rates of erosion in a sand-land for all the transect, the eroded fields, contributing areas only, and those localities directly affected by erosion were 1:6:26:110, and for clay-land they were 1:59:128:4183. The mean rates for the respective transects were 0.63 and 0.008 m^3 ha^{-1}.

The frequency of erosion was greatest, about once a year, where irrigation was used to grow field vegetables, and fields were eroded more often (over half the fields twice or more). In most localities fields that did erode eroded every two to four years, but very few fields were rilled every year.

Amounts of sediment deposited in a catchment can be compared with amounts of topsoil eroded. Volumes of sediment stored within or transported out of 367 catchments relate well statistically to topsoil textures in the eroded fields. Most sediment was retained within the catchment when soils were sandy, the regression equation expressing this being:

$$\text{sediment stored in catchment} = 15.8 + 0.71\% \text{ sand in topsoil } (R^2 = 27.0\%)$$

whereas the amount of soil transported out of the catchment was better correlated to the soil's clay content:

$$\text{sediment stored in catchment} = 83.0 - 1.56\% \text{ clay in topsoil } (R^2 = 19.9\%)$$

Erosion as a Problem

Results such as these showed that in particular localities water erosion was a problem and needed to be tackled because it was often associated with flooding and damage to property (Boardman, 1995; Evans, 1996) and with the delivery of fine sediment to streams. The latter impact helps explain the recent increase in water quality problems related to pesticides and nutrients in drinking water supplies and to the silting up of instream salmon and trout spawning beds, which is so damaging to fisheries (Environment Agency, 1998).

Understanding Erosion

The field-based assessments of erosion not only answered the question "Is erosion a problem?" but also greatly aided in the understanding of erosion processes. For example, rainfalls that caused erosion were often neither of large amounts nor of great intensities (Evans, 1990a; Chambers and Garwood, 2000) but rare storms could cause exceptional damage (Boardman, 1988, 1993). Erosion on slopes occurred dominantly below convexities and for it to occur fields had to have a relief of more than 5 m for erodible soils and 10 m for less erodible soils, and slopes had to be steeper than 3% (Evans, 1990a). Erosion increased as slopes and relief became greater. The extent of channel erosion was related to topsoil texture. Where soils were lighter in texture and contained less clay, erosion on slopes as a percentage of the total erosion (erosion on slopes plus valley floor) was greater. Erosion was more widespread in valley floors in localities where topsoils contained more clay (Fig. 2a) or where winter cereals predominated (Fig. 2b), though the two factors of topsoil clay content and area of land drilled to winter cereals were often related.

Predicting Erosion

From these results it was possible to predict the soil-landscape associations most vulnerable to erosion (Evans, 1990b); these predictions have been corroborated by later work (SSLRC, 1993; MAFF, 1997) and now form the basis for extension work and advice (MAFF, 1999). However, although it has proved possible on a regional basis to relate

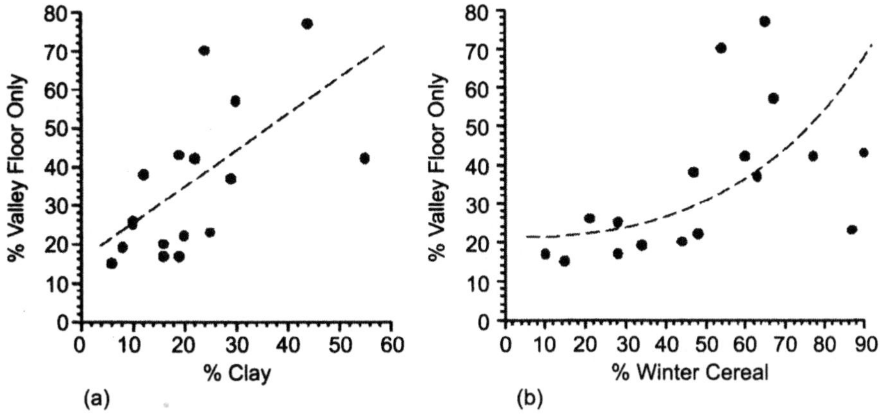

Fig. 2 Channel erosion on slopes and valley floors in relation to topsoil clay content and extent of winter cereals in 17 monitored localities in England and Wales. (a) Erosion in valley floors only, as a percentage of all erosion, and mean topsoil clay content in the locality. (b) Erosion in valley floors only, as a percentage of all erosion, and percent of eroded fields drilled to winter cereals.

amounts eroded to morphological factors (Evans, 1990a), it has not proved possible to predict rates of erosion for individual fields from some of the factors known to have caused or influenced erosion (Evans, 1998).

Factors that could explain the poor predictions are the following:

1. Poor quality information, not only on the individual storms that caused the erosion, but also on the extent of ground cover at the time. This insufficiency was inherent in the design of the project, which was to assess the size of the erosion problem rather than to investigate what the controlling factors were.
2. The highly skewed nature of the erosion rate data and the small range of median and mean values.
3. It is unlikely that an erosion event occurs when all the factors that could maximize erosion are in place, i.e., intense erosive rain on a bare, highly erodible soil on a steep slope. Generally, there is a complex interplay between these factors and it is difficult to separate out the factors and relate them individually to erosion rate.
4. Highly erosive, intense rainfalls are rare in the UK, and the positively skewed statistical distribution of erosion rates may well reflect the distribution pattern of erosive rains more than any other factor (Boardman and Favis-Mortlock, 1999).

Farmers' Perceptions of Erosion

It is the low rates of erosion, as exemplified by the median values for the localities and crop types, that the farmer sees in the field. Also, the same field often does not erode every year. So the volumes of soil eroded are small and hardly affect the land owner or manager, for the rills neither interfere with drilling, spraying or harvesting operations nor affect yields. The actual areas of ground in a field affected by erosion and deposition are generally very small (Fig. 3). A field-based assessment of erosion therefore helps explain why farmers in general do not perceive erosion as a problem in England and Wales. It is only when a major erosion event happens that the need for soil conservation becomes obvious; this is as true in Australia (Gardiner, 1992) as it is in England.

Comparisons with Other Field-based Assessments of Erosion

There have been a number of other field-based erosion monitoring schemes (for references see Evans, 2002). The statistical distributions of erosion rates, where these are given, are all positively skewed. Where sufficient numbers of fields have been sampled in studies, mean values can be compared (Fig. 4). Rates are not greatly different for localities in the UK, except where soils are more erodible (and often grow a wider range of crops), and these latter rates of erosion are similar to those of the erodible loessial soils of western Europe. Rates are low in Sweden, where fields were predominantly drilled to autumn-sown cereals. Rates are high on land cleared of tropical forest and put down to cultivation in Paraguay, where rainfall is high (c. 2,000 mm) (ERM/RSAC, 1998, 1999), where potato fields mostly have been left bare over winter on Prince Edward Island, Canada, and have suffered snowmelt runoff and erosion (Edwards et al., 1998), and especially so on steep slopes in tropical northern Thailand (Turkelboom and Trébuil, 1998).

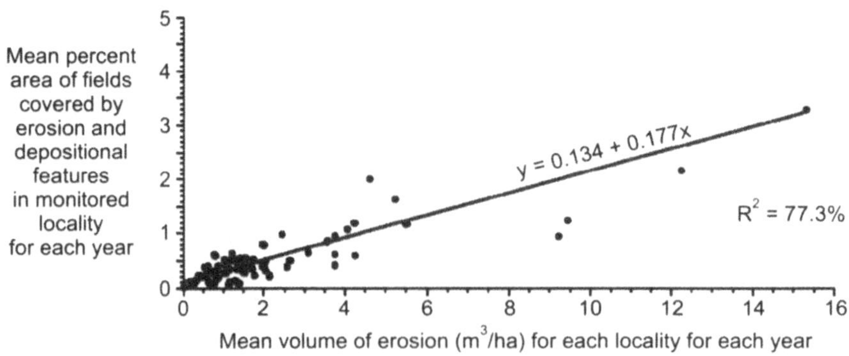

Fig. 3 Area of field affected by erosion as a function of volume of soil eroded.

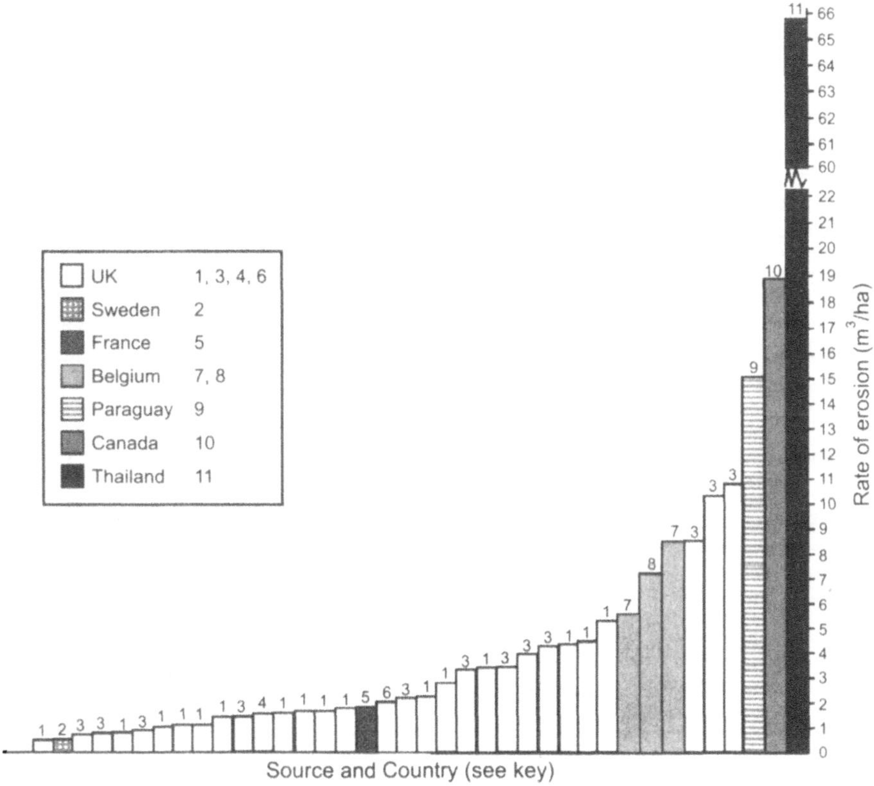

Fig. 4 Mean rates of erosion as measured in various localities. See Evans (2002) for references.

There is more information on maximum rates of erosion (Fig. 5). This is because erosion has been surveyed in some localities after a rare storm. A wide range of values has been recorded, most (28) being below 20 m^3 ha^{-1}, and few (6) above 80 m^3 ha^{-1}, the remainder (18) falling between these values. Maximum values reflect rainfall intensities and amounts falling in rare storms.

CONCLUSIONS

Field-based assessments of erosion work are quick and easy to do and, compared with plot experiments, inexpensive. They are more realistic than those carried out using predictive formulae based on plot experiments. They improve understanding and give good quality data on which government can base its actions.

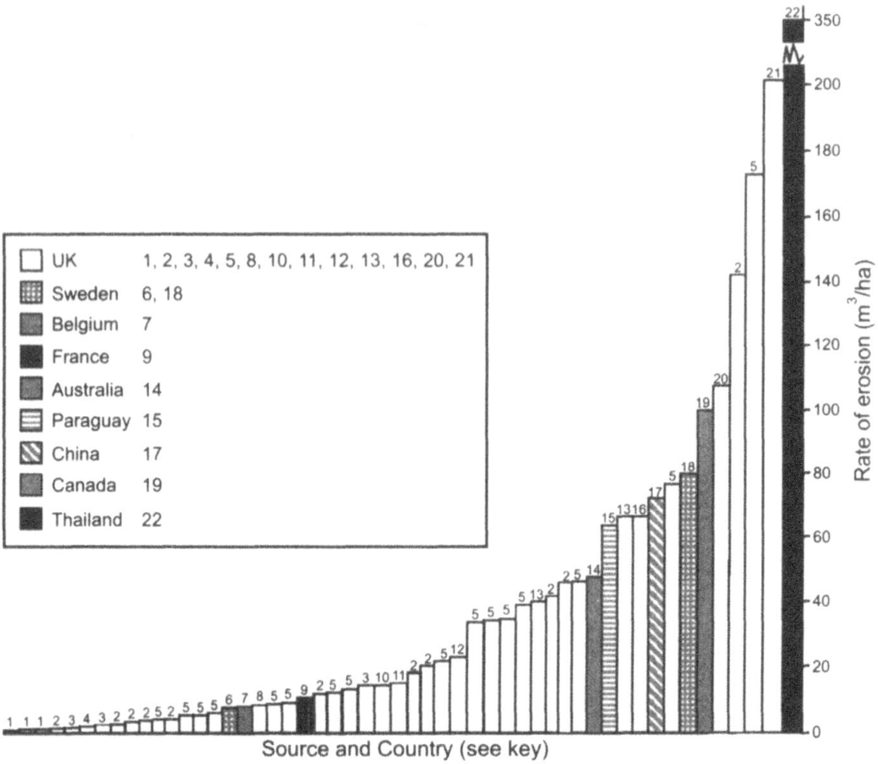

Fig. 5 Maximum rates of erosion as measured in various localities. See Evans (2002) for references.

Such field-based techniques could form the basis for a standardized method to assess erosion and could be used for monitoring and evaluating soil conservation and development projects or programmes. They could also be used to validate models to predict erosion.

Assessments of erosion in the field address the problem of scale because it is not difficult to relate data from the catchment within a field to that for the whole field, the river catchment, and the soil association, and extrapolate from the latter to the region or state. Field-based assessments can also address the problem of the lack of information on the delivery ratio of sediment from a catchment and help remedy "the lack of knowledge and uncertainties regarding the impact of deposition of sediment on yield and soil quality" (Lal, 1979: 1010).

In general, field-based estimates of erosion are likely to be (much) lower than those predicted from plot-based models such as RUSLE. It may not be possible, because of the interplay of factors, to accurately predict rates of erosion, except within homogeneous landscapes with similar weather patterns.

References

Boardman, J. 1988. Severe erosion on agricultural land in east Sussex, UK, October 1987. Soil Tech. 1: 333-348.

Boardman, J. 1990. Soil erosion on the South Downs: A review. pp. 87-105. *In:* J. Boardman, I.D.L. Foster and J.A. Dearing. [eds.] Soil Erosion on Agricultural Land. John Wiley and Sons, Chichester.

Boardman, J. 1993. The sensitivity of downland arable land to erosion by water. pp. 211-228. *In:* D.S.G. Thomas and R.J. Allison. [eds.] Landscape Sensitivity. John Wiley and Sons, Chichester.

Boardman, J. 1995. Damage to property by runoff from agricultural land, South Downs, southern England, 1976-93. Geogr. J. 161: 177-191.

Boardman, J. 1996. Soil erosion by water: problems and prospects for research. pp. 489-505. *In:* M.G. Anderson and S.M. Brooks. [eds.] Advances in Hillslope Processes, Vol. 1. John Wiley and Sons, Chichester.

Boardman, J. 1998. An average soil erosion rate for Europe: myth or reality? J. Soil Water Conservat. 53: 46-50.

Boardman, J. and D. Favis-Mortlock. 1999. Frequency-magnitude distributions for soil erosion, runoff and rainfall - a comparative analysis. Zeitschrift fur Geomorphologie N.F., Suppl.-Bd, 115, 51-70

Chambers, B.J. and T.W.D. Garwood. 2000. Monitoring of water erosion on arable farms in England and Wales. Soil Use Manag. 16: 93-99.

Dumanski, J. 1999. Monitoring progress towards sustainable land management. Paper given at ISCO conference, Purdue University, West Lafayette, Indiana, May 25, 1999.

Edwards, L. et al. 1998. Measurement of rill erosion by snowmelt on potato fields under rotation in Prince Edward Island (Canada). Can. J. Soil Sci. 78: 449-458.

El-Swaify, S.A. with an international group of contributors. 1999a. Sustaining the Global Farm - Strategic Issues, Principles and Approaches. International Soil Conservation Organization (ISCO), and Department of Agronomy and Soil Science, University of Hawaii at Manoa, Honolulu, Hawaii.

El-Swaify, S.A. 1999b. An action agenda from the discussion groups. Paper given at ISCO conference, Purdue University, West Lafayette, Indiana, May 28, 1999.

ERM/RSAC. 1998. Satellite optical and SAR data integration for soil erosion susceptibility mapping, monitoring and land management. Feasibility report to the British National Space Centre. Environmental Resources Management/ Remote Sensing Applications Consultants, London and Alton.

ERM/RSAC. 1999. Satellite optical and SAR data integration for soil erosion susceptibility mapping, monitoring and land management. Quarterly report to the British National Space Centre. Environmental Resources Management/ Remote Sensing Applications Consultants, London and Alton.

Environment Agency. 1998. The State of the Environment of England and Wales: Fresh Waters. The Stationery Office, London.

Evans, R. 1990a. Water erosion in British farmers' fields. Prog. Phys. Geogr. 14: 199-219.

Evans, R. 1990b. Soils at risk of accelerated erosion in England and Wales. Soil Use Manag. 6: 125-131.

Evans, R. 1993. On assessing accelerated erosion of arable land by water. Soils and Fertilizers 56: 1285-93.

Evans, R. 1995. Some methods of directly assessing water erosion of cultivated land - a comparison of measurements made on plots and in fields. Prog. Phys. Geogr. 19: 115-129.

Evans, R. 1996. Soil Erosion and Its Impacts in England and Wales. Friends of the Earth, London.

Evans, R. 1998. Field data and erosion models. pp. 313-327. *In:* J. Boardman and D. Favis-Mortlock. [eds.] Modelling Soil Erosion by Water. Springer, Berlin.

Evans, R. 2002. An alternative way to assess water erosion of cultivated land - field-based measurements and analysis of some results. Appl. Geogr. 22: 187-208.

Evans, R. and J. Boardman. 1994. Assessment of water erosion in farmers' fields in the UK. pp. 13-24. *In:* J. Rickson. [ed.] Conserving Soil Resources: European Perspectives. CAB International, Wallingford.

Evans, R. and D. McLaren. 1994. Monitoring Water Erosion of Arable Land. Friends of the Earth, London.

Gardiner, T.J. 1992. 'Colenso' soil conservation demonstration - a successful farm case study. pp. 480-483. *In:* P.G. Haskins and B.M. Murphy. [eds.] People Protecting Their Land, Vol. 2. Proceedings 7th ISCO Conference, International Soil Conservation Organization, Sydney.

Herweg, K. 1996. Field Manual for Assessment of Current Erosion Damage. Soil Conservation Research Programme, Ethiopia and Centre for Development and Environment, University of Berne, Switzerland.

Laflen, J.M. and E.J. Roose. 1998. Methodologies for assessment of soil degradation due to water erosion. pp. 31-55. *In:* R. Lal, W.H. Blum, C. Valentine and B.A. Stewart. [eds.] Methods for Assessment of Soil Degradation. CRC Press, Boca Raton.

Lal, R. 1997. In Discussion of 'Degradation and resilience of soils'. Phil. Trans.Roy. Soc. Lond. B. 352: 1008-1010.

MAFF. 1997. Controlling Soil Erosion. An Advisory Booklet for the Management of Agricultural Land. MAFF Publications, PB3280. Ministry of Agriculture, Fisheries and Food, London.

MAFF. 1999. Controlling Soil Erosion. A Manual for the Assessment and Management of Agricultural Land at Risk of Water Erosion in Lowland England. MAFF Publications, PB4093. Ministry of Agriculture, Fisheries and Food, London.

Reyes, M.R. et al. 1999. Comparing GLEAMS, RUSLE, EPIC and WEPP soil loss predictions with observed data from different tillage systems. Paper given at ISCO conference, Purdue University, West Lafayette, Indiana, May 28, 1999.

Risse, L.M., M.A. Nearing, A.D. Nicks and J.M. Laflen. 1993. Error assessment in the Universal Soil Loss Equation. Soil Sci. Soc. Am. Proc. 57: 825-833.

Scherr, S. 1999. The Cost of Environmental Protection versus Restoration. Paper given at ISCO conference, Purdue University, West Lafayette, Indiana, May 27, 1999.

SSLRC. 1993. Risk of Soil Erosion in England and Wales by Water on Land under Winter Cereal Cropping. Soil Survey and Land Research Centre, Silsoe.

Stocking M.A. and N. Murnaghan. 2001. Field Assessment of Land Degradation. Earthscan Publications Ltd., London.

Turkelboom, F. and G. Trébuil. 1998. A multiscale approach for on-farm erosion research: application to northern Thailand highlands. pp. 51-71. *In:* F.W.T. Penning de Vries, F. Agus and J. Kerr. [eds.] Soil Erosion at Multiple Scales. CAB International, Wallingford.

Watson, A. and R. Evans. 1991. A comparison of estimates of rates of soil erosion made in the field and from photographs. Soil Tillage Res. 19: 17-27.

14

Monitoring Erosion Using Microtopographic Features

Eelko Bergsma[1] and Abbas Farshad[2]

[1]Agricultural Engineer, formerly of the Soil Science Division, International Institute for Geo-information Science and Earth Observation (ITC), Enschede, The Netherlands. E-mail: bodem@wanadoo.nl

[2]Soil Scientist, ESA, International Institute for Geo-information Science and Earth Observation (ITC), Enschede, The Netherlands. E-mail: farshad@itc.nl

ABSTRACT

Experience has shown that rain erosion-induced microtopographic features can be grouped into seven types: original or resistant clods, eroding clods, flow paths, prerills, rills, depressions and possibly basal cover. The features represent the erosion that has occurred during a period previous to observation. Features are recorded per 25 cm on lines of 12.5 m along the contour. An indicator of erosion intensity can be derived from the erosion feature distribution. The indicator is calculated as the percentage of eroded clods plus two times the percentage prerill and rill area. It shows significant to highly significant correlation with measured soil loss. This opens the possibility of evaluating cropping systems and conservation practices for their protective effect against erosion. The erosion intensity can be compared for sites that represent the situation with and without conservation practices. The method can be used to monitor the development of erosion during a rain shower, a rainy season, or a series of years by recording the presence of the features at time intervals. Case studies in Colombia, Nepal, Tanzania and Thailand demonstrate the use of the microtopographic features for monitoring erosion intensity.

MICROTOPOGRAPHIC FEATURES AND SOIL EROSION

Rainfall erosion leaves behind microtopographic features on the soil surface. The possibly bewildering variation in soil surface microrelief is often called "random roughness". However, careful study of the erosion-induced microrelief has led us to distinguish seven features of microtopography (Table 1). A detailed description can be found in Bergsma (2001). The distribution of these features changes as the erosion proceeds, and it can be monitored. The aim of the monitoring is not to estimate soil loss, but to register the erosion intensity under different cultivation systems in locally representative conditions.

The method, which has so far been applied in Thailand, Nepal, Colombia, Tanzania and the Netherlands for various cultivation systems, uses the accumulated effect of the rain erosion in an erosive period previous to observation as it is expressed in the microtopographic erosion features (Bergsma, 1992, 2001; Bergsma and Farshad, 2003).

Recording the different types of features allows the determination of an indicator, hereafter referred to as *erosion intensity indicator*. The method can be used equally well for fields of annual or perennial crops, forests, plantations and grassland.

Using the indicator of erosion intensity, different types of land use can be judged in a comparative way. In other words, the erosion hazard of cultivation systems can be compared. Thus, the recording of microtopographic features can give immediate information on the relative resistance to erosion of areas within a soil and water conservation (SWC) project, for instance those with different practices on otherwise comparable sites. Perhaps of greater practical importance for the evaluation of SWC projects is recording the features on comparable sites

Table 1 Microtopographic features used to describe the soil surface (abbreviations in parentheses are used in the following tables and figures).

Microtopographic feature	*Brief description*
Original or resistant clods (res)	Resistant forms and those created by recent tillage
Eroded clods or surfaces (ero)	Forms rounded by splash and disintegration
Flow paths or surfaces (flo)	Flat areas of shallow unconcentrated flow, often with braiding pattern of lag sediment
Prerills (pre)	Shallow channels of concentrated flow, up to 3-5 cm deep
Rills (ril)	Micro-channels deeper than prerills and up to 20 cm deep
Depressions (dep)	Small low areas, enclosed by clods
Vegetative matter (veg)	Basal cover of plants and litter

outside the project area, as this will give information about the difference between the situation with and without conservation.

RELATED STUDIES ON SOIL MICROTOPOGRAPHY AND EROSION

Several erosion studies have paid attention to soil microrelief and some researchers recognized components of the microrelief.

Merritt (1984) identified four stages of micro-rill development:

1. sheet flow
2. flow line development
3. micro-rills
4. micro-rills with headcuts

There appears to be some correlation of the stages of micro-rill development of Merritt with the microtopographic features used in the method presented here. Stages 1 and 2 of Merritt, the sheet flow and the flow line development, could correspond with the *flow areas* of the present method; stage 3 could correspond with the development of *prerills*. The headcuts of stage 4 may begin as cross-scarps in the bed of prerills, which then take part in the development of rills.

Auzet (2001) discusses parameters for erosion prediction that can be derived from soil surface characteristics: vegetation cover, stone cover, crust development and surface roughness. They were insufficient to describe the influence of soil surface structure and microrelief on total erosion. A conclusion is that lack of knowledge of erosion processes and their interactions could be partly compensated for by describing in a simple way the soil surface characteristics that relate to types of processes.

Comparing the soil surface characteristics of Auzet with the microtopographic features (Table 1), the *vegetative cover* could be partly similar to *basal cover*. *Stone cover* would be classified as *resistant clods*. *Crusts* would most often be *flow surfaces*. *Surface roughness* would not be random, but has proved to be composed of seven distinguishable components.

Unexpected results may be found by using the present method of microtopographic features. Mainam (1991) showed that contour bunds of medium height have more erosion than bunds of low height in an area in northern Thailand. The higher the bunds, the more erosion occurs by overflowing in the heavy rains. Van Dijk (2001) suggests there is a large influence of oriented roughness that results from tillage. Overland flow may be more frequent in cases where up-and–down-slope tillage has been practised, but it is more diffuse and less erosive than the concentrated flow resulting from the tillage that diverts flow sideways.

Poesen (1988) gave an overview of the mechanisms of incipient rilling and gullying in the Belgian loess region. It gave attention to the development of surface (and subsurface) erosion features, whereby one feature is often a stage in the development of another.

Stages of change of the soil surface structure as caused by rain have been observed by Valentin (1985). The area of observation was in Niger near Agadez, at the southern fringe of the Sahara. Three soils were involved in his study: a pebble-covered desert pavement, a sandy soil, and a clayey alluvial soil. Stages that were observed in the development of the soil surface topography are the following:

1. *Before the rain:* the clods have sharp edges.
2. *During the rain that is absorbed by the soil:* the shapes of the clods are cratered, grains are washed free, swelling takes place, and small particles are moved downwards.
3. *At the beginning of the overland flow:* clods have become smooth. When the soil becomes saturated some deposition at the foot of clods takes place, there is exposure of resistant parts; some flow paths develop.
4. *When there is overland flow:* the surface of the clods consists of micro-aggregates, slaked material, and crusts. Surface flow erodes the clods sideways, and remnants of crusts occur in deposits.

In the observations made by Valentin (1985) the clods with a sharp-edged shape observed before the rain are similar to *original or resistant clods* of the present method (Table 1). After the rain starts, initially clods with cratered shape occur, and at the beginning of overland flow clods have become smooth and resistant parts are exposed. These forms will correlate with *eroding clods* and some remaining *resistant clods*. In the presence of overland flow, lateral erosion of the clods occurs; this correlates with the formation of *flow paths*.

Imeson and Kwaad (1990) found periods to be distinguished in the evolution of the structural elements starting from freshly tilled topsoil:

1. A short period with freshly tilled soil.
2. A period in which rainfall-induced processes lead to stepwise degradation of soil structure and a stepwise decline of various soil physical processes.
3. A period in which a continuous crust is present at the soil surface and in which no further change in soil physical properties takes place, except by biological activity.

In a study by Andrieux et al. (2001) soil surface features were described by criteria that can be observed at the field scale. Relating the surface criteria of Andrieux et al. with the microtopographic features in the method presented here (Table 1), the *surface seal* would probably be a *flow surface*. *Roughness* does not correspond with any feature, because roughness is described by its component features in the present method. *Grass cover* and *crop residue* would be similar to *basal cover* in the present method. *Stones* would be classified as *resistant material*. *Topsoil structure* is not considered as such.

In other research on the influence of soil surface on erosion, the microrelief height distribution receives attention. It plays a role in studies of storage capacity and overland flow generation. In several cases, stereoscopic methods have been used (Borselli et al., 2001; Ciarletti et al., 2001; Farres and Merel, 2001; Farres and Poesen, 2001).

On a very detailed scale, surface relief has been recorded by laser scanning of soil surface profiles. Areas of a few square metres could thus be modelled very accurately in three dimensions (Huang et al., 1988). Applications were the determination of surface storage capacity and the connectivity of overland flow paths (Abedini et al., 1997). Improvements on the laser device completed in 2001 allow measurement of 0.5 × 4 m area in about 7.5 minutes with 0.5 mm accuracy (Darboux and Huang, 2003).

On equally detailed scale, digital photogrammetry has been used to study sediment transport (Stojic et al., 1998) or to construct an elevation model of the soil surface; a model of 8 sq m area can be made in 10 minutes (Wegmann et al., 2001).

Studies of the change in the soil surface microrelief during erosion conducted for instance by Jester et al. (2001) and Torri et al. (2001) show that the decay of surface relief that occurs in defined conditions of rainfall and soil wetness is associated with a tendency of roughness increase when channel flow occurs. Like Andrieux et al. (2001), these authors do not work with components of the roughness.

To be able to realistically conclude about flow depth and (micro)channel occurrence, one has to look at what really happens in the field (Jetten, 2001). The method presented here uses direct field observation of the three-dimensional components of the microtopography and field stereophotos for reporting purposes.

The Field Manual for Assessment of Current Erosion Damage (Herweg, 1996) describes eroded parts of fields with their land use and management, seen in the local erosion toposequence. Five general microrelief classes are used, going from fresh clods to a smooth surface.

The *Handbook for Field Assessment of Land Degradation* (Stocking and Murnaghan, 2001) uses field indicators of erosion to derive an estimate of soil loss. Three indicators are related to the general surface microtopography of fields: pedestals, armour layer, and rock exposure. The handbook stresses the viewpoint of the land user and the socio-economic-political conditions in which the land user has to work.

Table 2 shows a brief comparison of these two methods and the method using soil surface topography.

METHOD OF RECORDING FEATURES

In a field to be studied, the general eroded part indicates the place of highest erosion hazard. There, a measuring tape (of, for instance, 2.5 m length) is stretched along the contour, so that the features made by flow are met across the tape during recording. The tape has alternate coloured intervals of 25 cm. For each interval the dominant microtopographic feature type is recorded. The recording uses 50 intervals, following a contour line. Each tape interval represents 2% of the area and the percentage distribution of the seven features is determined. The procedure is repeated twice along parallel lines, situated at one or two metres above or below the first observation line.

The procedure is not difficult to learn and does not take much time to apply. A short video shows the recognition of the features under natural rainfall (Bergsma, 2003). In a case of erosion plots bordering each other, up to 24 records have been made in one day. It is more efficient and

Table 2 Characteristics of three methods of monitoring land degradation.

Assessment of current erosion damage	*Microtopographic erosion features*	*Field assessment of land degradation*
The method estimates soil loss from recent storms by rill volumes, as a part of the site description. It is not a means of predicting soil loss. Site description is repeated at intervals of one or more years to monitor erosion damage.	Microtopographic erosion features are recorded along the contour. An indicator of erosion intensity is derived, which correlates with measured soil loss. A comparison of erosion intensity can be made for sites of different land use or conservation practices.	It aims to identify underlying causes and effect of conservation and rehabilitation on land of individual users. The degree of erosion is derived from various estimators of soil loss and expressed in t/ha.
It is not a method for assessing the general status of land degradation.	It is not a method for assessing the general status of land degradation or soil loss in t/ha.	*The Handbook* gives guidelines to assess land degradation, its causes and fitting remedies.

Fig. 1 Measuring tape with coloured intervals of 25 cm, stretched along the contour for recording microtopographic erosion features; the site is a maize stubble field on loess, in Zuid Limburg, the Netherlands.

stimulating to have a team of two persons in the field. The feature recording can be done on any type of land use, be it annual or perennial crops, grassland, forest, orchard or plantation. The method was used in several doctoral studies (Turkelboom, 1999: 87-90; de Bie, 2000: 143-164).

INDEX OF EROSION INTENSITY DERIVED FROM THE MICROTOPOGRAPHIC FEATURES

A record of the different types of microtopographic erosion features allows the determination of an indicator of erosion intensity. Using this indicator, the intensity of erosion of different types of land use and cultivation systems can be compared. The indicator of the erosion intensity is derived from the most serious erosion features, which are rills, prerills and flow paths. The indicator is calculated as the percentage flow area plus two times the percentage prerill and rill area. The unequal weight allotted to features tries to represent the relative importance of the features in the erosion process that causes soil loss.

The indicator values appeared to correlate with measured soil loss in previous research cases (Table 3 and Bergsma, 2001).

Table 3 Measured soil loss and erosion intensity derived from microtopographic features.

Location and date	*Number of erosion plots, as treatments × replications*	*Spearman rank correlation coefficient*		
		All individual plots	*Number of plots excluded †*	*Plots grouped per treatment*
Chiang Dao, Thailand, August 1994	5 × 4 and 2 × 1	0.39	3: 0.76*** 4: 0.79***	0.69 † 1 treatment: 0.85*
Doi Tung, Thailand, July 1997	5 × 4 and 3 × 1 of which 8 plots studied	0.59	1: 0.84*	0.90** † 1 treatment: 0.94**

*** = significance level of 0.001
** = significance level of 0.01
* = significance level of 0.05

† = excluded for reasons of faulty plot management, deposition within plot, and one derived but unlikely extreme erosion intensity in 1997

The comparison of the measured soil loss and the indicator of erosion intensity has led at times to recognition of faulty soil loss measurements as well as faulty feature observations.

The feature method is not meant to detect erosion-governing factors with the aim that a (better) prediction of soil loss may be attempted. The feature method aims at a comparison of the actually observed erosion intensities that are characteristic of local land use types. It provides a comparison about the relative resistance to erosion of these land use types. This in turn will support the evaluation of SWC projects. Rural extension workers and land use planning officers can benefit from these elementary data for their recommendation of certain cultivation practices.

EXAMPLE OF MONITORING EROSION TO EVALUATE THE EFFECT OF CULTIVATION PRACTICES

In an area of about 30 km^2 near the village of Lom Kao, north of the city of Phetchabun, Thailand, the erosion development in five major land use types was monitored using microtopographic features over a period of two months, starting roughly at the beginning of the rainy season. The sites were comparable in rainfall erosivity, general topography and soil (Table 4; basic data from Woldu, 1998).

The microtopographic erosion features were recorded with repetitions situated in the upper, middle and lower part of the field. Records of features were made after each rain in June and July, and once in August. The last record had only two repetitions, both made in the middle part of the field with one metre between the lines. These records of feature distribution are shown in Fig. 2. This pattern of change in

Table 4 Characteristics of five fields near Lom Kao, Thailand.

Field code	*Location*	*Parent material, soil, position, steepness*	*Cultivation system (the critical factor for erosion intensity is in italics)*
1A	Hill-land, 2 km E of Lom Kao.	Parent material: Shale. Soil: fine silty over skeletal Typic Paleustults and Typic Haplustults (association), moderately deep to deep, well drained, Ap coarse strong subangular blocky, pH 4, CL over B-horizon, pH 3.5, C. Position: straight mid-slope. Steepness: 28%.	Four years fallow after forest, now one-year-old mixed orchard with tamarind, spacing 6 m × 6 m (some 8 m × 8 m), tillage once by tractor, up and down slope, weeding by tractor *only around trees, elsewhere weeds and rough soil surface*.
1B		Parent material: Shale. Soil: Lithic Ustorthents, Ap and part of B lost by erosion, topsoil gravelly silty clay, bedrock here at 20 cm, at 50-60 cm elsewhere. Position: straight mid-slope. Steepness: 26%.	20 years maize cultivation, then five years fallow, now ploughed for tamarind planting, up and down the slope; there is high soil animal activity and *weed growth*.
2	Hill in Piedmont, 12 km N of Lom Kao, 1 km NW of Ban Kok Kathon.	Parent material: Andesite. Soil: Pachic Haplustolls, moderately deep to deep, moderately well drained, Ap Silty clay, 0-2% gravel, weak fine to medium subangular blocky, pH 7.5. Position: straight mid-slope. Steepness: 28%.	Maize, tillage once by tractor, up and down slope, plants in rows along contour till 20° off, spacing 40-50 cm between plants, 50-80 cm between rows, 5-9 plants per m^2, *some empty spaces*, germination uneven because of lack of moisture and poor seed, weeding once when plants are at knee height, 150 kg NPK applied half at planting and half after weeding.

(Table 4 Contd.)

(Table 4 Contd.)

3 A	Hill in northern Piedmont, 14 km N of Lom Kao, 1.6 km N of Ban Kok Kathon.	Parent material: Andesite. Soil: Ustic Dystropepts (eroded), Ap 10-15 cm, pH 6.0, profile is moderately deep, well drained, topsoil clayloam, 'cloddy' strong medium subangular blocky over B, clay. Position: straight mid-slope. Steepness: 32%.	Maize, tillage by heavy tractor, one time, up and down slope, crop residue hinders ploughing and *is removed* beforehand, seedbed preparation and weeding by hoe, plants in rows roughly along the contour, rows spaced 70-85 cm, plants spaced 35-60 cm, on average 9 plants per m^2, germination uniform, stand denser than at Site 2, fertilizer applied at planting and when plants are at knee height.
3 B		Parent material: Sandstone. Soil: Kanhaplic Haplustults, Ap 15 cm, topsoil sandy clayloam, fine and loose topsoil structure, strong medium subangular blocky. Position: straight mid-slope. Steepness: 28%.	Sweet potato, tillage by heavy tractor along the contour, plants spaced 15-20 cm on convex ridges made by tractor, spaced 125 cm, 60 cm high, *running down the slope*, soil was pulverized by plowing and ridge making (over-worked), lowering the grade of soil structure.

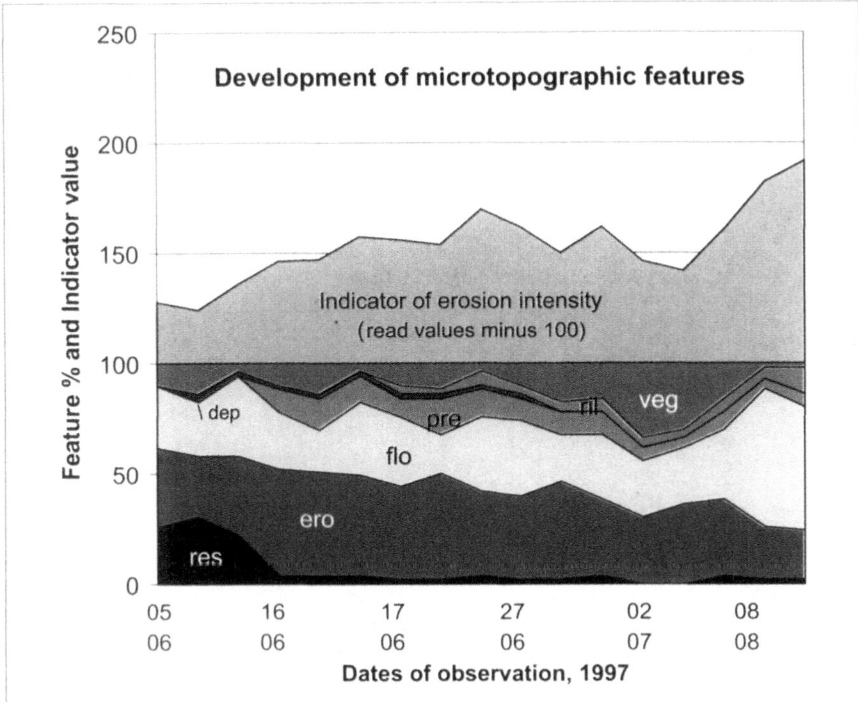

Fig. 2 Microtopographic features during a two-month period (Site 2, Woldu 1998). Recording is repeated two times in August and three times at other dates. See Table 1 for abbreviations.

microtopographic features under successive rains is often found in the development of erosion over time.

On Site 2, the *area of flow paths* remains rather constant during the first five rains but increases strongly in August after harvest and the removal of residue. *Prerills* start after the second rain, and their importance remains rather constant till July, when basal cover increases. *Rills* occur after the third rain. They continue to cover a small part of the area and increase in August after basal cover disappears. The *basal cover* provided by the crop and the weeds remains low and constant during four rains. It increases in July because of crop residue and weed growth. It disappears after burning in August.

Table 5 shows percentage cover by features on two dates. On July 2 much basal vegetation exists on some of the five fields. A month later the fields have less basal cover, harvest has taken place, residues have been burned or have rotten, and land preparation by tillage has turned residues

Table 5 Relative erosion intensity on 2 July 1997 and 8 August 1997.

Field	*Features on 2 July 1997*							*Indicator of erosion intensity*	
	res	*ero*	*flo*	*pre*	*ril*	*dep*	*veg*	*Value*	*Rank*
1A	4	40	20	1	2	1	32	26	1
1B	0	39	21	4	0	1	35	29	1
2	3	38	28	10	5	1	15	58	3
3A	1	39	24	19	15	0	2	92	4
3B	3	69	2	7	18	0	1	52	2
Field	*Features on 8 August 1997*							*Indicator of erosion intensity*	
	res	*ero*	*flo*	*pre*	*ril*	*dep*	*veg*	*Rank*	*Order*
1A	3	24	31	18	-	-	22	67	1
1B	2	13	61	20	-	-	4	101	4
2	2	23	59	14	-	-	2	87	3
3A	5	18	49	25	3	-	-	105	4
3B	2	40	36	1	21	-	-	80	2

under. Field 1B has a relatively stronger increase in erosion intensity between the two dates. This reflects the different influence of the crop growth and the post-harvest situation.

Field 3A of maize is 4-6% steeper than other sites. At the later date a number of rills has changed into prerills under the influence of splash in the uncovered stage. The sweet potato field 3B has cultivation ridges and furrows that run down the slope but are not formed by erosion.

The data of all five fields studied by Woldu (1998) show a tendency of three stages in the development of erosion that has been found generally in our studies and that was also noticed by Valentin (1985) in the soil surface relief. There is a first stage of rapid change, a second stage of rather constant feature distribution, and a last stage of rapid change again.

In the case of Site 2 of Woldu, the transitions are around the dates of 16/6 and 2/7. The three phases can be seen in Figs. 2 and 3.

Development of Microtopographic Features and Final Erosion Intensity

Erosion intensity derived from microtopographic features was studied on 22 fields in the area of Kao Khor, Thailand (Mainam, 1991) for various site conditions and crop management practices. It was found that the development of microtopographic erosion features during the observation period has characteristics that allow prediction of the final

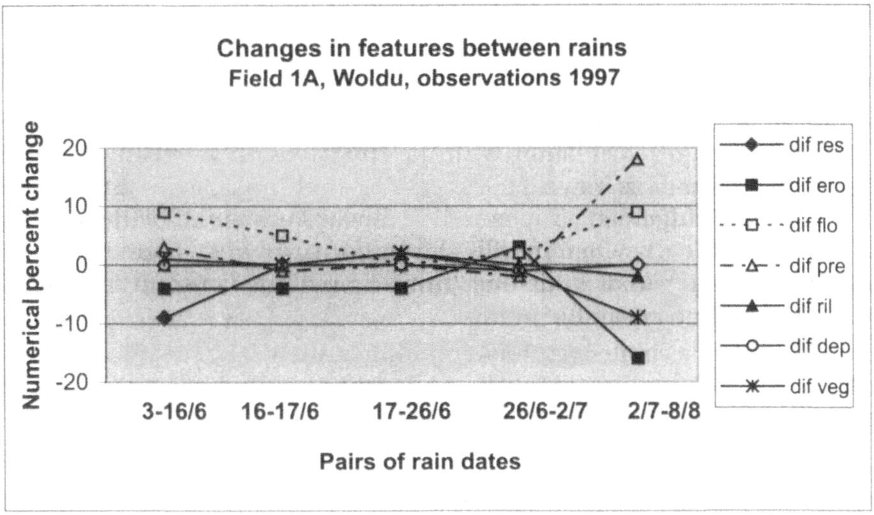

Fig. 3 Example of three phases in erosion feature development.

Table 6 Correlation between the development of certain microtopographic features and the erosion intensity at the end of the observation period, 29/4-24/5, 1990 (data of Mainam, 1991).

	Duration of fall in res	*Amount of fall in ero*	*Duration of rise of ril*	*Period up to rill formation*	*Period up to pre/rill formation*
Rs	0.76	0.71	0.82	0.86	0.91
t value	5.2	4.5	6.3	7.7	9.8
Sign. level	0.0005	0.0005	0.0005	0.0005	0.0005

Duration and Period are expressed in number of rains
Rs = Spearman rank correlation coefficient
t-value = Student's *t*
Sign. level = Statistical significance level

res = resistant/ original clods
ero = eroding clods and surfaces
pre = prerills
ril = rills

erosion intensity. Table 6 shows correlations between the Erosion Intensity Indicator on the final date of observation and the development of certain microtopographic features up to that date. Such a correlation holds for the period of crop influence. In the post-harvest stage, other large changes in erosion features may occur, such as are shown by field 1B (Table 5).

The length of period up to rill development. This characteristic of erosion development has the great advantage that it is rather simple to use as an indicator in the field. The start of rill formation is easier to observe than the continuation of an increase in rill area. Correlation with the final erosion intensity was high.

Period up to pre/rill formation. The period up to prerill and rill formation shows a stronger correlation with the final erosion intensity than rill formation alone. It is, however, less easy to apply and carries greater risk of subjective judgement. The period, as always measured by the number of rainstorms, ends when prerills start to form, or when there is a rise in prerill formation that is distinct from a previous constant level. Rills usually form later than the prerills.

The use of a single microtopographic feature will be practical in some cases, but the erosion intensity index gives a more comprehensive measure of erosion resulting from a previous period.

ACCURACY OF MEASURED SOIL LOSS AND PRECISION OF THE INDICATOR OF EROSION INTENSITY

At some erosion stations, the eroded bedload is kept in the collector furrow while the suspended load overflows a barrier and is allowed to pass. Bedload is measured on a hand-held balance that has for instance a 9 kg maximum load. A correction is made for the moisture content. This method is judged to be more accurate than sampling from a barrel that has received the sediment, because after stirring the water and sediment, the sand-size particles will settle too quickly to be represented well in a sample taken from the barrel. Zöbisch et al. (1996) stress the influence of the sampling procedure on the accuracy of soil loss data using barrels. In an experiment with five operators who sampled the same eroded volume, only three results agreed with each other within acceptable limits.

In the recommended procedure, there are three replications, made on lines one or two metres apart. A difference between the three feature records is caused by the natural variation of erosion in the field and an amount of error in the recording. To obtain an estimate of the error in the recording of the features, 10 repetitions of 50 feature readings were made on the same observation line of 12.5 m. The site was an arable field with maize stubble (autumn, 1998), in a loess region in Zuid Limburg, the Netherlands. It was concluded that recording the erosion-induced microtopographic features needs prior practice to rehearse the application of the details of their classification criteria. And from the data of the experiment a maximum observation error of 5 was found in the value of the erosion intensity indicator, average of three repetitions.

CONCLUSIONS

The following conclusions can be drawn from the present study:

- Erosion-induced microtopographic features show the accumulated effect of erosion over a period since the previous management of the entire soil surface. An erosion intensity indicator can be derived from the recorded features. In this way the erosion intensity on various sites can be compared. The comparison can be made for a certain moment or through monitoring the erosion development by repeated observations.
- In conservation planning, the use of the erosion intensity indicator is a rapid and simple way to compare the effect of land use systems. Their comparative erosion hazard can thus be judged. In a SWC project, partial areas may be compared as in a lay-out of alternative practices. The effect of a project as a whole may be judged from recording the features on comparable sites outside the project area as this will give information about the difference in erosion between the with-project and without-project situations.
- For practical purposes, the expected relative erosion intensity under land use types is indicated by the delay in rill development from the start of a rainy season or seeding time. The delay is expressed in the number of showers.
- In trying to characterize the expected erosion of an arable land use system by the development of microtopographic features, it is important to be aware of three stages in the development of the features, namely large changes in the microtopography after land preparation, minor changes during crop maturing, and large changes after harvest.

ACKNOWLEDGEMENTS

Access to observation sites in Zuid Limburg is by courtesy of the Experimental Farm Wijnandsrade. Provision of soil loss data and facilitating the research in Zuid Limburg is by courtesy of Drs F.J.P.M. Kwaad, at the time working at the Physical Geography and Soil Science Laboratory, University of Amsterdam.

References

Abedini, M.J., W.T. Dickinson and R.P. Rudra. 1997. Integration of GIS tools and laser-scanned DEM with implications for rainfall-runoff modeling. J. Am. Soc. Agr. Eng. (ASAE) 1997, No. 973029, 15 pp.

Andrieux, P., A. Hatier, J. Asseline, G. de Noni and M. Voltz. 2001. Predicting infiltration rates by classifying soil surface features in a Mediterranean wine-growing area. International Symposium, The Significance of Soil Surface Characteristics in Soil Erosion. September 2001, Strasbourg, France. COST 623 Workshop, Soil Erosion and Global Change. Organized by RIDES, network of research groups.

Auzet, A.V. 2001. Descriptors of soil surface characteristics for infiltration/runoff and erosion assessment. International Symposium, The Significance of Soil Surface Characteristics in Soil Erosion. September 2001, Strasbourg, France. COST 623 Workshop, Soil Erosion and Global Change. Organized by RIDES, network of research groups.

Bergsma, E. 1992. Features of soil surface micro-topography for erosion hazard evaluation. pp. 15-26. *In:* H. Hurni and K. Tato. [eds.] Erosion, Conservation and Small-scale Farming. Selection of papers presented at the 6th International Soil Conservation Organisation (ISCO) conference, Ethiopia and Kenya, 1989.

Bergsma, E. 2001. Erosion intensity evaluated from micro-topographic soil erosion features, and its correlation with conservation practices, presence of fertiliser, and the erosion development between contour hedges; data of Doi Thung, northern Thailand. *In:* D.E. Stott, R.H. Mohtar, G.C. Steinhardt. [eds.] Sustaining the Global Farm. Selected papers from the 10th International Soil Conservation Organization (ISCO) Meeting, 1999, West Lafayette, Indiana.

Bergsma, E. 2003. Erosion by Rain - its subprocesses and diagnostic microtopographic features. A video/CD-Rom, 33 min., ITC. For use in education, research and consulting; with handout.

Bergsma, E. and A. Farshad. 2003. Multiple use of erosion-induced microtopographic features. International Symposium, 25 years of Assessment of Erosion. Ghent, Belgium.

Borselli, L., M.P.S. Sanchis, M.S. Yañez and D. Torri. 2001. Dynamics and properties of ponding areas. International Symposium, The Significance of Soil Surface Characteristics in Soil Erosion. September 2001, Strasbourg, France. COST 623 Workshop, Soil Erosion and Global Change. Organized by RIDES research groups.

Ciarletti, V., P. Biossard, L.M. Bresson, M. Zribi, L. Bennaceur and M. Chapron. 2001. Tridimensional investigation of the soil roughness evolution along time by using a rainfall simulator and a stereovision device. International Symposium, The Significance of Soil Surface Characteristics in Soil Erosion. September 2001, Strasbourg, France. COST 6223 Workshop, Soil Erosion and Global Change. Organized by RIDES, network of research groups.

Darboux, F. and C.H. Huang. 2003. An instantaneous-profile laser scanner to measure soil surface microtopography. Soil Sci. Soc. Am. J. 67: 92-99.

de Bie, C.A.J.M. 2000. Comparative performance analysis of agro-ecosystems. Doctoral thesis, Wageningen University, The Netherlands. Chapter 11: Soil erosion indicators for maize-based agro-ecosystems in Kenya.

Farres, P.J. and A.P. Merel. 2001. An assessment of the use of terrestrial photogrammetry as a method for the monitoring of natural soil surface changes over an extended period. International Symposium, The Significance of Soil Surface Characteristics in Soil Erosion. September 2001, Strasbourg, France. COST 623 Workshop, Soil Erosion and Global Change. Organized by RIDES, network of research groups.

Farres, P.J. and J. Poesen. 2001. Ridge furrow surface forms: a laboratory experimental study combining the use of low altitude terrestrial photogrammetry and soil micromorphology. International Symposium, The Significance of Soil Surface Characteristics in Soil Erosion. September 2001, Strasbourg, France. COST 623 Workshop, Soil Erosion and Global Change. Organized by RIDES, network of research groups.

Herweg, K. 1996. Field Manual for Assessment of Current Erosion Damage. Soil Conservation Research Program, Ethiopia and Centre for Development and Environment, University of Berne, Switzerland.

Huang, C.H., I. White, E. Thwaite and A. Bendeli. 1988. A noncontact laser system for measuring soil surface topography. Soil Sci. Soc. Am. J. 52: 350-55.

Imeson, A.C. and F.J.P.M. Kwaad. 1990. The response of tilled soils to wetting by rainfall and the dynamic character of soil erodibility. *In:* J. Boardman, I.D.I Foster and J.A. Dearing. [eds.] Soil Erosion on Agricultural Land. John Wiley, pp. 3-14.

Jester, W., A. Klik, G. Hauer and C.C. Truman. 2001. Interrill wash and splash erosion as affected by surface roughness and rainfall intensity. International Symposium, The Significance of Soil Surface Characteristics in Soil Erosion. September 2001, Strasbourg, France. COST 623 Workshop, Soil Erosion and Global Change. Organized by RIDES, network of research groups.

Jetten, V. 2001. The effect of surface roughness on the hydraulic radius. International Symposium, The Significance of Soil Surface Characteristics in Soil Erosion. September 2001, Strasbourg, France. COST 623 Workshop, Soil Erosion and Global Change. Organized by RIDES, network of research groups.

Mainam, F. 1991. Evaluating the effect of agricultural practices and crops for soil conservation on steep lands by monitoring features of soil surface microtopography - case study of Khao Kho area, Thailand. MSc thesis, ITC, International Institute for Aerospace Survey and Earth Sciences, Enschede, The Netherlands. 133 pp.

Merritt, E. 1984. The identification of four stages during micro-rill development. Earth Surface Processes and Landforms 9: 493-96.

Poesen, J. 1988. A review of the studies on the mechanisms of incipient rilling and gullying in the Belgian loam region. Proceedings of the International Symposium on Erosion in S.E. Nigeria, 1988. Federal University of Technology, Owerri, Nigeria.

Stocking, M.A. and N. Murnaghan. 2001. Handbook for the Field Assessment of Land Degradation. Earthscan Publications Ltd., London; Sterling, Virginia, USA.

Stojic, M., J. Chandler, P. Ashmore and J. Luce. 1998. The assement of sediment transport rates by automated digital photogrammetry. Photogrammetric Engineering & Remote Sensing, 64(5): 387-395.

Torri, D., L. Borselli, M.P.S. Sanchis and M.S. Yañez. 2001. Splash-induced soil surface dynamic. Paper presented at the International Symposium, The Significance of Soil Surface Characteristics in Soil Erosion. September 2001, Strasbourg, France. COST 623 Workshop, Soil Erosion and Global Change. Organized by RIDES, network of research groups.

Turkelboom, F. 1999. On-farm diagnosis of steepland erosion in northern Thailand - Integrating spatial scales with household strategies. Doctoral thesis, Faculty of Agricultural and Applied Biosciences, Catholic University of Leuven, Leuven, Belgium.

Valentin, C. 1985. Organisations pelliculaires superficielles de quelques sols de région subdésertique - dynamique de formation et conséquences sur l'économie en eau. Editions de l'ORSTOM, Collection études et théses, Paris.

van Dijk, P. 2001. Effects of conservation tillage on random and oriented surface roughness and the implications for depression storage on the hillslopes. International Symposium, The Significance of Soil Surface Characteristics in Soil Erosion. September 2001, Strasbourg, France. COST 623 Workshop, Soil Erosion and Global Change. Organized by RIDES, network of research groups.

Wegmann, H., D. Rieke-Zapp and F. Santel. 2001. Digitale Nahbereichs-photogrammetrie zur Erstellung von Oberflächenmodellen für Bodenerosionsversuche. Wissenschaftl. Techn. Jahrestagung der DGPF, October 2000, J. Albertz and S. Dech, eds.

Woldu, H.D. 1998. Assessment of the effect of present land use on soil degradation - a case study in the Lom Sak area, central Thailand. MSc thesis, International Institute for Aerospace Survey and Earth Sciences, Enschede, the Netherlands.

Zöbisch, M.A., P. Klingspor and A.R. Odour. 1996. The accuracy of manual runoff and sediment sampling from erosion plots. J. Soil Water Conservat. May-June 1996, 231-233.

15

The Walnut Gulch Experimental Watershed — 50 Years of Watershed Monitoring and Research

Mary H. Nichols
Southwest Watershed Research Center, 2000 E Allen Rd., Tucson, AZ 85719, USA.
E-mail: mnichols@tucson.ars.ag.gov

ABSTRACT

The Walnut Gulch Experimental Watershed of the United States Department of Agriculture—Agricultural Research Service was established in 1953 with the broad objectives to (1) determine the effects of conservation projects on water yield and sediment movement and (2) quantify flood runoff from semi-arid rangeland watersheds. The 150 km^2 watershed was instrumented with rain gauges and runoff-measuring flumes arranged in a pattern of nested subwatersheds. Data collected during the past 50 years have been analyzed to characterize precipitation in regions dominated by convective thunderstorms and to study and model subsequent flood wave movement, transmission losses, and water yield from complex watersheds. The effects of topography and various soil, vegetation, and surface cover complexes on water and sediment movements have been studied at spatial scales ranging from plots to watersheds. The initial research objectives have expanded to include remote sensing, nutrient cycling, and development of decision support systems. The comprehensive database has been used to characterize baseline conditions and variability inherent in semi-arid rainfall and runoff. The data have also been used to develop rainfall, runoff, and erosion prediction technologies. This chapter focuses on hydrologic and erosion research and includes a description of instrumentation and monitoring sites, a description of major research findings, and a summary of lessons learned from measurement and field experiences associated with both long-term and short-term projects.

INTRODUCTION

Soil erosion and land degradation across the United States became a national crisis in the 1930s. Although the problems were serious, there was a lack of technology to address the problem, and a lack of basic data to develop new technology. In response, erosion control experiment stations and watershed research programs were initiated to collect experimental data and develop new soil and water conservation (SWC) technologies. These data were needed to quantify rainfall, runoff, and erosion and to understand their relation to land management. In addition, erosion control demonstration projects were an important mechanism for transferring SWC information to land users and managers.

In the southwestern US, implementing upstream SWC programmes was problematic because of the potential affect on downstream water yields. Prior appropriation water laws existed in most of the western states in the US and a concern about reducing downstream water yield was paramount. The work of the Southwest Watershed Studies Group began 1 July 1951 with broad research objectives to determine whether conservation practices would affect water yields and sediment movement and to evaluate flood runoff from semi-arid rangeland watersheds. The objectives were driven by fears among water users that range conservation work would deplete irrigation water supplies. In 1954, the research and personnel of the Southwest Watershed Studies Group were transferred to the newly formed United States Department of Agriculture (USDA) Agricultural Research Service (ARS); in 1961, the Southwest Rangeland Watershed Research Station was established with headquarters in Tucson, Arizona. In the 1990s, the name was changed to the Southwest Watershed Research Center (SWRC).

Today, nine scientists at the SWRC in Tucson conduct research to understand and quantify semi-arid watershed processes and to develop technology for the sustainable management of natural resources. The mission of the SWRC is to understand and model the effects of changing climate, land use, and management practices on the hydrologic cycle, soil erosion processes, and watershed resources; to develop remote sensing technology and apply geospatial analysis techniques; to develop decision support tools for natural resource management; and to develop new technology to assess and predict the condition and sustainability of rangeland watersheds.

The objectives of this chapter are to (1) describe measurement and monitoring on the USDA-ARS Walnut Gulch Experimental Watershed, (2) describe major research findings and technology transfer, and (3) present an overview of monitoring lessons learned from 50 years of experimental research.

MEASUREMENT AND MONITORING ON THE USDA-ARS WALNUT GULCH EXPERIMENTAL WATERSHED

The Walnut Gulch Watershed was selected as a site for research by the USDA by a team of scientists and engineers (Renard and Nichols, 2003) who traveled throughout Arizona, New Mexico, and southern Colorado to examine, screen, and select watersheds suitable for long-term hydrologic, range management, and erosion research. Several criteria were developed for watershed selection with primary focus on the physical attributes of the watershed while incorporating the social impacts of the proposed research. Suitable research watersheds would range in size from 65 to 194 km^2 and would include a secondary tributary to a main channel that furnished irrigation water. The watersheds should receive 250-400 mm of annual precipitation. Vegetative cover would include range grasses (blue grama, *Bouteloua gracilis*; black grama, *B. eriopoda*; and their associates), with little or no cultivated land. Vegetative cover would not be deteriorated beyond recovery. The watershed would contain no closed basins, minimal water would be lost to deep percolation, and the watershed would be in a sediment-producing area. Research efforts required that the sites be accessible during stormy weather and contain sufficient bedrock in the channel at or near the surface upon which to build gauging stations. The cooperation of ranchers within the area was a very important consideration in the selection of research watersheds. In addition, the chosen watersheds would be situated within a major drainage area in which controversy over water supplies existed or had the potential to develop. After investigating several locations, the team identified the Walnut Gulch Experimental Watershed (WGEW) in and adjacent to Tombstone, Arizona, as the site to establish an intensively monitored research watershed (Fig. 1).

Monitoring objectives were clearly specified during initial planning of the WGEW with priority given to studying flood hydrology, which required measuring rainfall and runoff. Measurement methods and instrumentation were less clearly defined, because in the early 1950s very little was known about the spatial variability of air mass thunderstorm

Fig. 1 USDA-ARS Walnut Gulch Experimental Watershed location map.

rainfall and the high-velocity, short-duration runoff events that characterize semi-arid regions. In addition, available instrumentation, including runoff gauges and sediment samplers, were developed for monitoring perennial flow in humid climates and did not perform adequately in steep gradient ephemeral channels.

The WGEW is operated by the SWRC as an outdoor laboratory supporting multidisciplinary research with current research emphasis on hydrology, erosion and sedimentation, global climate change, CO_2 fluxes, remote sensing, and decision support systems. The long-term monitoring network at the WGEW is a critical component of research conducted by the SWRC.

The 150 km^2 WGEW (http://www.tucson.ars.ag.gov) is located in the semi-arid transition zone between the Sonoran and Chihuahuan Deserts in the Southwestern US. The main Walnut Gulch channel is a normally dry tributary to the San Pedro River. The San Pedro River, which originates in Sonora, Mexico, and flows north into Arizona, is generally ephemeral with a perennial section associated with bedrock near the surface. Thunderstorm rainfall during the summer "monsoon" season produces most of the surface runoff. Average annual precipitation on the WGEW ranges from 300 mm at the lower end (1,275 m asl) of the watershed to

340 mm at the upper end (1,585 m asl). Precipitation during July, August, and September accounts for approximately two-thirds of the annual total (Osborn, 1983; Nichols et al., 2002), and results in nearly all of the surface runoff.

CORE MONITORING NETWORK

Work to instrument the WGEW began in 1953. Rainfall monitoring was initiated with 11 weighing bucket rain gauges distributed throughout the WGEW. Because of the limited aerial extent of thunderstorms, it quickly became apparent that the rain gauge network was inadequate to monitor spatially varied precipitation. Rain gauges were added to the network and today rain gauges are installed at 88 sites with a density of 1.7 rain gauges km^{-2}.

Operation, maintenance, data collection, and reduction associated with the original analog data-recording network at WGEW were costly and labor intensive. By the early 1990s, the mechanical rainfall and runoff sensors were becoming increasingly obsolete. In 1996, the SWRC began a multi-year effort to fully re-instrument the WGEW with electronic sensors and digital dataloggers. Each rain gauge consists of a weighing rain gauge retrofitted with a precision, temperature-compensated load cell that outputs a voltage in response to the weight of water collected in the gauge. Voltages are stored in a datalogger and are telemetered to the Tombstone field office every 24 hours. The electronics and dataloggers are stored in a metal cylinder below ground at each rain gauge. This reduces vandalism and damage from lightning strikes and helps to minimize temperature variations.

Measuring runoff and sediment in ephemeral alluvial channels is especially complex. Flows are infrequent but reach high velocities and carry heavy sediment loads. Initially, options for measuring runoff were limited to structures such as V-notch weirs that were commonly used in perennial systems. However, on the WGEW, conditions of highly variable flow with heavy sediment loads quickly filled behind weirs and in the throat of flumes, resulting in the loss of hydraulic control through the measurement section. Several weirs and flumes on the WGEW failed or provided inadequate measurements, and new measurement structures needed to be designed. Following the initial structural failures at Walnut Gulch, a project was begun with personnel of the ARS Hydraulic Structures Laboratory in Stillwater, Oklahoma (Gwinn, 1964, 1970), to develop a new "Walnut Gulch Supercritical-Flow Measuring Flume". Based on the early field experiences and scale model work, supercritical flume designs evolved (Smith et al., 1981) to measure the flow in "flashy" ephemeral streams. Significant advances were made in developing measuring structures for high-velocity, sediment-laden flows (Brakensiek et al., 1979).

Within the WGEW, runoff monitoring stations were located according to a nested design so the relationship between the rainfall and runoff of successively larger subwatersheds could be analyzed. Runoff is monitored within the WGEW channel network upland, at the outlet of small watersheds (50-200 ha), and within upland subwatersheds (2-20 ha). From 1964 through 1967, 11 Walnut Gulch supercritical runoff-measuring flumes (Fig. 2) were constructed along the main stem of Walnut Gulch and major tributaries. There are also 10 instrumented stock ponds that collect water and sediment. These ponds are instrumented with stilling wells and floats to measure water depth, and periodic topographic surveys are completed to measure sediment accumulation. Four of the ponds have sharp crested weirs in the spillway to measure outflow. Based on the success of the Walnut Gulch supercritical flumes in the large channels, a small supercritical flume was designed, tested, and named the Santa Rita Critical Depth Flume. The Santa Rita Critical Depth Flume is widely used to measure runoff rates generally less than 2.8 m^3 sec^{-1}. Runoff instruments have been upgraded and analog output is produced in

Fig. 2 Walnut Gulch supercritical runoff-measuring flume.

parallel to a digital data stream at each measurement site. Water level recorders have been retrofitted with linear potentiometers and voltage outputs are stored in a datalogger and transmitted to the field office.

Sediment in channel runoff is sampled with a traversing slot sampler (Fig. 3) that was designed in response to limitations of alternative sampling methods (Renard et al., 1986). When flow depth is greater than 0.06 m, the traversing slot travels across the outlet of the flume and diverts depth-integrated samples to evenly spaced, stationary slots below the flume exit. Water and particles smaller than the 13 mm slot are directed into sample bottles. The samples are dried and weighed to quantify sediment concentration.

Every 24 hours, each of the 125 instrumentation sites is automatically and sequentially contacted via radio. Stored data are transmitted to the Tombstone field office and the data are stored temporarily. Data are then transferred to the SWRC file server located in Tucson where the data are processed and archived.

In addition to the core rainfall, runoff, and sediment data, recent research focusing on global change and remote sensing has expanded the monitoring network to include meteorologic stations, carbon flux instruments, and soil moisture sensors. *As a result, the WGEW is the most highly instrumented semi-arid experimental watershed in the world. It also has one of the largest published collections of satellite- and aircraft-based imagery with coordinated ground observation in the world.*

Fig. 3 Santa Rita critical depth flume and traversing slot sediment sampler.

RESEARCH FINDINGS AND TECHNOLOGY TRANSFER

Fifty years of data collection, analysis, and interpretation have resulted in an extensive array of publications, research accomplishments, and technologies. A critical early accomplishment of work on the WGEW was the development of instruments to monitor the hydrologic and erosion cycle in semi-arid areas (Renard et al., 1993). The need for specialized instrumentation led to the design and construction of the largest pre-calibrated structure for measuring runoff in semi-arid regions in the world. The precipitation and runoff monitoring network and the innovations and research results have allowed researchers to prepare a water balance for Walnut Gulch that is typical of semi-arid rangeland watersheds (Renard et al., 1993). In addition to a general water balance, thunderstorm rainfall characteristics have been quantified and rainfall models have been developed (Osborn et al., 1980; Osborn, 1983). The temporally continuous, spatially distributed WGEW precipitation database was used to develop the first depth-area-intensity relationships for semi-arid convective airmass thunderstorms. Research to quantify the role and magnitude of transmission losses during runoff in ephemeral streams has resulted in simulation models that have been incorporated in watershed-scale runoff models (Lane, 1983).

SWRC scientists and data collected from the WGEW have played a critical role in the development and transfer of several national ARS simulation modeling efforts. Major contributions have been made to the development of RUSLE (Revised Universal Soil Loss Equation) (Renard et al., 1991), CREAMS (a field-scale model for Chemicals, Runoff, and Erosion from Agricultural Management Systems) (Knisel, 1980), EPIC (Erosion/Productivity Impact Calculator), SPUR (Simulation of Production and Utilization of Rangelands) (Hanson et al., 1992), WEPP (Water Erosion Prediction Project) (Nearing and Lane, 1989; Laflen et al., 1991), and KINEROS (Kinematic Runoff and Erosion Model) (Woolhiser et al., 1990). Many of these models are being used outside of the US. They are being used in combination with collected data to improve the scientific understanding of semi-arid watershed processes.

A range of research projects has been conducted during the last 50 years to improve the condition of deteriorating rangelands, and to improve SWC practices. In the late 1800s, the condition of rangelands in the southwest began to rapidly deteriorate in response to droughts and grazing pressure. Range renovation experiments conducted at the SWRC have resulted in new information on the relationship between grass seeding and precipitation patterns (Cox and Jordan, 1983), and mechanical and chemical treatments for brush control (Cox et al., 1983).

Research projects to improve rangeland condition also resulted in the development and evaluation of land imprinting systems for brush management and seedling establishment (Dixon and Simanton, 1977). The land imprinting system consists of a conservation plow for imprinting land surfaces with complex geometric patterns. The land imprinter forms rainwater-irrigated seedbeds that help to ensure successful seed germination, seedling growth, and a subsequent cover of vegetation. Water harvesting research on the WGEW has contributed to technology used worldwide in arid and semi-arid areas. Research conducted to improve methods and materials for collecting and storing precipitation led to a stepwise guide for the design, selection of materials, installation, and maintenance of water-harvesting systems (Frasier and Myers, 1983). These larger-scale applied research projects are complemented by plot-scale experiments to understand basic runoff and erosion processes.

An important research tool for conducting plot-scale experiments is the rainfall simulator. Experiments designed at the WGEW using the rotating boom rainfall simulator have produced the world's largest database of rangeland hydrology and erosion measurements (Simanton et al., 1986). The simulator is used to conduct experiments under controlled conditions. Rainfall, runoff, infiltration, and sediment measurements have been used both for basic process studies and to develop, evaluate, and parameterize point-to-hillslope scale runoff and erosion models. Recently, a new computer-controlled variable intensity rainfall simulator, called the Walnut Gulch Rainfall Simulator, was developed to quantify the relationship between rainfall intensity and steady state infiltration (Paige et al., 2003).

MONITORING LESSONS LEARNED

Clearly, monitoring on the scale of the WGEW is well beyond the scope, scale, and objectives of most short-term projects. However, short-term monitoring should be conducted and interpreted in the context of available longer-term characteristics of rainfall and runoff. Collecting adequate data to evaluate soil conservation and watershed development projects in regions where extended periods of above or below average precipitation are the norm can take several years or even decades. As an example, an experiment in the 1970s on the WGEW to convert a shrub-covered landscape to grass through ripping and seeding resulted in a short-term reduction in sediment yield, but rainfall during subsequent years was insufficient to establish the grasses in the longer term. The landscape is currently covered with shrubs and the short-term reduction in sediment yield has not significantly altered the long-term average sediment yield.

Data from long-term research programmes play a critical role in interpreting data collected as part of short-term monitoring and evaluation efforts by providing (1) baseline conditions against which to interpret, (2) temporal data from which variability can be quantified, and (3) a framework within which to develop and test new measurement methods. Data collected during short-term projects should be interpreted in the context of baseline conditions and temporal trends. In semi-arid regions the spatial variability of watershed characteristics such as soils, geology, drainage patterns, and cover can be dramatic over short distances, adding to the difficulty in developing general conclusions from individual project evaluations. Simulation models can be used to design conservation projects, as well as to design experiments. Modelling can be conducted prior to implementing field work to select sites for treatment, to select specific practices, and to understand the impacts of temporal and spatial variability in hydrologic and ecosystem variables on soil erosion.

Research experiments and the associated monitoring, measuring, and observations require enormous contributions of time, labour, and money. Access to data beyond the life of an individual project will maximize the return on this investment. To ensure that future users have access to the data, it is important that thought be given to data management before data collection begins. Data management is a critical component of measuring and monitoring. An efficient data management system consists of a framework to store, organize, archive, and retrieve data associated with each project. Efficient access to quality-controlled data allows a broader audience of users. In addition to the data, institutional knowledge becomes increasingly important as the length of data collection increases. Research programmes with long-term data collection histories face a significant challenge as institutional knowledge is lost through changes in personnel. A well-designed database can be used to capture and store some of this knowledge.

SUMMARY

The multidisciplinary research program has addressed SWC in semi-arid regions by quantifying hydrologic and erosion processes, developing simulation models and decision support tools, and incorporating new technologies into research. Sufficient data to quantify the variability in semi-arid rainfall and runoff and their affect on water supply, water quality, and energy fluxes from rangeland watersheds requires a long-term monitoring program. Such data are being used in combination with data collected to quantify upland and channel erosion and sedimentation processes to quantify landscape evolution patterns and the sustainability of rangeland ecosystems with respect to land use and management. The ecosystem responses of grass and shrub communities to short-term rainfall variability as well as long-term changes in global climate are being evaluated. New technologies such as satellite-based sensors to monitor temporal changes in forage production, soil moisture, and evaporation are being incorporated in the monitoring network. These core monitoring efforts are providing data for developing computer-aided decision-making tools that incorporate simulation models, databases, and expert opinion for improving semi-arid watershed management.

Long-term research programmes are critical to the conservation of semi-arid lands. A critical component of this research is instrumentation for monitoring, measuring, and collecting data. Data collected on the WGEW are of national and international importance and make up the most comprehensive semi-arid watershed dataset in the world. Research continues at the WGEW today in cooperation with local ranchers, federal agencies, universities, and international scientists interested in understanding semi-arid watersheds.

References

Brakensiek, D.L., H.B. Osborn and W.J. Rawls (coordinators). 1979. Field Manual for Research in Agricultural Hydrology. USDA Agriculture Handbook 224. 500 pp.

Cox, J.R. and G.L. Jordan. 1983. Density and production of seeded range grasses in southeastern Arizona (1970-1982). J. Range Manag. 36(5): 649-652.

Cox, J.R., H.W. Morton, J.T. LaBaume and K.G. Renard. 1983. Reviving Arizona's rangelands. J. Soil Water Conservat. 38(4): 342-345.

Dixon, R.M. and J.R. Simanton. 1977. A land imprinter for revegetation of barren land areas through infiltration control. Hydrology and Water Resources in Arizona and the Southwest. Office of Arid Land Studies, University of Arizona, Tucson, 7: 79-88.

Frasier, G.W. and L.E. Myers. 1983. Handbook of Water Harvesting. USDA-ARS, Agriculture Handbook No. 600, 45 pp.

Gwinn, W.R. 1964. Walnut Gulch supercritical measuring flume. Trans. Am. Soc. Agr. Eng. 10(3): 197-199.

Gwinn, W.R. 1970. Calibration of Walnut Gulch supercritical flumes. Proc. Am. Soc. Civil Eng. 98(HY8): 1681-1689.

Hanson, J.D., B.B. Baker and R.M. Bourbon. 1992. SPUR II Model Description and User Guide: GPSR Technical Report Number 1. USDA-ARS, Great Plains Systems Research Unit, Ft. Collins, Colorado.

Knisel, W.G. 1980. CREAMS: A Field-scale Model for Chemicals, Runoff and Erosion from Agricultural Management Systems. USDA Conservation Research Report No. 26, 640 pp.

Laflen, J.M., L.J. Lane and G.R. Foster. 1991. WEPP A new generation of erosion prediction technology. J. Soil Water Conservat. 46(1): 34-38.

Lane, L.J. 1983. SCS National Engineering Handbook, Section 4, Hydrology, Chapter 19: Transmission Losses. USDA-SCS, Washington, D.C, 32 pp.

Nearing, M.A. and L.J. Lane. 1989. USDA Water Erosion Prediction Project: Hillslope Profile Model Documentation. NSERL Report No. 2, USDA-ARS-NSERL, West Lafayette, Indiana.

Nichols, M.H., K.G. Renard and H.B. Osborn. 2002. Precipitation changes from 1956-1996 on the USDA-ARS Walnut Gulch Experimental Watershed. J. Am. Water Resour. Assoc. 38(1): 161-172.

Osborn, H.B., L.J. Lane and V.A. Myers. 1980. Rainfall/watershed relationships for Southwestern thunderstorms. Trans. ASAE 23(1): 82-87, 91.

Osborn, H.B. 1983. Precipitation Characteristics Affecting Hydrologic Response of Southwestern Rangelands. USDA-ARS Agricultural Reviews and Manuals: ARM-W-34, January.

Paige, G. B., J.J. Stone, J.R. Smith and J.R. Kennedy. 2003. The Walnut Gulch Rainfall Simulator: A Computer-Controlled Variable Intensity Rainfall Simulator. Appl. Eng. Agr. 20(1): 1-7.

Renard, K.G., J.R. Simanton and C.E. Fancher. 1986. Small watershed automatic water quality sampler. Proceedings of the 4th Federal Interagency Sedimentation Conference, Las Vegas, Nevada, Vol. 1, pp. 51-58.

Renard, K.G., G.R. Foster, G.A. Weesies and J.P. Porter. 1991. RUSLE Revised Universal Soil Loss Equation. J. Soil Water Conservat. 46(1): 30-33.

Renard, K.G., L.J. Lane, J.R. Simanton, W.A. Emmerich, J.J. Stone, M.A. Weltz, D.C. Goodrich and D.S. Yakowitz. 1993. Agricultural Impacts in an Arid Environment: Walnut Gulch Studies. Hydrol. Sci. Tech. 8(1-4): 145-190.

Renard, K.G. and M.H. Nichols. 2003. History of small watershed research in non-forested watersheds in Arizona and New Mexico. Proceedings of the 1st Interagency Conference on Research in the Watersheds, Oct. 27-30, 2003 Benson, Arizona, pp. 296-301.

Simanton, J.R., C.W. Johnson, J.W. Nyhan and E.M. Romney. 1986. Rainfall simulation on rangeland erosion plots. pp. 11-17. *In:* L.J. Lane. [ed.] Erosion on Rangelands: Emerging Technology and Data Base. Proceedings of the Rainfall Simulator Workshop, Jan. 14-15, 1985, Tucson, Arizona. Society for Range Management, Denver, Colorado.

Smith, R,E., D.L. Chery, K.G. Renard and W.R. Gwinn. 1981. Supercritical Flow Flumes for Measuring Sediment Laden Flow. USDA-ARS Technical Bulletin No. 1655, 72 pp.

Woolhiser, D.A., R.E. Smith and D.C. Goodrich. 1990. KINEROS, A Kinematic Runoff and Erosion Model: Documentation and User Manual. USDA-ARS, ARS-77, 130 pp.

16

Geomatics: An Effective Technology for Monitoring and Evaluation of Soil Conservation and Watershed Development Projects

Abdelaziz Merzouk[1], Ferdinand Bonn[2] and Mounif Nourallah[3]

[1]Professor of Soil and Water Conservation at Hassan II Institute of Agronomy and Veterinary Medicine, freelance consultant in SCWM to FAO, IFAD, World Bank and other international agencies, Rabat, Morocco. E-mail: merzouk@mtds.com

[2]Professor and Chairholder, Canada Research Chair in Earth Observation, University of Sherbrooke, freelance consultant in environmental assessment, Sherbrooke, Quebec, Canada (deceased).

[3]Professor of Agronomy and presently Country Portfolio Manager for the Near East and North Africa Division, International Fund for Agricultural Development (IFAD), Rome, Italy.

ABSTRACT

Geographic Information Systems, in conjunction with the related geomatic technologies of Remote Sensing, Global Positioning Systems, telecommunication or internet-intranet for telegeoprocessing and information sharing, have operational applications that can streamline Soil Conservation and Watershed Development (SCWD) planning, implementation, and monitoring and evaluation (M&E). By the late 1990s, these technologies developed to a point that they can provide a valuable and cost-effective contribution to SCWD project engineering. Their potential is even greater for M&E activities focusing on project achievements and impact. In the present era with its extraordinary progress and innovation in computer technology, the current developments in geomatics are no longer technology-driven. It is rather the limitation of awareness and technology transfer that should be overcome in the field of SCWD, mostly in developing countries. The purpose of this chapter is to briefly review the recent developments in geomatics applications in SCWD projects with focus on the M&E environment. While the examples chosen relate mainly to Morocco, taken herein as a typical developing country with major SCWD problems and programmes, most applications and suggestions apply more generally.

INTRODUCTION

In the past three decades, soil conservation and watershed development (SCWD) have made great advances in western as well as developing countries. However, everyone involved in the endeavour for sustainable development recognizes that it was a rough road on all fronts: scientific, technological, political, socioeconomical and financial. During this period, the Earth Summit of Rio de Janeiro 1992 was a turning point on an international level because the approved Agenda 21 programme of action served to move all these programmes to the forefront in many countries. Every country has its own SCWD story, and one can find many similarities, due generally to the strong involvement and investment made by various United Nations development and financing institutions and agencies (FAO, UNEP, UNDP, World Bank, IFAD, UNESCO and WFP), international and regional scientific and professional societies, as well as the private sector. Readers of this chapter can draw a parallel between their own country's experience and that of Morocco as a typical developing country striving to establish and sustain a national SCWD programme and as an example illustrating how many countries are engaged in the modernization of SCWD tools to gain more time and precision and to lower the cost of implementation for all stages of the project cycle, from identification and diagnosis to monitoring and evaluation (M&E). Figure 1 shows the map of Morocco.

Increasing human pressure on water and land resources forced Morocco to launch several ambitious SCWD projects in the early 1970s. However, much of the original political will and strong technical enthusiasm were not matched with substantive commitment of public funds and farmer participation. An international watershed management conference was held in Morocco in 1988 (MAMVA, 1988) with the participation of FAO and the World Bank to assess the progress and look at the lack of funding of SCWD projects. It was shown at the time that out of 50 large dams, endangered by premature silting up, less than eight had

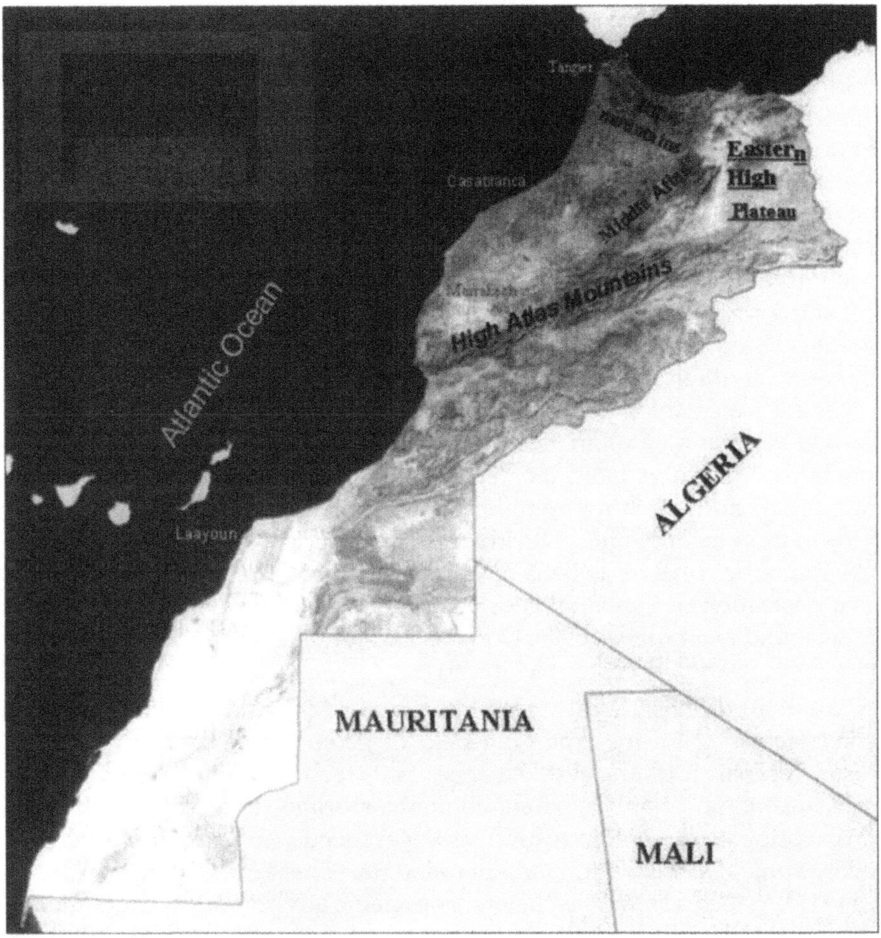

Fig. 1 The upper part of Morocco.

a watershed management plan. The process of SCWD project design and planning took an average period of four years for a medium-size watershed (150,000 ha). As for the project impact assessment, besides the data and assumptions regarding the control of sediment yield and the costs avoided, there was no useful information on project impact upstream of the dams. All conference participants, including those coming from various Mediterranean countries, stressed two urgent needs: (1) to improve the project design and implementation methods and (2) to form concrete ideas on the tangible results to be expected from the actions and investments involved before any national or international financing institution could embark on any programme.

In 1992, Morocco began preparation of a national watershed management plan and made a new start for the organization and methodology of SCWD project design and planning. With the active contribution of UNDP/FAO, six regional centres for watershed studies and management were created, each with a multidisciplinary team trained and equipped with the latest in remote sensing (RS) and Geographical Information System (GIS) computer technology. The rationale behind the project was to help the country make up for the accumulated delay in SCWD project design by the modernization of planning procedures and tools. Knowing that most of the SCWD projects were located in remote and marginal upland areas with no baseline surveys of their resources, most of the efforts were invested in the reconnaissance, integrated survey, and preparation of the comprehensive project development plan. Less attention was given to M&E activities and methods. Ten years later, the centres met their goals and gave concrete demonstrations of how beneficial and essential it was to adopt new information technologies. During this same period, most of the scientific disciplines involved in SCWD have actively worked on the evaluation and demonstration of the benefits of using the new information technologies (Kovar and Nachtnebel, 1996; Douglas and Johnson, 2001). Figures 2 and 3 show rangelands in eastern Morocco.

Much emphasis was put on the input of RS and GIS technologies, considered part of the emerging field of geomatics, a modern scientific term referring to the art, science and technologies related to the measurement, analysis, management, storage and display of the descriptions of geographically referenced information, termed geographical information or spatial data. These data concern earth-physical features as well as the anthropogenic environment. The principal disciplines embraced today by geomatics include the mapping sciences, GIS, RS and surveying, Global Positioning Systems (GPS), environmental visualization, geodesy, and photogrammetry. To date, the role of GIS, RS and GPS has expanded substantially in the reconnaissance, integrated analysis, and planning phases of SCWD projects.

It is impressive to see how close the geomatic developers got to the end-users in order to adjust the products and solutions to their real needs and problems. Most of the GIS, RS or GPS application software has used the field of SCWD for demonstration, tutorial and examples of applications for products (Arcview, ArcInfo, Idrisi, Erdas, SPANS, ILWIS, MapInfo, ATLAS, PCI/Easi-Pace, Geomatica, and Envi). At the time of writing, most of the geomatics software is running on cheaper and more powerful PCs as well as on desktop or palm-unit products that can be used by most SCWD project staff with little or no computer training.

Fig. 2 Local forage, halfe (*Strippa tenacissima*) in 3.5 M ha in eastern Morocco.

Fig. 3 Contour furrows of *Atriplex* for grazing.

The present chapter does not intend to provide precise guidelines for all geomatic technologies and their potential applications in M&E. This contribution should be a reminder to all technicians and professionals involved in participatory M&E that geomatics can make their job efficient and more precise. It offers an effective technology for building up and maintaining a valuable knowledge and information management system. Some M&E handbooks and guidelines developed recently by United Nations institutions (IFAD, 2003a; FAO, 1996) and scientists involved in natural resources management projects have recommended the use of geomatics (Merzouk and Mejjati Alami, 2001). However, they did not stress the potential and benefits of geomatics, and therefore it is still difficult to make provision for funds or to even consider using geomatics in M&E. Recent experience showed that when the use of geomatic tools is not well prescribed and justified in M&E, they failed to perform and become looked upon as a gimmick if not just a waste of funds (MADREF, 2002; IFAD, 2003b). Fortunately, there are enough examples of successful geomatics applications in every country to prevent this perception, and the following sections will review some.

WATERSHED SPATIAL INFORMATION SYSTEM (WSIS) AS A BASIS FOR M&E

An integrated and participatory approach to SCWD requires the build-up of a Watershed Spatial Information System (WSIS) as a comprehensive computerized system to manage, access, and analyse large amounts of spatial data and other information. More often soil conservation activities represent the major component in an integrated rural development, rangeland development, forest development, or natural resources management project. To make progress towards local community participation and directly impact the farming environment, many UN organizations with governments have adopted a village or a micro-zone approach for implementation. However, they kept routing soil conservation activities and their hydrological and environmental impact to wider watershed levels (MADREF, 2000). The new approach does not mean that the watershed approach should be discarded; rather, it is to be enriched with flexibility in working at various scales with more capacity for integration and aggregation. Therefore, the soil conservation project area will not always fit a major watershed demarcation and the user can modify the WSIS with a project spatial information system. Many UN and international organizations and academics have produced guidelines, computer software and handbooks describing the information needs, and possible sources and methods for acquiring and analysing information for

SCWD projects. It is important to stress that the information needed for the planning is the basis for M&E (IFAD, 2003a). It is usually put into four major data categories: physical and land use data, human resources data, socioeconomic profile, and man-made environment. The last group relates to infrastructure and existing facilities and their distribution. Most of these data sets are spatially distributed and they usually relate to some time-based pattern of change, such as land use and vegetative cover, soil qualities, water supply and uses, productivity, and various other socioeconomic and built infrastructure.

There is no universal menu for the WSIS database construction, as it should be guided by user needs. The above data sets represent the requirements of an almost perfect database for SCWD. Unfortunately, in developing countries, natural resource surveys are not always available for SCWD projects as they are located in remote mountainous areas or marginal agricultural and pastoral lands. The Moroccan government was reluctant to commit the expenditure of public funds for building up a natural resources database. In Morocco, less than 20% of the country has a reconnaissance map and database for soil and water resources (Merzouk, 1994; MADREF, 1996) and there are no updated land use maps. This is because traditional ground reconnaissance survey and monitoring methods are laborious, expensive, slow, localized, and subjective. It is impossible to cover whole regions by traditional methods in the short period between the identification and the preparation or the appraisal of SCWD project. Therefore, the SCWD project planning team very often compromises on the requirements of a perfect database, and WSIS involves practical solutions. Many of these practical solutions are based on assumptions, extrapolation of the few existing data, simulation models, or professional expertise with a comforting promise that the project M&E will progressively complete the information and the database. Unfortunately, once the project starts, the promise often gets forgotten.

Experience has shown that when the baseline survey is missing or incomplete, the project M&E system will not perform. During the last decade, Morocco experienced failure of the M&E system in the case of the rangeland development project in the eastern zone as well as in other integrated rural development projects that had a major soil conservation component. IFAD has reported many examples of such failure in M&E and published guidelines on M&E for impact (IFAD, 2003a,b). In the case of the Al Haouz rural development project, a SCWD in a High Atlas Mountain Watershed funded by IFAD in 2001 (MADREF, 2000), the government allocated all the required public funds for building up a comprehensive and up-to-date database and GIS during the first year of the project. This involved various contracts with the private sector and universities to carry out detailed resource and land use surveys using geomatics technologies and to build up the project spatial information system based on the project logical framework and the M&E system and indicators. The project now has a raw model for its comprehensive spatial information system, but more importantly knowledge of how to use it for project implementation and participatory M&E. Besides the database storage and analysis capabilities, the WSIS contains a relational database package that allows spatial modelling on widely available software such as Access or dBase. Land use dynamics and changes, hydrology/erosion and productivity models are examples of environmental models that can be programmed within the GIS spatially distributed framework (Anys et al., 1994). Developments in this area have been going on worldwide, with initiatives such as ANSWERS (Areal Non point Source Watershed Environmental Response Simulation, Beasley et al., 1980) for small catchments and SWAT (Soil and Water Assessment Tool, Arnold et al., 1993) for larger ones. The latter is already included as a module in popular GIS software (Arcview).

SPATIAL DATA COLLECTION, ANALYSIS AND COMMUNICATION: A CHALLENGING TASK

It has been repeatedly recommended that the project choose spatial data related to indicators that are simple, measurable (quantitatively or qualitatively) and well understood by the parties in charge of the data collection, including the project beneficiaries, and are also suitable for monitoring progress as well as impact. The monitoring and analysis of the project effects will be based on the comparison of the data acquired during the implementation with that in the pre-project baseline information. Most of the indicators used involve spatially distributed and time-based information that require reliable and regular data collection and analysis. Unfortunately, with more focus on actions for implementation of project activities, the field staff has often failed to conduct the very demanding task of data collection, extrapolation, and analysis required for M&E. The lack of time, technical understaffing, and field difficulties in data collection are some of the reasons the project management units give to explain this deficiency in the monitoring of physical progress.

Subcontracting and partnerships with research institutions, usually far from the field, did not improve the M&E in many countries in North Africa. The project management and the stakeholders must do as much data collecting, collating and analysing as possible. Monitoring, for example, the water level and storage behind an earth dam is an indicator of great interest to the local water users' association and the design engineers and is essential for impact assessment. In many soil conservation projects, M&E systems were prescribed and initiated for monitoring soil moisture regime or other soil quality indicators in treated and untreated agricultural or rangeland plots, but did not last. More successfully, a regional applied research project (HYDROMED) was initiated in 1996 in six Mediterranean countries to modernize data collection, analysis and modelling tools in connection with M&E of soil and water conservation structures (small earth dams and bench terraces). From water level recorders installed in wells, streams, canals and reservoirs to soil water content and tension meters and water quality monitoring probes, electronics advances and miniaturization presently offer solutions with fascinating automatic measurement and autonomy in data recording and logging. The project succeeded in introducing this new approach to the field data measurement and logging in North Africa and the Middle East (Albergel et al., 2001).

To facilitate the data sharing and to reduce the cost of field transportation in some ongoing SCWD projects in Tunisia and Morocco, the HYDROMED project introduced the experience of data satellite transmission to the field office, headquarters, or any other partner of the project. With the widespread coverage and simplification of data transmission by commercial satellites, the cost of equipping the field data loggers with modems and frequent connections is much less than the cost of field transportation and labour of data collection. The satellite data transmission connecting field measurement devices allows direct and real time geo-processing and monitoring with the help of GIS modelling. More extension and simplification will lead to broader use of these technologies. New technologies stimulate the motivation and performance of the M&E staff, who feel isolated in remote areas and hindered by a lack of involvement in the use of the data and the information collected. More recent developments include intelligent sensor webs in the field, which have the capability to respond or call an alarm in specified circumstances (Wood et al., 2002).

REMOTE SENSING IN M&E

Aerial photo interpretation has always been a standard source of information for SCWD projects. Unfortunately, its high cost in funds and acquisition time never allowed its use for repetitive mapping and monitoring of watershed surface conditions. This is why watershed specialists were early users of data and imagery acquired by space-borne sensors, i.e. satellite remote sensing. Remote sensing provides synoptic and repetitive coverage of earth resources at relatively low cost per surface area. Most importantly, it provides users with data and imagery in numerical format readily pre-processed for direct image analysis and mapping and easily integrated into computer-based watershed geographical information. There are numerous books, proceedings and manuals that have been published since the late 1980s, theorizing and developing RS applications in the SCWD field (FAO, 1989, 1993, 1996; Bonn, 1996). Table 1 shows the spatial, spectral and temporal scales as well as some major M&E indicators observed using the imagery from current satellites.

It is important to know that SCWD projects can obtain satellite images today through a simple internet purchase order from a number of international suppliers. During the last 20 years, many developing countries have established national RS centres to promote and assist in developing applications. The cost and the time of acquisition and processing has been reduced drastically during the last five years; RS has been simplified, demystified, and brought for direct use by field office technicians in the SCWD projects.

Another development in RS that marked the end of the 1990s was a new generation of satellites such as Ikonos and Quickbird offering sub-meter spatial resolution imagery. These new products have the spatial precision and panchromatic spectral range of traditional aerial photography but add the multispectral view in the visible and near infrared spectral range and allow the direct use of the digital nature of the information generated in GIS. This enhanced the thematic mapping capacity with RS for natural resource inventory at the village and micro-zone level (Cissé et al., 1999). However, the cost of these very high resolution data is greater than that of the previous generation of satellite data.

The benefit of RS in M&E lies in the capacity to rapidly obtain up-to-date data or information at a relatively low cost. Spatial RS provides a valuable and cost-effective contribution to monitor progress. It helps detect and assess land use and land cover changes (LULCC). Remote sensing information obtained from the sub-meter spatial imagery allows the detection of meaningful change and monitors progress of individual small soil conservation works and we can say that LULCC detection is the core application of RS in M&E of SCWD projects. The first principle of SCWD is the wise use of the land, and most project cost-benefit analyses are based on the comparison of the pre-project baseline information on land use and the information obtained as the implementation progresses. At regular intervals, the project implementation progress is shown on the map and compared to pre-project as well as to projected land use maps. The computer program will automatically locate the changes and provide

Table 1 Spatial (m) and spectral resolutions of land imaging satellite systems and some M&E indicators available.

*SATELLITE**	*PAN*	*VNIR*	*SWIR*	*TIR*	*Radar Resolution/ Band*	*Swath km*	*G. cover Repeat. Days*	*Possible mapping scale of some M&E physical indicators available from RS*
	Medium resolution with frequent global coverage							Mapping at 1/250,000 – 1/50,000
IRS-1 C,D/ India	5	23	70			70, 142	48, 24	Land use, irrigation drainage networks, ice snow cover, vegetation species and surface cover and characteristics, ground water recharge and discharge areas, soil quality parameters like salinity, erosion and stoniness, large soil conservation works, and gully control, water harvesting structures, and many other indicators plus their temporal change detection.
IRS-P6/ India		6.2-23	70			24, 124	125, 24	
Spot 4/ France	10	20	20			120	26	
Spot 5/ France	2.5-5	10	20			120	26	
CBERS/ China-Brazil	20, 80	20	80	160		120	26	
Landsat 7/ US	15	30	30	60		185	16	
TERRA (ASTER) Japan-US	15	30	90			60		
ALOS-1/ Japan		10	10	20		70, 70	48	
	High resolution, small area coverage							Mapping at 1/50,000 – 1/10,000
Ikonos-2/ Space Imag.	1	4				12	247	Land use at village and farm level, irrigation drainage networks and sub-meter single soil conservation structures (check dams, terracing, tillage work, etc., vegetation characteristics and density, soil quality parameters like salinity, erosion and stoniness, construction, and any physical work on the land, very precise temporal change detection for all the above parameters and many others.
QuicBird-2/ EarthWatch	0.6	2.5				16	185	
OrbView/ Orbimage	1	4				8	NSS	
EROS B1-6/ ImageSat	1	4				16	185	
SPIN-2/ Russia	2, 10					180, 200	NSS	
Cartosat 1/ India	2.5					30	99	

(Table 1 Contd.)

(Table 1 Contd.)

	RADAR, visibility under cloudy conditions		Mapping at 1/100,000 – 1/50,000
Radarsat 1/ Canada	8.5/C	50	Land morphology, soil moisture, land use, snow water equivalent, vegetation and characteristics, ground water recharge and discharge areas, soil roughness and erosion, major infrastructure roads, irrigation schemes and change detection.
Radarsat 2/ Canada	3/C	50	
ERS-2 ENVISAT/ ESA	10, 30/C	100	
ALOS/ Japan	10/L	70	

PAN, panchromatic; VNIR, Visible and near infrared (IR); SWIR, Short wave IR; TIR, Thermal IR

* More information on land imaging satellites is available from the ASPRS guide to land satellites compiled by W.E. Stoney (2004) and posted at http://www.asprs.org/asprs/news/satellites/satellites.htm

the relevant statistics, animation and scenarios needed for reporting and evaluation.

Detection of LULCC using RS information has been a very useful exercise to relate the changes to the driving forces behind many land or water quality degradation processes. It helped justify numerous SCWD projects by showing and simulating the consequences of uncontrolled land degradation processes in no-project situations (Merzouk and Dhman, 1998). This same approach and methods are used for monitoring and evaluating LULCC. In all projects launched during the last four years in Morocco in cooperation with IFAD and the World Bank, provision was made for sufficient funds to finance the RS application in M&E (MADREF, 2000, 2002).

GIS APPLICATIONS FOR M&E

GIS is considered the core of geomatics and is commonly defined as "A system of computer hardware, software, and procedures designated to support the capture, management, manipulation, analysis, modelling and display of spatially referenced data for solving complex planning and management problem" (Maguire et al., 1991). This definition shows the list of powerful capabilities that back the claim that GIS can serve as an effective tool in SCWD project planning and implementation (FAO, 1996). The technology is even more relevant for the M&E activity, which involves the detailed and integrated analysis of a complex set of geographical information. Understanding and quantifying the economic and environmental impact of SCWD activities requires more operational, thus more complex and voluminous, spatial modelling that only GIS technology can make possible.

GIS can be the dynamic centre for almost every aspect of M&E. It is the best computer tool for integrating, analysing, retrieving, and updating spatial data and producing new thematic maps and data in cost-effective ways. Combined with RS, GIS technology has undergone explosive growth over the last decade, which extended its use in the SCWD projects. The increasing adoption of computer technology by SCWD specialists was facilitated by the high performance and competitiveness of the hardware and software industries, which succeeded in putting on the market new generations of more powerful and affordable equipment and software. Progress on this front was able to overcome the constraints of shrinking budgets allowed to M&E activities in SCWD projects.

Can GIS, alone or combined with other geomatic technologies, support the M&E of SCWD projects? Researchers and the private sector developing GIS technologies worked hard to evaluate and demonstrate the effectiveness of GIS as a natural resource management planning tool. The positive answer to the above question is supported by the increasing flow of scientific publications, reports of applied research programmes, and special GIS symposiums. The RS/GIS Committee of the American Society of Rangeland Management (ASRM), for example, could not handle the overwhelming volume of information from GIS/RS specialists to rangeland professionals and therefore decided to increase their journal's capacity. The committee believed that this was just the beginning and decided to use recent advances in computer, internet and compact disk technologies to facilitate this flow. ASRM is not the only professional society that strove to demonstrate the potential applications of RS/GIS in SCWD projects. The same insight, enthusiasm and concrete results are posted by watershed, soils, hydrology, agricultural development, water resources and many other interest groups.

Regrettably, in the field, decision-makers on SCWD projects did not show the same understanding and enthusiasm. In the mid-1990s, rangeland development and integrated agricultural development projects in Morocco and Tunisia, with heavy emphasis on soil conservation, have invested substantial funds in GIS/RS hardware and software and trained technicians abroad in order to initiate the GIS/RS-based M&E system (MADREF, 2002). Two years later, there was no continuity, which led to failure of the M&E system in both projects. Fortunately, this has changed, and with more awareness and continued training of the field staff as well as the central decision-makers, the gap has shrunk and there are ongoing projects funded by IFAD, World Bank and many other international donors and lending institutions that are implementing effective applications of GIS/RS in M&E of SCWD projects.

GIS technology is making many types of M&E work more efficient and improving employee effectiveness. Some of these applications are reviewed in the following paragraphs.

Developing M&E Database

Building the M&E system starts with the organization of a required database founded on the project objective, approach and the recommended indicators and methods for data collection and analysis. This will yield a large amount of data. The GIS provides the full capabilities of a database management system (DBMS). In fact, most GIS software uses the widely available desktop database packages such as dBase or Access, provides easy instructions for building up M&E database conceptual models, and allows the M&E DBMS to directly input existing data. Most DBMS software provides built-in analysis capabilities and offers more capacity for customizing. However, the most important feature of the GIS-based DBMS is the ability to link the database table to geographic features (maps) by a unique identifier (coordinate values).

This suitability can be exemplified in an SCWD project initiated in 2001, in the N'fis watershed (High Atlas, Morocco). A systematic pre-appraisal survey of the 420 villages yielded a very useful but voluminous database on resources, infrastructures, and farmer evaluation of the limitations and potential for integrated development. By adding the village coordinates, the obtained database made the nucleus of the project GIS that was progressively enriched with all the existing data and information. The DBMS facilitates the communication by providing instant display and analysis of the spatial dimensions of the problems by linking the database to maps and multimedia-supported information.

Mapping Capabilities

Monitoring and evaluation of SCWD involves a great deal of thematic and dynamic cartography. It is usually data related to location attribute, a thematic description, and a chronological attribute. Conventionally these types of information have been reported in thematic map or tabular form but at a very slow pace in field offices (MADREF, 2002). It was even more difficult to report dynamic changes in land cover and land use or any land or water quality parameters.

GIS tools allow the project field staff to carry out accurate mapping and engineering drawings with ease and efficiency. The GIS mapping capabilities include digitizing, which can be done easily now on screen, geo-referencing, graphic and image display and generation, interactive editing and transformation, and plotting. A typical M&E mapping exercise was developed for a watershed development project in Tunisia (ORSTOM, 1997). It uses detailed topographic map or geo-referenced satellite images and a GPS to locate different physical achievements on the field as the implementation progresses. In the office, the project staff can update the progress matrices and produce achievement maps and carry out all mapping needs. Obtaining information on the implementation progress for a given time period by sub-watershed, village, or administrative unit and by type of activity used to be weeks of manual work and is now just a click away. The digital nature of the maps produced allows better storage security and facilitates transfer and exchange with all partners in the project.

Geo-computation and Spatial Modelling in M&E for Impact

GIS analytical capabilities and spatial-temporal modelling, which is possible today with desktop and user-friendly software, are particularly suitable for the M&E of SCWD projects. Monitoring and evaluation of project impacts require a great deal of understanding and integrated modelling of the processes involved. It can be a simple function such as the productivity of an irrigated land or rangeland, the population migration function in response to some project activities, or degradation function of land or water quantity and quality (e.g. erosion, salinization, pollution, drought).

The advantage of using this technology is the speed, reliability and power of spatially integrated, multi-parameters and modelling of complex processes that can be performed by the project staff. The geomatics firms in their quest for new market niches became involved and added new capabilities in their software. Detection of LULCC using various sources of inputs (aerial photos, satellite imageries, GPS point-data collection, tele-transmitted data) is now a routine function in many kinds of GIS software and is the same for all major watershed analysis and modelling.

A typical example of spatio-temporal modelling for impact in SCWD is the evaluation of land use change resulting from project actions on the water balance and quality. Using multi-temporal land use maps and a spatially distributed hydrological model, the GIS system compares runoff volume, providing precise quantitative estimation on the watershed water storage capacity and its distribution (Merzouk and Dhman, 1998). The need for integrated spatial analysis and modelling is derived from the project objectives and can be outlined from the logical framework. For M&E-specific needs as far as integrated analysis and modelling are involved, the project preparation and appraisal phases usually outline information needs.

Specialized computer programs have been developed by the World Bank and the FAO's Investment Centre to carry out crop, farm modelling, and agricultural project analysis (FARMOD, COSTAB, COSTBEN/ DISPLAY, RURALINVEST). These software programs can be downloaded from the World Bank website and easily geo-referenced and integrated in the project GIS with the M&E corresponding indicators. While there cannot be a universal model or software for this, there are several ones that can be used with minimal adaptation and extrapolation. This is the case, for example, of IMPEL (Integrated Model to Predict European Land use), which was developed by the European Union (De la Rosa et al., 1996). The model is built using a bottom-up approach based on extrapolation of the models from the field or farm to broader regions using spatially distributed environmental parameters.

GIS-assisted Knowledge-based Approach to M&E

Communicating M&E information and findings in proper ways and in a timely fashion to all partners is crucial to the success of the M&E system in improving project implementation and impacts. The main purpose in communicating information and results is to encourage ownership by all partners or impose accountability. Preliminary results need to be discussed with many partners and feedback and final findings need to be shared as soon as possible. Many M&E systems did not have a clear communication strategy. Data and report transfer and circulation that relied on paper and traditional mail services were slow and difficult. Fortunately, we have now entered an era in which the dissemination of information from one place to another has virtually become instant (with email communication, SCWD project websites, discussion forums, data-satellite transmission). GIS technology can play a major role as a platform for information and knowledge networking. GIS with its database management capabilities has improved collation and storage of M&E information during the last decade, and it started recently to play an important role in communication of information and results of SCWD projects.

The GIS technology is well suited to bridge with electronic networking for sharing of information and knowledge. In fact, during the last five years, we have seen a marked shift in emphasis from data collection and processing to technology integration and information packaging and networking. Geomatics firms are retooling to meet these new demands and new entrants from related disciplines, such as engineering, environmental sciences, and agricultural development, are getting much of the attention. Internet-based knowledge systems are being developed for M&E that can process remotely sensed or measured data and GIS-derived products to improve project implementation and results.

The integration into GIS of multimedia and internet technology is also ensuring a solid linkage between research, technology and production in the environment of M&E. This ease in communication and information sharing is bridging the gap between knowledge and practice at the local, regional and international levels. IFAD is setting a leading example of knowledge networking for their projects with more emphasis on the M&E results of their rangeland management projects (IFAD, 2003b). Figure 4 illustrates the key elements of recently established geomatic-based M&E system of rangelands development in eastern Morocco (IFAD, 2003c).

Fig. 4 Illustration of five key elements of a geomatic-based GIS for M&E of a rangelands development project: (**A**) Geomatic technologies encourage participatory monitoring and mapping with representatives of range users' associations in Eastern Morocco. (**B**) PC-based GIS and RS computer system allows greater spatial data analysis and modeling. (**C**) GPS-guided field monitoring and data collection. (**D**) Land use/cover monitoring and mapping is facilitated with RS and 3D digital data analysis. and (E) computer screen showing GIS capacity to display spatial data (digital map) and the corresponding database and to output the status of the M&E indicators.

CONCLUSIONS

The objective of this chapter has been to describe the growing status of geomatics within the M&E of SCWD projects. Striking advances have been made in the last two decades in the development and implementation of geomatics technologies such as GIS, RS, GPS, and telemetry. In concert with remarkable developments in computer technology and software engineering, these modern technologies have shown enormous potential in improving the cost and enhancing the manner in which we collect, store, analyse, display and communicate information and data for planning, implementation, and M&E of SWCD projects.

Easy availability of data in the temporal domain, mainly from high-precision RS and electronic measurements, provides a new dimension and generates a large volume of useful information for M&E. Geomatics technology, mainly the GIS environment, is integrating various automated methods for data capture, both spatial and non-spatial, so as to minimize manual effort and ensure a higher degree of accuracy and standardization. Today Ikonos, Quickbird and other satellite systems, which are providing data with sub-meter resolution and Palm technology, make it possible to map natural resources and SCWD field physical progress at a detailed level of villages or cluster of villages.

Some geomatics technologies, mainly GIS and RS, are becoming commonplace in watershed development in western and in developing countries, as was illustrated in the Moroccan case. Unhappily, early attempts of geomatics integration into M&E systems of SCWD projects were not notably successful or were aborted. For a time, decision-makers as well as field technicians ignored the major advances being made in making the technology useful and user friendly. Now we can say that that phase is over; many projects funded by IFAD and the World Bank are making full use of these technologies in the M&E system. Today the expectations are high for the role of geomatics in M&E of SCWD projects and with greater awareness of programs and successful examples, use of this technology will become routine.

References

Albergel, J., S. Nasri, R. Ragab, J. Khouri, F. Morino, R. Berndtsson and A. Merzouk. 2001. Les lacs collinaires dans les zones semi-arides du pourtour méditerranéen. International Cooperation with Developing Countries (INCO-DC), Contract Number: ERBIC 18 CT 960091, Final Report, IRD/INGREF, Tunis, 120 pp. + 6 annexes.

Arnold, J.G., P.M. Allen and G. Bernhardt. 1993. A comprehensive surface-groundwater flow model. J. Hydrol. 142: 47-69.

Anys, A., F. Bonn and A. Merzouk. 1994. Remote sensing and GIS based mapping and modeling of water erosion and sediment yield in a semi-arid watershed of Morocco, Geocarto International 9(1): 31-40.

Beasley, D.B., L.F. Huggins and E.J. Monke. 1980. ANSWERS: a model for watershed planning. Trans. Am. Soc. Agr. Eng. 23(4): 938-944.

Bonn, F. 1996. Précis de télédétection. Volume 2: applications thématiques. Sainte-Foy, PUQ/ AUPELF, 663 pp.

Cissé, G., P. Odermatt, M. Tanner and L.Y. Maystre. 1999. Use of GPS and GIS software to assess seasonal home gardening area variations in the Ouagadougou (Burkinao Fasso) urban setting. Sécheresse 2(1): 123.

De la Rosa, D. 1996. MicroLEIS 4.1: Microcomputer-based land evaluation information system. Integrated system for land data transfer and agro-ecological land evaluation. Software + Documentation, CSIC-IRNAS, Sevilla (http:// www.irnase.csic.es/ microlei.htm).

Douglas, E. and X.X. Johnson. 2001. Remote Sensing/ GIS Symposium. J. Range Manag. 54(2): 2001.

FAO. 1989. Remote Sensing Applications to Water Resources. RSC Series No. 50. Rome.

FAO. 1993. Remote sensing and its application to investment projects identification and preparation. RSC Series No. 65. Rome.

FAO. 1996. Computer-assisted Watershed Planning and Management. FAO Conservation Guide No. 28/1. Rome.

Herbert, C. 1995. Un SIG pour le développement Rural: Cas de l'ODSYPANO_Tunisie. (GTZ) Gmbh. Tunis.

IFAD. 2003a. Managing for Impact in Rural Development: A Guide for Project M&E. International Fund for Agricultural Development (IFAD), Rome. CD-ROM edition, available also online, www.ifad.org.

IFAD. 2003b. Effective Project M&E Systems in SADC Countries: Lessons Learned by Theme. URL: www.ifad.org/evaluation/public_html/doc/Ile/1003mese.html.

IFAD. 2003c. Projet de développement des parcours et de l'élevage dans l'Oriental (PDPEO–II) Phase II. Rapport d'évaluation, No. 03/010 IFAD-MOR, Rome.

Kovar, K. and H.P. Nachtnebel. 1996. Application of geographical information systems in hydrology and water resources management. Proceedings of the HydroGis'96 Conference, Vienna, Austria, 16-19 April 1996. IAHS Publication No. 235. Wallingford, UK.

MADREF. 1996. Plan national d'aménagement des bassins versants. Rapport général. Administration des Eaux et Forêts, Rabat.

MADREF. 2000. Projet intégré de développement rural dans les zones montagneuses de la Province d'Al Haouz. Ministère de l'Agriculture du Développement Rural et des Eaux et Forets. DPA de Marrakech, Marrakech.

MADREF. 2002. Projet de développement des parcours et de l'élevage dans l'Oriental (Maroc). Rapport de la phase de l'evaluation. Document préparé par la SCET-Maroc, Rabat.

Maguire, D.J., M.F. Godchild and D.W. Rhind. 1991. Geographical Information Systems. John Wiley and Sons, New York.

MAMVA. 1988. Aménagement des bassins versants. Actes du Séminaire International sur l'aménagement des bassins versants, 17-18 January 1988, Rabat.

Merzouk, A. 1994. Utilisation de la télédétection spatiale dans l'étude et l'inventaire des sols en zone semi-aride.*In:* F. Bonn. [ed.] Télédétection de l'Environnement dans l'Espace Francophone. PUQ, Sainte-Foy. pp. 21-30.

Merzouk, A. and H. Dhman. 1998. Shifting land use and its implication on sediment yield in the Rif Mountains (Morocco). Adv GeoEcology 31: 333-340. Reiskirchen.

Merzouk, A. and M. Mejjati Alami. 2001. Using geomatics as support for land use change assessment and watershed management planning in Morocco: The Tleta case study. *In:* International Symposium: Arid Regions Monitored by Satellites from Observing to Modelling for Sustainable Management. Marrakech, Morocco, 12-15 November 2001. University Cadi Ayyad, Marrakech.

ORSTOM. 1997. Besoin de cartographie des espaces aménagés du projet pilote de développement du bassin versant du l'oued Mellegue: Rapport de mission en Tunisie. Document ronéo, 20 pp. Paris, France.

SCOT CONSEIL. 1997. Prestations d'assistance technique et de conseils dans le domaine de la cartographie des parcours par images satellite dans le cadre du PDEEO, commune d'El Ateuf et O. M'Hamed, Rapport final, Marché N° 57/94/ DPA/52/DP, 150 pp., Ramonville, France.

Wood, M.D., I. Henderson, T.J. Pultz, P.M. Teillet, J.G. Zakrevsky, N. Crookshank, J. Cranton and A. Jeena. 2002. Integration of remote and in situ data: prototype flood information management system. Proceedings of the 2002 IEEE Geoscience and Remote Sensing Symposium (IGARSS 2002) and the 24th Canadian Symposium on Remote Sensing, Toronto, Ontario, Volume III, pp. 1694-1696, also on CD-ROM.

17

Use of Environmental Radionuclides to Monitor Soil Erosion and Sedimentation in the Field, Landscape and Catchment Level before, during and after Implemention of SWC Measures

Felipe Zapata

Soil scientist, Joint FAO/IAEA Programme, International Atomic Energy Agency, P.O. Box 100, Wagramerst. 5, A-1400 Vienna, Austria. E-mail: f.zapata@iaea.org

ABSTRACT

Effective soil conservation programmes can counter soil erosion losses and associated sediment load. Recent initiatives based on selected land use and management are aiming at preventing or mitigating erosion losses in agricultural lands. In this context, there is an urgent need to obtain reliable soil erosion data to underpin the development of scientifically sound land use policies and selection of effective soil conservation measures. The quest for alternative techniques for assessing soil erosion to complement existing methods and to meet new requirements has directed attention to the use of fallout or environmental radionuclides, in particular ^{137}Cs, as tracers for documenting soil redistribution within the landscape at several temporal and spatial scales. Advantages and limitations of the ^{137}Cs technique are highlighted. Significant developments that have been made to refine and standardize the ^{137}Cs technique as well as to exploit its application in a wide range of studies are reported. This information has proven valuable to perform evaluations in the context of soil conservation projects and design policy responses to arrest soil erosion. The FAO/IAEA programme through research networks and other mechanisms is involved in further development and applications of fallout radionuclides to assess the effectiveness of soil conservation measures.

INTRODUCTION

Soil erosion and its associated sediment deposition are natural landscape-forming processes but they can be accelerated by human intervention, thus becoming a serious threat to the sustainable intensification of agricultural production as well as a major problem for sustainable watershed management and conservation of the natural resource base (OECD, 2004). Recent reports highlight the seriousness of soil degradation, in particular soil erosion at the regional and global level (UNEP, 2000). Current concern for both on-site and off-site problems associated with accelerated soil loss is generating an urgent need for obtaining reliable quantitative data on the extent and actual rates of soil erosion worldwide. Such data are required for a more comprehensive assessment of the magnitude of the problems to obtain a better understanding of the main factors involved, to validate existing and new prediction models, and to provide a basis for developing scientifically sound land use policies and selecting effective soil conservation measures and improved land management strategies, including assessment of their economic and environmental impacts (OECD, 2004). On the other hand, it is also important to recognize that effective soil conservation programmes can successfully counter soil erosion losses and sediment production. An approach commonly used by industrialized countries is the removal of cropland susceptible to erosion from production and use of improved soil conservation measures (OECD, 2004). Similarly, conservation agriculture to intensify crop production focuses on the adoption of selected land use and management practices to mitigate or control soil erosion (FAO, 2003). In this context, there is an urgent need for direct measurement of soil erosion, which can be done using erosion plots, surveying methods, and nuclear techniques.

The choice of one or another method will basically depend on the objectives of the study and the availability of resources. Existing classical techniques such as erosion plots and surveying methods for monitoring soil erosion are capable of meeting some of these requirements but they have a number of important limitations in terms of the representativeness of the data obtained, their spatial resolution, their potential to provide information on long-term rates of soil erosion and associated spatial patterns over extended areas, and the costs involved. The quest for alternative techniques for soil erosion assessment to complement existing methods and to meet new requirements has directed attention to the use of fallout or environmental radionuclides, in particular ^{137}Caesium, as tracers for documenting rates and spatial patterns of soil redistribution within the landscape (IAEA, 1995). The objectives of this chapter are (1) to highlight advantages and limitations of the use of environmental radionuclides, in particular the ^{137}Cs technique in soil erosion and sedimentation investigations, (2) to report on progress made on the applications of the ^{137}Cs technique in soil erosion and deposition studies at several scales, and (3) to present recent developments on the use of environmental radionuclides in the assessment of the effectiveness of soil conservation technologies.

METHODS

It is postulated that an integrated approach to assess soil erosion and sedimentation rates at the landscape level should be applied before the implementation of any soil conservation measure to establish a baseline for comparison purposes and to obtain an assessment of the magnitude of the erosion and sedimentation problems. Thereafter, the efficacy of soil conservation measures can be evaluated either using a time series analysis to assess changes over a period of time or by relatively comparing the soil erosion and deposition rates of the soil conservation treatment to a reference treatment. Besides the objectives of the study and the availability of resources, the choice of a particular method for measuring soil erosion would require a thorough knowledge of the advantages and limitations of the technique to be used. A number of workers have explored the potential for using environmental radionuclides, in particular fallout ^{37}Cs (bomb fallout) from the atmospheric nuclear weapons tests of the 1950s, to obtain rates and patterns of soil erosion and deposition for a number of purposes. Specialists in the fields of hydrology and soil geography have commonly been the main users of these techniques. The key assumptions and requirements of the ^{137}Cs technique have been fully described in many publications. The main advantages and limitations of the ^{137}Cs technique to be considered in the context of soil erosion and sedimentation studies are summarized below (Walling and Quine, 1993; Zapata, 2002).

Advantages and Limitations of the ^{137}Cs Technique

The key advantages of the ^{137}Cs technique in soil erosion studies can be listed as follows:

- Estimates are based on contemporary sampling and provide a retrospective assessment of medium-term (30-40 years) rates of soil redistribution.
- Estimates can be obtained on the basis of a single site visit, thus there is no need for long-term monitoring.
- Resulting estimates of soil redistribution rates are integrated medium-term, average data and less influenced by extreme events.
- Estimates are of individual points within the landscape and information on rates and spatial patterns can be assembled at several scales. Information can be scaled up and down.
- Sampling does not require significant disturbance of the landscape or study area.
- The results are compatible with recent developments in physically based distributed modelling and the application of Geographical Information Systems and geostatistics to soil erosion and sediment yield studies.
- Rates of soil redistribution represent integrated effects of all landscape processes resulting in movement of soil particles under defined land use and management.
- The technique provides information on both erosion and deposition in the same watershed and therefore net rates of sediment export.
- There are no major scale constraints, apart from the number of samples to be analysed.
- From the interpretation of the data, the technique permits quantification of processes such as soil tillage redistribution and soil loss and deposition associated with water (sheet) erosion.
- In some areas, it has been successfully applied to obtain wind erosion estimates.
- There are applications in fingerprinting suspended sediment sources and estimating rates of over-bank floodplain accretion.

It is also important to recognize that the technique inevitably has some limitations:

- A multi-disciplinary team is needed for successful application of the technique. This is a particular limitation in countries with limited scientific staff and expertise.

- Specialized laboratories with sample preparation and costly gamma counting equipment are required.
- The technique requires quality assurance and control of low-level gamma spectrometry measurements.
- The technique is effectively limited to documentation of sheet and general surface lowering, but it can be applied to study rill and gully erosion in cultivated areas and to obtain wind erosion estimates.
- It is an indirect approach that depends on the link between measured soil redistribution and observed ^{137}Cs redistribution.
- There are uncertainties associated with the selection of the conversion models used to estimate erosion and deposition rates from the ^{137}Cs measurements.
- It is unable to document short-term changes in erosion rates such as those related to changes in land use and management practices.
- The technique needs further standardization of the protocols for its worldwide application and specialized application in some areas, i.e. areas affected by Chernobyl fallout.

While some of the limitations are inherent in the technique, several others have been successfully addressed in the framework of the research networks reported below. The significance of these limitations, in particular the cost, must be evaluated in the context of the particular advantages of the ^{137}Cs technique and the important limitations of other techniques.

Other Potential Environmental Radionuclides

The ^{137}Cs technique has mostly monopolized recent work in this field, and there have been few attempts to use other fallout radionuclides such as ^{210}Pb and ^{7}Be, despite increasing evidence of considerable potential for their use in soil erosion investigations, both individually and as a complement to ^{137}Cs. The progress made in the standardization of the ^{137}Cs technique is paving the way to their use as described below.

Unsupported or Excess ^{210}Pb

Unlike ^{137}Cs, fallout ^{210}Pb (Pb-210, half-life 22.2 years) is of natural origin and represents a product of the ^{238}U decay series derived from the decay of gaseous ^{222}Rn, the daughter of ^{226}Ra. ^{226}Ra occurs naturally in soils and rocks and diffusion of a small proportion of the ^{222}Rn from the soil introduces ^{210}Pb into the atmosphere. Fallout ^{210}Pb is commonly termed unsupported or excess ^{210}Pb when incorporated into soils or sediments in order to distinguish it from the ^{210}Pb produced *in situ* by the decay of ^{226}Ra

(Walling and He, 1999b). Although ^{210}Pb has been widely used for dating lake sediment cores, its potential for estimating soil erosion has been essentially little exploited to date. Because the deposition of fallout ^{137}Cs and ^{210}Pb exhibits different temporal patterns, there is potential to use the two radionuclides as independent indicators in soil erosion studies to provide additional information on the erosion history of a site. Two approaches have been postulated for the conjunctive use of ^{210}Pb and ^{137}Cs in soil erosion investigations (Wallbrink and Murray, 1996a; He and Walling, 1997). The similar behaviour of fallout ^{210}Pb in soils to ^{137}Cs makes it an alternative to ^{137}Cs in soil erosion investigations in areas where ^{137}Cs measurements prove to be inapplicable, for instance, in areas where a significant amount of Chernobyl-derived fallout was received or where the level of bomb-derived ^{137}Cs fallout was too low or heterogeneous (Walling and He, 1999b). Further studies are required to explore the potential for using unsupported ^{210}Pb measurements in these environments.

^{7}*Beryllium*

^{7}Be is also a naturally occurring radionuclide produced by the bombardment of the atmosphere by cosmic rays causing spallation of nitrogen and oxygen atoms in the troposphere and stratosphere. In this case, this radionuclide is extremely short-lived (half-life of 53.3 days) relative to ^{137}Cs and ^{210}Pb, and it offers potential for investigating soil erosion dynamics over much shorter time scales. ^{7}Be is typically concentrated in the upper 5 mm of the soil profile and it is therefore capable of providing good discrimination between sediment derived from the uppermost soil surface and that derived from depths greater than 10 mm, where concentrations will be effectively zero. This radionuclide has been used in erosion plot experiments to distinguish sediments mobilized by sheet erosion and rill erosion, and to demonstrate the progressive development of rills during the course of a single simulated runoff event (Burch et al., 1988; Wallbrink and Murray, 1996b). Monitoring the spatial distribution of the ^{7}Be activity across small plots immediately after storm events could also provide a basis for investigating local erosion patterns in response to microtopographic controls (Blake et al., 1999). There is also scope to couple measurements of ^{7}Be activity with equivalent measurements of ^{137}Cs and unsupported ^{210}Pb to provide increased capability for sediment source discrimination. Combined measurements of ^{137}Cs and ^{7}Be have been used to derive estimates of medium- and short-term rates of soil redistribution within the study field (Walling et al., 2000).

RESULTS AND DISCUSSION

Use of the ^{137}Cs Technique in Soil Erosion and Sedimentation Studies

The joint implementation of two closely linked IAEA-networked research projects greatly contributed to further development of the ^{137}Cs technique by bringing together 26 research groups and promoting exchange of knowledge and sharing of experiences among participants (Zapata, 2001). Much of the initial work focused on validating the approaches for the application of the ^{137}Cs technique in a range of environments. Standardized methods and protocols were successfully developed and documented in a handbook for further dissemination and application (Zapata, 2002). The ^{137}Cs technique worked well in a range of environments and scales in the course of these projects. The data assembled by the contributors using standardized protocols for the application of the ^{137}Cs technique have provided directly comparable and representative information on soil erosion rates in a wide range of environments worldwide. As the efficacy and value of the technique were increasingly recognized, the scientific teams participating in the projects exploited its potential in a wide range of studies at several scales (Zapata, 2003a). The following investigations illustrate some of these developments and practical applications:

- The relationship between soil redistribution and topographic controls (Pennock, 2003).

Fig. 1 Sheet (laminar) erosion on cropland after corn harvest in the Cerrado region of Brazil (*Source:* Dr. Urquiaga, Embrapa-Agrobiologia, Seropedica, RJ, Brazil).

- The relationship between soil quality parameters and soil erosion as inferred from ^{137}Cs redistribution (Li et al., 2000).
- The role of tillage in soil redistribution (Schuller et al., 2003, 2004).
- Development of improved conversion models for estimating soil erosion (Walling and He, 1999a).
- Tracing sources of suspended sediment in river basins (Wallbrink, et al., 1999).
- The assessment and management of erosion or sedimentation risks in view of the restoration of fish populations in a river basin (Bernard and Lavardiere, 2000).
- Estimating wind erosion in several environments (Chappell et al., 1998; Chappell and Warren, 2003).
- National reconnaissance survey of soil erosion in Australia (Loughran and Elliott, 1996).

The unique information on estimates of soil redistribution rates and their spatially distributed patterns provided by the ^{137}Cs measurements offer a great potential for an improved understanding of the relationships between soil loss or gain and other parameters such as soil quality, soil carbon, nutrient redistribution, and the fate of agrochemicals and other contaminants in the agricultural landscapes (Pennock and Frick, 2001; Ritchie and McCarty, 2003). The IAEA is promoting further development and applications of the ^{137}Cs technique and other environmental radionuclides in soil erosion and sedimentation research through several mechanisms, in particular national and regional Technical Co-operation projects (Zapata, 2003a). In March 2006, Ritchie and Ritchie recorded 3,644 publications at the URL http://hydrolab.arsusda.gov/cesium. Recent expanding applications include the following:

- The relationship between on-site soil erosion and off-site impacts (sediment delivery and nutrient export) at several scales (Wallbrink et al., 2003).
- The impact of forest harvesting operations on soil redistribution (Wallbrink et al., 2002).
- Validation of distributed erosion and sediment yield models within catchments (Walling et al., 2003).
- Tracing sediment mobilization and delivery in river basins as an aid to catchment management (Walling, 2004) and tracing suspended sediment sources in catchments and river systems (Walling, 2005).
- Estimating land surface erosion from radionuclide data in lake and reservoir sediments (Zhang and Walling, 2005).

Using Environmental Radionuclides in Soil Conservation Projects

The recent developments in the use of the ^{137}Cs technique in soil erosion and sedimentation studies obtained through the implementation of two research networks have paved the way to both extending the approach to other environmental or fallout radionuclides (FRNs) and exploiting the essentially unique advantages provided by the technique to assess soil redistribution to pilot-test interventions that conserve the natural resource base and protect the environment. Further developments in the combined and/or selective application of FRNs are pursued through the joint implementation of two new research networks on "Assessment of the effectiveness of soil conservation technologies using fallout radionuclides" and "Isotope techniques for sediment source characterization" during the period 2003-2007 with an overall goal of achieving sustainable agricultural production and watershed management. The specific objectives are to further develop the FRN methodologies, in particular the combined use of ^{137}Cs, ^{210}Pb and ^{7}Be for measuring rates and patterns of soil redistribution at several time and spatial scales, and to use these techniques to assess the impact of changes in land use types and evaluate the effectiveness of soil conservation measures tailored to local conditions and resources. These new and refined methods will allow systems of land use and management, and the effectiveness of specific soil conservation technologies, to be rapidly evaluated in a cost-effective manner. In total 28 scientists, from Argentina, Australia, Austria, Brazil, Canada (2), China (2), Chile, Japan, Morocco, Pakistan, Poland, Romania, Russia, Switzerland, Turkey, UK, USA and Vietnam, one FAO representative, and seven observers attended the first and second co-ordination meeting of the projects (Zapata, 2003b; Bernard, 2004). The linkage of the networks to the World Overview of Conservation Approaches and Technologies (WOCAT) consortium will ensure proper implementation of the soil conservation component through the following activities: (1) applying methodologies for the standard characterization of soil conservation measures; (2) gathering available information on soil and water conservation (SWC) from WOCAT databases; (3) conducting impact assessment (socioeconomic evaluation) in selected case studies; (4) inputting the information generated using FRNs into the existing databases and making these data available to the WOCAT consortium and the scientific community, in general; and (5) establishing co-ordination to conduct collaborative research within the framework of WOCAT and the National Centre for Competence for Research Programme on Research Partnerships for Mitigating Syndromes of Global Change (http://www.nccr-north-south.unibe.ch) in the Horn of Africa and Central Asia.

A representative from WOCAT secretariat is participating in the network and the co-ordination meetings.

STANDARD CHARACTERIZATION OF THE SOIL CONSERVATION TECHNOLOGIES IN STUDY

At the first co-ordination meeting of the projects, it was agreed that the SWC technologies in study should be described in a standard and comprehensive way in order to fully document them, share knowledge and experiences, and disseminate information about their effectiveness (Zapata, 2003b; Bernard, 2004). For this purpose it was recommended that the standard terminology (e.g. SWC categorization) and procedures of the WOCAT programme be used. Detailed information on this categorization is available at www.wocat.net. Results expected from this work will be the comprehensive compilation of existing knowledge about these technologies in a way allowing comparison, identification of knowledge gaps, clarification, and pros and cons: e.g. environmental and socioeconomic efficiency, inputs and outputs. It also should be noted that when using an existing methodology such as WOCAT there are always some shortcomings because it has not been developed specifically for the user's own purpose. However, the main advantage is the possibility of sharing knowledge and experience with others. Users should try to use the existing methodology as much as possible and provide feedback to WOCAT for improvement, making additions and modifications as needed. Information about FRN experiments not covered by WOCAT should also be recorded. There is a plan to create a link between WOCAT and the FRN database.

THE USE OF ENVIRONMENTAL RADIONUCLIDES IN SOIL CONSERVATION PROJECTS: APPROACHES AND CHALLENGES

At the first co-ordination meeting of the projects, several issues relating to the application of environmental radionuclides in soil conservation projects were discussed. The key focus of the Soil Conservation Co-ordinated Research Project (CRP) is to develop novel applications of environmental radionuclides or FRNs for assessing the impact of soil conservation measures on soil erosion and sedimentation. ^{137}Cs and excess ^{210}Pb and ^{7}Be have already been identified as the primary radionuclides for use in the investigations (Zapata, 2003b; Bernard, 2004).

Approaches

The following potential approaches involving these radionuclides were identified, although it was emphasized that this list should not be seen as exhaustive. Furthermore, it is to be hoped that additional approaches might be developed within the framework of the CRP.

1. Use of detailed measurements of ^{137}Cs depth distributions and inventories to establish changes in soil redistribution rates over the period covered by bomb fallout in response to a major change in land management. A recent study made by Schuller et al. (personal communication) usefully exemplified the potential of this approach. It should be noted, however, that the change in land use and management has to date back to many years to apply this approach successfully.
2. Combined use of Chernobyl and bomb ^{137}Cs to investigate changes in soil redistribution rates associated with the post-Chernobyl period. This approach was likely to be limited primarily to those areas of Europe where Chernobyl and bomb fallout inventories were of similar magnitude (Golosov, 2003).
3. Time series measurements of ^{137}Cs inventories aimed at quantifying changes in soil redistribution rates during specific periods (Lobb, personal communication). It was clear that the application of this approach would be constrained by the need for periods of sufficient duration between measurement campaigns to ensure that the change in inventory was greater than the uncertainty associated with the sampling and laboratory measurement procedures. Its application was, therefore, likely to be restricted to areas with relatively high erosion rates. This approach requires an initial measurement campaign to provide a baseline against which to compare future measurements. This meant that in general it was unlikely to be possible to obtain results from this approach during the CRP period. However, there are a few detailed data sets in existence, which could be used to provide an existing baseline.
4. Use of ^{7}Be for short-term comparisons and measurements. The successful use of ^{7}Be measurements to quantify erosion rates during individual events or for short periods had been reported by Wallbrink and Murray (1996b) and Blake et al. (1999). The approach clearly involved a number of complexities and would require careful application. In addition an appropriate experimental design would be required to ensure that meaningful results were obtained. For example, in determining the

effectiveness of soil conservation measures it will be necessary to have a control area for comparison purposes.

5. Use of ^{210}Pb measurements. There is still potential for using ^{210}Pb measurements to estimate soil redistribution rates, a potential that should be explored more fully (Walling and He, 1999b). The concerns raised for ^{137}Cs in point 3 above also apply for ^{210}Pb.
6. Conjunctive use of ^{137}Cs, ^{210}Pb and ^{7}Be. The studies reported by Wallbrink and Murray (1996a), He and Walling (1997), and Walling et al. (2000) had demonstrated the potential for using two or three of these radionuclides in combination to derive information on change in soil redistribution rates through time.
7. Source fingerprinting procedures. Recent unpublished studies reported by Wallbrink in Australia and Onda in Japan had clearly demonstrated the potential for using fingerprinting techniques for discriminating and identifying suspended sediment sources. Coupled with an appropriate experimental design this approach could afford a powerful tool for assessing changes in sediment sources resulting from the introduction of soil conservation measures. This would be valid in cases where the sites with conservation measures and the control area exhibit contrasting fingerprint characteristics.
8. Explore the potential to use FRNs to assess or predict losses of carbon, nutrients, pesticides, etc. from the watershed (Pennock and Frick, 2001; Ritchie and McCarty, 2003).

Challenges

A number of technical challenges in applying and developing the approaches outlined above were also identified. These are outlined below.

1. Global location and environmental conditions: The environmental conditions under which FRNs are currently applied vary considerably in climate, soil, topography, and land use. One of the most significant consequences is the variety of FRN deposition histories and current inventory levels (which can be quite low or highly variable). Uncertainties still exist as to the inventories to be expected in some areas of the world (e.g. ^{137}Cs inventories in tropical areas, excess ^{210}Pb inventories in coastal areas with prevailing onshore winds, and high levels of supported ^{210}Pb relative to excess ^{210}Pb in some areas). In some circumstances low FRN levels can be dealt with by analysing samples for longer durations (requiring more detectors, longer studies, or alternate sampling strategies – bulking) or using more efficient detectors (requiring more capital investment), but practical limits may exist.

2. Reference sites: Reliable reference sites can be difficult to establish because of the high degree of variability in local precipitation and the difficulty of identifying undisturbed stable sites. In this respect, it is essential that researchers take the utmost care to establish the reference inventory of FRN in their study area. Finding suitable reference sites is a particular problem in mountain areas and areas with intensive cultivation. In large-scale studies, the use of a single value for reference inventory may be inappropriate.
3. Fate of FRNs: There are still some uncertainties associated with the behaviour of fallout FRNs in the soil and related environments, e.g. plant interception, preferential soil sorption mechanisms, and plant uptake. These uncertainties require further elucidation.
4. Use of ^{7}Be: Although many projects proposed to use ^{7}Be, it should be noted that its use involves considerably more complexity than the traditional ^{137}Cs method (e.g. the need to monitor fallout inputs throughout the year). Further work and testing of this approach is required before it can be widely adopted with confidence.
5. Quality assurance, quality control, and technical support: Several issues were identified under this broad topic: (a) need for training to establish a minimum skill set in field procedures, analytical procedures and data interpretation, (b) access to standards for calibration, (c) sample exchange programmes, and (d) analytical support for some CRP members.
6. New technologies: Gamma spectrometry undergoes regular "advances" (new hardware such as *in situ* detectors and software), which may or may not afford new opportunities for the use of FRN in soil erosion studies. These advances need to be rigorously tested to assess their value for use in soil conservation projects.
7. Appropriate measurement technologies: This issue was raised with regard to the analysis of ^{210}Pb, e.g. appropriate low level Hyperpure Germanium (HPGe) detectors and the use of alpha spectrometry as an alternative method.
8. Conversion models: Further development and validation of existing (and development of new) conversion models is required. Uncertainties associated with these models need to be better understood (e.g. grain size effects). There is a need for standardization of methods, although with the development of new approaches the potential for standardization was bound to be reduced.
9. Source fingerprinting: Although basic procedures are now well established, there is a need to explore the application of new fingerprint properties capable of providing improved

discrimination between potential sources and to develop rigorous statistical and numerical procedures for establishing the relative importance of the sources considered. Again, an appropriate and rigorous experimental design was essential to ensure the generation of meaningful results.

10. Climatic changes (extreme events or storms): It is important that any attempt to quantify changes in soil redistribution rates resulting from changes in land management should be able to distinguish them from those changes arising from climate change alone (frequency and intensity of extreme events or storms). Such considerations should be incorporated in the experimental design of the study.
11. Integrated approach to soil erosion and sedimentation studies: It is relevant that these links should be established and further strengthened to ensure that the results obtained from the project were applicable at both the field and watershed scales. This would also promote the integration of studies focusing on either on-site or off-site effects.

CONCLUSIONS

Developments in the refinement and harmonization of the ^{137}Cs technique have been achieved through the implementation of two networked research projects. The developed methods and protocols provide a standardized framework for the application of the technique. This is an essential prerequisite to obtain directly comparable and representative information on soil erosion rates in a wide range of environments and to understand the influence of the main factors affecting soil redistribution in the landscape. The efficacy and value of the technique has been increasingly recognized in other studies, and a wide range of applications demonstrates its potential for an improved understanding of the relationships between soil loss or gain and other parameters such as soil quality, soil carbon, nutrient redistribution, and the fate of agrochemicals and other contaminants in the agricultural landscapes. As there is an urgent need to arrest soil erosion and associated land degradation worldwide and there is still considerable scope for the wider application of ^{137}Cs and other FRNs, the FAO/IAEA programme is implementing mechanisms to bring these recent developments into the agricultural ecosystems to promote sustainable land management. In this context two research networks are being implemented aiming at the development of novel applications of FRNs such as ^{137}Cs and excess ^{210}Pb and ^{7}Be for assessing the impact of soil conservation measures on soil erosion and sedimentation. The application of FRN technologies in soil conservation projects entails many challenges arising from the influence of site-specific factors and their interactions, the occurrence of other degradation processes, and the need for an integrated approach to assess both on-site and off-site effects at the watershed level, even if a specific soil conservation measure is studied. Substantial progress in the approaches and methodologies has been achieved by addressing several of the issues mentioned above. The linkage to WOCAT with the collaborating projects is essential in order to document, evaluate, monitor and disseminate worldwide the data generated on the effectiveness of soil conservation measures. This information will also be useful to perform socioeconomic evaluations and design scientifically sound land use and management policy responses to prevent or mitigate soil erosion and associated problems.

References

Bernard, C. 2004. Report of the Second Research Co-ordination Meeting of the Project "Assess the effectiveness of soil conservation techniques for sustainable watershed management and crop production using fallout radionuclides." IAEA-311-D1-RC-888.2, Vienna. URL: http://www-naweb.iaea.org/nafa/swmn/crp/d1_5001.html

Bernard, C. and M.R. Lavardiere. 2000. Using ^{137}Cs as a tool for the assessment and the management of soil erosion/sedimentation risks in view of the restoration of the Rainbow Smelt (*Osmerus mordax*) fish population in the Boyer river basin. Acta Geologica Hispanica 35(3-4): 321-327.

Blake, W.H., D.E. Walling and Q. He. 1999. Fallout beryllium-7 as a tracer in soil erosion investigations. Appl. Radiat. Isot. 51: 599-605.

Burch, G.J., C.J. Barnes, I.D. Moore, R.D. Barling, D.H. Mackensie and J.M. Olley. 1988. Detection and prediction of sediment sources in catchments: use of Be-7 and Cs-137. pp. 146-151. *In:* Proc. Hydrology and Water Resources Symposium, ANU, Canberra, February 1988.

Chappell, A., A. Warren, M.A. Oliver and M. Charlton. 1998. The utility of ^{137}Cs for measuring soil redistribution rates in southwest Niger. Geoderma 313-337.

Chappell, A. and A. Warren. 2003. Spatial scales of ^{137}Cs-derived soil flux by wind in a 25 km^2 arable area of eastern England. Catena 52: 209-234.

Food and Agriculture Organization of the United Nations (FAO). 2003. Intensifying crop production with Conservation Agriculture. http://www.fao.org/waicent/faoinfo/agricult/ags/AGSE/Main.htm

Golosov, V. 2003. Application of Chernobyl-derived ^{137}Cs for the assessment of soil redistribution within a cultivated field. Soil Till. Res. 69: 85-97.

He, Q. and D.E. Walling. 1997. The distribution of fallout ^{137}Cs and ^{210}Pb in undisturbed and cultivated soils. Appl. Rad. Isot. 48: 677-690.

International Atomic Energy Agency (IAEA). 1995. Use of nuclear techniques in studying soil erosion and siltation. IAEA-TECDOC-828.IAEA Publications, Vienna.

Li, Y., M.J. Lindstrom, J. Zhang and J. Yang. 2000. Spatial variability of soil erosion and soil quality on hillslopes in the Chinese Loess Plateau. Acta Geologica Hispanica 35(3-4): 261-270.

Loughran, R.J. and G.L. Elliott. 1996. Rates of soil erosion in Australia determined by the caesium-137 technique: a national reconnaissance survey. pp. 275-282. *In:* D.E. Walling and B.W. Webb. [eds.] Erosion and Sediment Yield: Global and Regional Perspectives. IAHS Publ. No. 236.

Organization for Economic Co-operation and Development (OECD). 2004. Agricultural impacts on soil erosion and soil biodiversity: Developing indicators for policy analysis (R. Francaviglia, ed.). Proc. from an OECD Expert Meeting. Rome, March 2003, 654 pp. URL: http://webdomino1.oecd.comnet/agr/soil_ero_bio.nsf/viewHtml/index/$file/Publ

Pennock, D.J. 2003. Terrain attributes, landform segmentation, and soil redistribution. Soil Till. Res. 69: 15-26.

Pennock, D.J. and A.H. Frick. 2001. The role of field studies in landscape-scale applications of process models: an example of soil redistribution and soil organic carbon modelling using CENTURY. Soil Till. Res. 58: 183-191.

Ritchie, J.C. and G.W. McCarty. 2003. ^{137}Cesium and soil carbon in a small agricultural watershed. Soil Till. Res. 69: 45-51.

Ritchie, J.C. and C.A. Ritchie. 2006. Bibliography of publications of ^{137}Cesium studies related to erosion and sediment deposition. http://hydrolab.arsusda.gov/cesium137bib.htm

Schuller, P., A. Ellies, A. Castillo and I. Salazar. 2003. Use of the ^{137}Cs technique to estimate tillage- and water-induced soil redistribution rates on agricultural land under different use and management in central-south Chile. Soil Till. Res. 69: 45-51.

Schuller, P., D.E. Walling, A. Sepulveda, R.E. Trumper, J.L. Rouanet, I. Pino and A. Castillo. 2004. Use of ^{137}Cs measurements to estimate changes in soil erosion rates associated with changes in soil management practices on cultivated land. Appl. Rad. Isot. 60: 759-766.

United Nations Environment Programme (UNEP). 2000. Global Environment Outlook 2000. Earthscan Publ. Ltd., London. UK.

Wallbrink, P.J. and A.S. Murray. 1996a. Determining soil loss using the inventory ratio of excess Lead-210 to Caesium-137. Soil Sci. Soc. Am. J. 60: 1201-1208.

Wallbrink, P.J. and A.S. Murray. 1996b. Distribution and variability of Be-7 in soil under different surface cover conditions and its potential from describing soil redistribution processes. Water Resour. Res. 32: 467-476.

Wallbrink, P.J., A.S. Murray and J.M. Olley. 1999. Relating suspended sediment to its original soil depth using fallout radionuclides. Soil Sci. Soc. Am. J. 63: 369-378.

Wallbrink, P.J., C.E. Martin and C.J. Wilson. 2003. Quantifying the delivery of sediment-P and fertilizer-P from forested, cultivated and pasture areas at the land use and catchment scale using fallout radionuclides and geochemistry. Soil Till. Res. 69: 53-68.

Wallbrink, P.J., B.P. Roddy and J.M. Olley. 2002. A tracer budget quantifying soil redistribution on hillslopes after forest harvesting. Catena 47: 179-201.

Walling, D.E. 2004. Using environmental radionuclides to trace sediment mobilization and delivery in river basins as an aid to catchment management. pp. 121-135. *In:* Proc. Nine Int. Symp. River Sedimentation, 18-21 October 2004, Yichang. China.

Walling, D.E. 2005. Tracing suspended sediment sources in catchments and river systems. Sci. Total Environ. 344: 159-184.

Walling, D.E. and Q. He. 1999a. Improved models for estimating soil erosion rates from cesium-137 measurements. J. Environ. Qual. 28: 611-622.

Walling, D.E. and Q. He. 1999b. Use of fallout lead-210 measurements to estimate soil erosion on cultivated land. Soil Sci. Soc. Am. J. 63: 1404-1412.

Walling, D.E. and T.A. Quine. 1993. Use of ^{137}Cs as a tracer of erosion and sedimentation: Handbook for the application of the ^{137}Cs technique. Report to the UK Overseas Development Administration, Exeter, UK.

Walling, D.E., Q. He and W. Blake. 2000. Use of ^{7}Be and ^{137}Cs measurements to document short- and medium term rates of water-induced soil erosion on agricultural land. Water Resour. Res. 35: 3865-3874.

Walling, D.E., Q. He and P.J. Whelan. 2003. Using ^{137}Cs measurements to validate the application of the AGNPS and ANSWERS erosion and sediment yield models in two small Devon catchments. Soil Till. Res. 69: 27-44.

World Overview of Conservation Approaches and Technologies (WOCAT). Detailed information available at the URL: http://www.wocat.net

Zapata, F. 2001. Final Report of the Co-ordinated Research Project on "Assessment of soil erosion through the use of the Cs-137 and related techniques as a basis for soil conservation, sustainable agricultural production and environmental protection". IAEA-311-D1-RC-629.4, IAEA Publications, Vienna.

Zapata, F. [ed.] 2002. Handbook for the Assessment of Soil Erosion and Sedimentation Using Environmental Radionuclides. Kluwer Acad. Publ., Dordrecht, The Netherlands.

Zapata, F. [ed.] 2003a. Field application of the Cs-137 technique in soil erosion and sedimentation. Special Issue Soil Till. Res. 69: 1-153.

Zapata, F. 2003b. Report of First Research Co-ordination Meeting of the Project "Assess the effectiveness of soil conservation techniques for sustainable watershed management and crop production using fallout radionuclides." IAEA-311-D1-RC-888.1, Vienna. URL: http://www-naweb.iaea.org/nafa/swmn/crp/d1_5001.html

Zhang, X. and D.E. Walling. 2005. Characterizing land surface erosion from ^{137}Cs profiles in lake and reservoir sediments. J. Environ. Qual. 34: 514-523.

18

Integrated Conservation Planning in a Small Catchment on the Chinese Loess Plateau

Rudi Hessel[1], Minh Ha Hoang-Fagerström[2], Jannes Stolte[1], Ingmar Messing[2] and Coen J. Ritsema[1]

[1]Soil Science Centre, ALTERRA Green World Research, Wageningen University and Research Centre, P.O. Box 47, 6700 AA Wageningen, The Netherlands. E-mail: rudi.hessel@wur.nl

[2]Department of Soil Sciences, Swedish University of Agricultural Sciences, P.O. Box 7014, SE-750 07 Uppsala, Sweden.

ABSTRACT

The Chinese government recognizes soil erosion as a major problem on the Chinese Loess Plateau and intends to combat the problem by land use change. The EROCHINA project aimed to define alternative land uses for the Danangou catchment, using monitoring and modelling of erosion, combined with scenario development and participatory planning. Agriculture is currently the main land use in the catchment and is also practised on steep slopes. However, it has severe environmental consequences, in particular extensive soil erosion. Measurements of soil loss showed that large amounts of soil are lost during storms. Simulations with the Limburg soil erosion model indicated that significant decreases in soil and water loss could be achieved by changes in land use. Simple conservation measures were predicted to be less effective but could nevertheless result in some reduction of runoff and erosion. The local people are highly aware of alternative conservation land uses suitable for different parts of the catchment. However, due to the poor living standard and the lack of off-farm employment, any effort to improve land use toward more sustainable practices would be difficult without support from the government. It is recommended that implementation of such conservation measures be carried out by active collaboration between farmers, extension workers, and researchers. Measurement and modelling of erosion can give valuable data to be used in such collaboration.

INTRODUCTION

Soil erosion on the Chinese Loess Plateau is a major problem because on site it causes loss of arable land, while off site it can cause silting up of rivers and reservoirs. Erosion rates are very high because of several factors. First, loess soil is highly erodible, especially when wet. Second, the area has a semi-arid climate with occasional heavy thunderstorms in summer. Third, large parts of the Loess Plateau have considerable relief, and slope angles are steep. Finally, vegetation cover is generally low. This is partly caused by the semi-arid climate with cold winters, but also at least partly by deforestation and grazing (Jiang Deqi et al., 1981). Because of these very high erosion rates, sediment contents in the Loess Plateau rivers can be extremely high, and concentrations of 1,000 g L^{-1} are reported frequently (Jiang Deqi et al., 1981; Zhang et al., 1990; Wan and Wang, 1994). These high concentrations have caused rapid sedimentation of the Yellow River along its lower reaches, which has raised the river bed to several metres above the surrounding landscape. Consequent breaching of the dikes could result in disaster. The Chinese government is well aware of the problems posed by the Yellow River and is determined to combat them. Since the Loess Plateau is the source of over 90% of all sediment entering the Yellow River (Douglas, 1989; Wan and Wang, 1994), major conservation efforts have been started in the area. In 1999, the Chinese government formulated ambitious new policies for the Loess Plateau. These policies aim to decrease erosion rates through changes in land use, the so-called "re-greening" of the Loess Plateau.

In 1997, the EROCHINA project was initiated to find ways to decrease erosion rates in a small catchment area on the Loess Plateau. Since it was a research project it did not actually implement conservation measures. The project used an integrated planning approach that combined measurements of erosion, land evaluation, and modelling of erosion using a Participatory Approach (PA). The PA was used to develop potential land use scenarios (LUS) using Participatory Rural Appraisal and Participatory Household Economy Analysis (PHEA) (Hoang-Fagerström et al., 2003a). A PA was chosen because it strengthens the feedback from farmers to scientists and creates a way for farmers to affect research directly (Francis et al, 1990). In this way, account can be taken of farmers' objectives and constraints, which should result in understanding of the factors that lead farmers to adopt conservation practices. By combining the farmer's knowledge of local conditions with scientific knowledge, the quality of research projects may be improved, and farmers can participate more actively in the decision-making process (Hoang-Fagerström et al., 2004).

The aim of this chapter is to give a brief description of the EROCHINA project, focussing on erosion monitoring, scenario simulations and PA. From this description some recommendations will be drawn for use of erosion monitoring and modelling in monitoring and evaluation of conservation projects.

STUDY AREA

Figure 1 is an elevation map of the Danangou catchment and shows the approximate position of the catchment (marked X) on the Loess Plateau (grey area on China map). Shading is used to bring out relief. The position of the weir used for measurements is indicated with a four-pointed star. The map of China was adapted from Pye (1987).

The Danangou catchment is a typical small (3.5 km^2) Loess Plateau catchment in northern China with steep slopes and a loess thickness close to 200 m. Elevation in the catchment ranges from 1,070 to 1,370 m and the catchment is deeply dissected by gullies, which have very steep slopes. The soils are mainly silt loams that are classified as Calcaric Regosols/ Cambisols in the WRB/FAO reference-system (Messing et al., 2003a). The climate is semi-arid, with occasional heavy thunderstorms in summer. At Ansai town, 5 km from the Danangou catchment, total average yearly rainfall was 513 mm over the period 1971-1998 (data from Ansai County Meteorological Station). Most of the rain (72%) falls in the period June-September, during which period a few heavy storms each year cause runoff and erosion. The main land uses in the Danangou catchment are wasteland (40%) and cropland (35%). Wasteland is mainly located on the steeper parts of gully slopes as well as on gully bottoms and was until recently used for grazing goats. This practice has ceased, since grazing was prohibited in September 1999. Cropland is located mainly on hilltops and on relatively gentle slopes at lower elevation. Nevertheless, slopes of up to 50% are not uncommon in croplands. The most common crops in the area are potato, millet, soybean, buckwheat and maize. Fallow land is mostly situated along the hilltops and woodland in the upper parts of some of the valleys. Some conservation measures are already being used, in particular terraces, check dams, tree and shrub planting, intercropping and crop rotation.

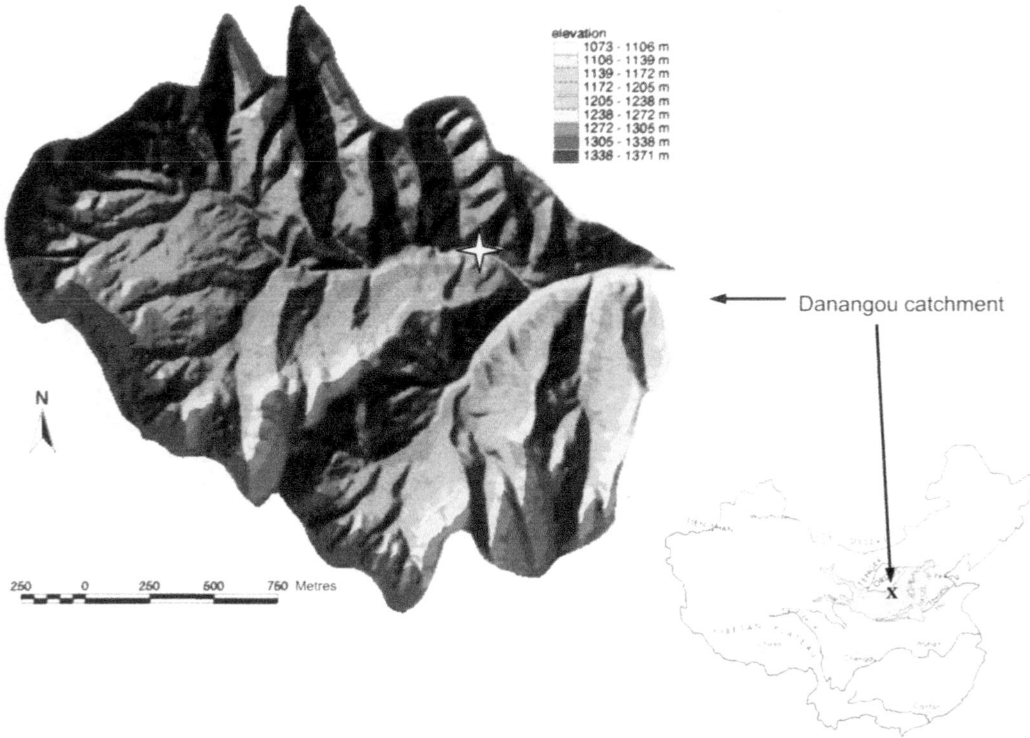

Fig. 1 Elevation map of the Danangou catchment, showing the approximate position of the Danangou catchment (marked X) on the Loess Plateau (grey area on China map). Shading is used to bring out relief. The position of the weir is indicated with a four-pointed star. The map of China was adapted from Pye (1987).

There are two villages in the catchment, Danangou and Leipingta. In 1998, 104 people (25 households) were living in Danangou and 102 people (21 households) in Leipingta. Land area per household is about 1.9 ha in each village, scattered in a number of small plots. The most productiv land in Danangou is alluvial land along the Yanhe River (outside the catchment), while in Leipinta it is terraced land with loess soils. Mules, donkeys and cows are kept to provide draught power and manure, while pigs, goats, and chicken are kept for meat, eggs, manure and cash. There are some orchards close to the villages. Household income is generated from farming, off-farm activities and government support, while the main expenditures are on some food, clothes, social fees, school fees and taxes.

PROJECT STRUCTURE AND METHODS

Figure 2 shows a flow chart of the EROCHINA project. The project consisted of two major parts, one dealing with erosion measurement and modelling, and the other with the development of alternative LUS. These scenarios were then evaluated for their effect on erosion using the Limburg soil erosion model (LISEM) (De Roo et al., 1996; Jetten and De Roo, 2001) and their effect on the economical situation of farmers by using PHEA (Hoang-Fagerström et al., 2003a,b).

Erosion Measurement and Monitoring

Soil loss from the upper catchment (2 km^2) was quantified using a weir that was built in 1998 (Van den Elsen et al., 2003). At the weir, water level and sediment concentration were measured on event basis. Soil loss from croplands was estimated from a 6 × 30 m erosion plot that was installed in 1999 (Hessel et al., 2003a). Input data for LISEM were collected every fortnight for all the major land use soil combinations that occur in the catchment (Liu et al., 2003; Stolte et al., 2003; Wu et al., 2003). Rainfall was measured with several tipping bucket rain gauges. After the input data was collected, the LISEM was calibrated (Hessel et al., 2003a).

Land Use Scenarios

Three main LUS were developed in the project (Chen et al., 2003). As can be seen from Fig. 2, these scenarios were based on the following:

1. Land evaluation, including a biophysical resource inventory in the area (Messing et al., 2003b), soil mapping, soil profile description and land use mapping. These data were used to develop suitability maps for the different types of land use (Chen et al., 2003).
2. Farmers' perceptions as found in PA studies (Hoang-Fagerström et al., 2003a,b; Messing and Hoang-Fagerström, 2001), including their

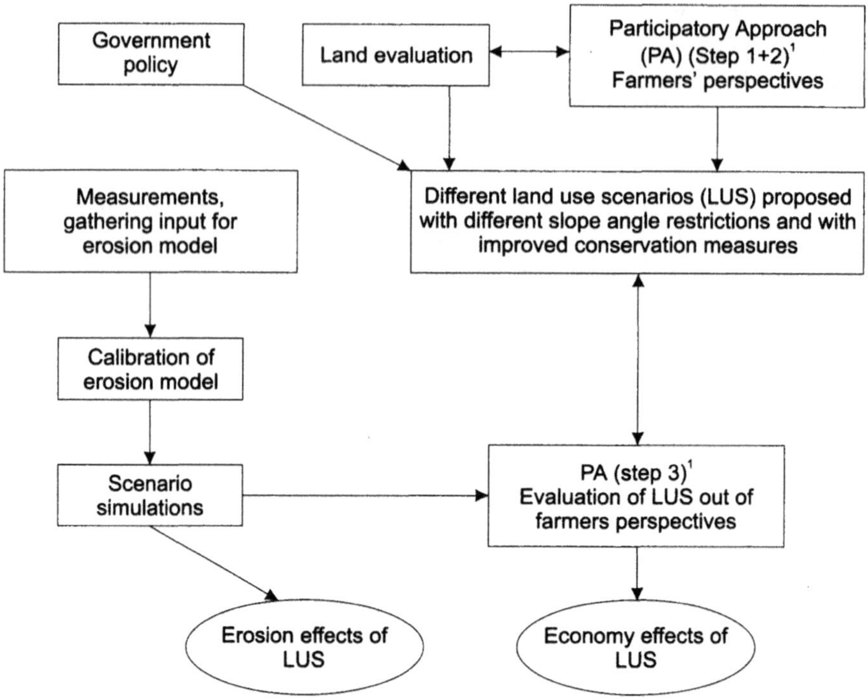

Fig. 2 Flow chart of EROCHINA project ([1]Steps 1, 2, 3 in Table 1).

views on soil workability, water availability and crop suitability. Their opinion about different types of conservation measures was also taken into account.

3. The plans of the authorities on re-greening the Loess Plateau. These policies aim to decrease erosion rates through changes in land use. In particular, they aim at a large decrease in cropland area so that all fields on steep slopes (above 15°) are changed from cropland to other uses. Grazing was also prohibited in 1999 to protect the steep slopes. The decrease in cropland should be accompanied by intensification of production on the remaining cropland and by an increase in woodland, shrubland and orchards (cash trees). The idea is that in the long term the income of the farmers should increase once they get better yields from the remaining cropland as well as income from fruit trees and other cash trees. Since it takes time before the new land use can start to benefit the farmers, the government is considering giving some compensation to the farmers to make the change economically feasible for them.

To evaluate the effects of the different LUS, the process-based distributed erosion model LISEM, using a storm scenario, was applied (Hessel et al., 2003b). Such models are quite sensitive to uncertainty in model input, but in the case of scenarios, the same uncertainty applies to all scenarios and one can therefore assume that the differences produced by simulations of different scenarios are consequences of changes in scenario, not uncertainty in inputs.

Participatory Work

The PA was carried out in three steps (Table 1). During Step 1, researchers learned to understand local people and their perspectives on local conditions, farming systems and livelihood strategies, and land use, using semi-structured interview techniques. Several key informants (e.g. village leaders) were interviewed, about 30% of the households in the catchment were studied, and two village meetings were held. During Step 2, the local

Table 1 Steps of the Participatory Approach.

	Step 1 (1998)	*Step 2 (1999)*	*Step 3 (2000)*
Objectives	Farmers' perspectives on general local conditions	Farmers' perspectives and perception on land use as a basis to establish different LUS	Evaluation of LUS from farmers' perspectives
Methods[1]	Village sketch Transect Seasonal calendar Timelines Problem/solution flows Ranking Venn diagram	Village model Transect Timelines Ranking Brainstorming	Village map Timelines Ranking Resource flows Questionnaire PHEA Bar charts
Outputs	1. Village profile (history, environmental, economical, social, and institutional patterns) 2. Farming systems components and livelihood strategies	1. Resource development and management 2. Current land use and farmers criteria for land suitability 3. General picture of current situation and visions of land use	1. Family composition and land area 2. Estimation of current household economy 3. Predict changes in household economy due to different LUS 4. Feedback on the findings to and from farmers

[1]Step 1: Conway and McCracken, 1990; Pretty, 1995; Hoang-Fagerström et al., 2003a. Step 2: Conway and McCracken, 1990; Pretty, 1995; MRDP, 1998; Hoang-Fagerström et al., 2003a; Messing and Hoang-Fagerström, 2001.

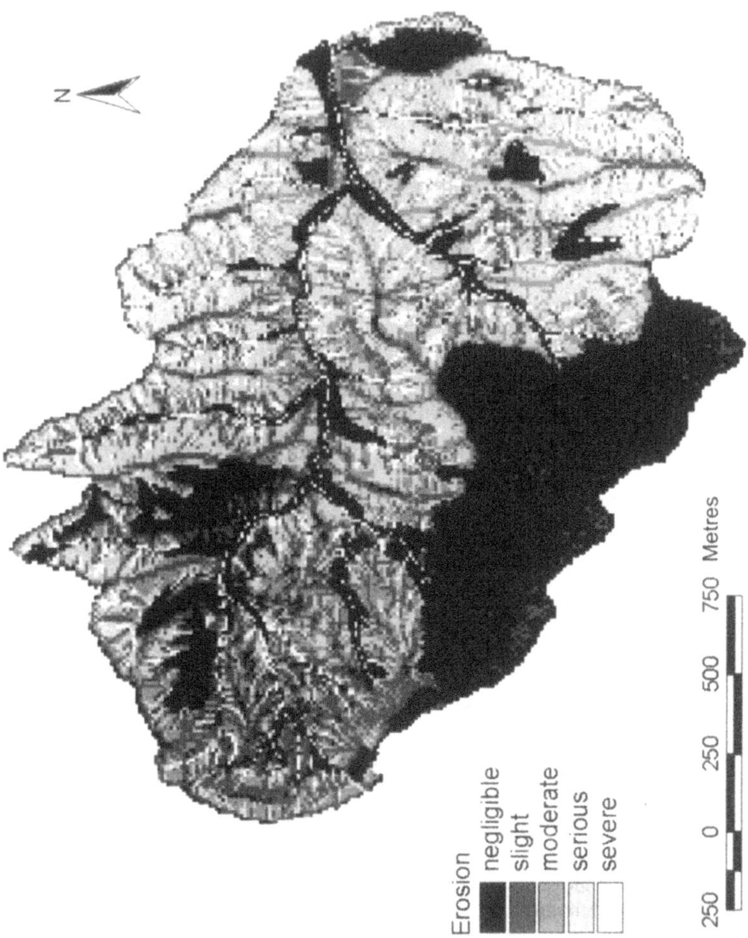

Fig. 3 Classified LISEM erosion map for the current land use and the 1 August 1998 storm.

people analysed their own conditions and alternative land uses with the help of researchers. This was carried out through six farmers' group meetings and one village meeting in Danangou. The meetings were organized on the basis of two principles: (1) The local people are planners, the village leaders are organizers, and the researchers are facilitators. (2) Gender mainstreaming is used in all steps, i.e. group meetings separately for male and female. During Step 3, farmer-researcher two-way feedback was carried out on LUS and their possible effects on erosion control and household economy. A PHEA was developed to predict potential changes in household economy.

It consisted of (1) the selection of representative farms using a formal interview and wealth ranking and (2) an in-depth study of the selected households (Hoang-Fagerström et al., 2003a,b). During the in-depth study, the inflows (cash income), within-farm flows (farm consumption and savings), and outflows (cash expenditure) of each selected household were analysed by researchers together with interviewed farmers. The balance between the inflows and outflows was estimated. The changes in farm production due to converting land with cut-offs at either 25° or 15° slopes, with or without intensification and with or without government support, were estimated.

The output of the whole PA, including LUS and their possible effects on erosion control and household economy, were then finally presented to the local farmers in order to get their feedback at a village meeting in each study village. At the village meeting, a poster exposition with simple visualization was used to facilitate researcher-to-villager feedback. The LUS were subsequently revised on the basis of farmer response and erosion modelling, until the predicted results were acceptable with respect to erosion and to household economy.

RESULTS

Measurement Results

Table 2 shows that the data on the different recorded events are consistent with the proposition that the more rain, the higher the discharge, and the more soil is lost. The results further suggest that about 11.5 mm of high intensity rain is needed to produce runoff at the weir. They also show that sediment concentrations were very high, indicating that erosion is a major problem in the catchment, both on field scale (sediment plot) and on catchment scale.

Table 2 Results of erosion monitoring.

Catchment outlet	*980801*	*980823*	*990720*	*000811*	*000829*
Event rainfall (mm)	15.1	13.0	14.1	11.6	17.8
Peak discharge (m^3 sec^{-1})	5.1	0.7	3.6	0.2	8.7
Discharge/rainfall (%)	12.6	3.0	12.1	0.9	16.3
Total soil loss (ton)	1,280	96	770	16	2,630
Soil loss (t ha^{-1})	6.2	0.5	3.7	0.1	12.7
Max sediment concentration (g L^{-1})	361	154	371	129	498
Sediment plot	*990710*	*990721*	*000707*	*000811*	*000829*
Event rainfall (mm)	10.6	5.0	16.3	12.0	16.7
Peak discharge (L sec^{-1})	0.18	1.20	0.12	0.03	9.4
Discharge/rainfall (%)	1.6	21.7	2.5	1.3	55.2
Total soil loss (kg)	24.8	162.1	27.9	1.3	659
Soil loss (t ha^{-1})	1.3	8.4	1.4	0.1	34.0
Average sediment concentration (g L^{-1})	736	748	336	39	357}

Scenario Simulations

Three main scenarios were developed: (1) base-line land use, (2) 25°, cropland only on slopes below 25 degrees, and (3) 15°, cropland only on slopes below 15 degrees. These scenarios were further divided into three sub-scenarios. A simple scenario was used that assumed that no further interventions were made. But the effects of biological interventions (such as mulching and improved fallow) and mechanical interventions (e.g. contour ridges) were evaluated separately in two sub-scenarios. These measures are relatively simple and inexpensive but labour-intensive and are therefore realistic given the local circumstances.

Table 3 shows the land use distribution for the three main scenarios. Compared to the base-line land use, the other scenarios have much more woodland/shrubland, while cropland area decreased according to the specified slope limits. It was assumed that one-fifth of the cropland will be fallow, since intensification of agriculture on the remaining cropland (e.g. use of fertilizers, improved fallow) might reduce the fallow area compared to the current situation. The decrease in cropland area for the 25° and 15° scenarios is accompanied by a gradual increase in orchard/ cash tree on slopes between 15° and 25°. The other land uses (including all slopes of more than 25°) remained unaffected for these two scenarios.

Figure 3 shows a LISEM erosion map for the base-line land use as affected by the 1 August 1998 storm. For a large area in the southern part of the catchment, no serious erosion was predicted. This was caused by the

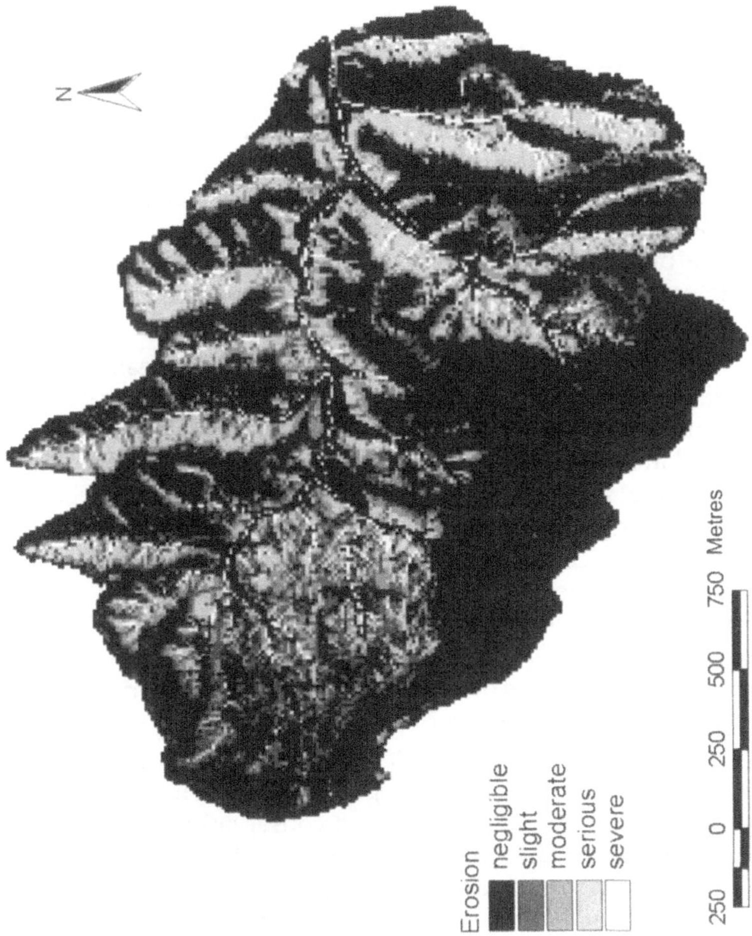

Fig. 4 Predicted erosion for the 15° scenario.

Table 3 Land use distribution scenarios.

Land use (% of catchment)	*Current*	*25°*	*15°*
Cropland	35.4	21.0	6.7
Orchard/cash tree	2.4	0.0	17.8
Woodland/shrubland	13.4	38.2	38.3
Wasteland	41.4	35.5	35.6
Vegetables	0.1	0.1	0.1
Fallow	7.3	5.2	1.6

fact that, according to the measured rainfall data, less rain fell in this part of the catchment during the 980801 event. Other areas with negligible erosion rates for the 980801 storm were mainly woodland areas. The hilltop areas generally had slight or moderate erosion rates, while the steeper parts of the slopes had serious or severe erosion rates. Apart from the effects of changing amounts of woodland, not much difference in erosion was found for the different land uses.

Figure 4 shows the results of the 15° scenario with the same storm. It is evident that the erosion in the catchment has decreased compared to the base-line situation (Fig. 3). The area with negligible erosion rates has more than doubled in size, while the other erosion classes have all decreased in area. Negligible erosion rates mainly occurred under woodland/shrubland. The large decrease in predicted erosion was therefore probably a direct consequence of the increase in woodland/shrubland area (Table 3). Figure 5 shows that soil loss from the catchment was much larger for base-line land use than for the 25° and 15° scenarios, while the differences in predicted soil loss between the 25° and 15° scenarios were relatively small. Differences between the 25° and 15° scenarios are mainly caused by cropland to orchard conversion.

Figure 5 shows that the relative decrease in soil loss was always larger than the relative decrease in discharge. Relatively small decreases are shown in discharge and soil loss for the conservation measures for all scenarios (ranging from 2 to 21%), whereas large decreases (ranging from 33 to 71%) are shown for the alternative LUS compared with base-line land use (Fig. 5). In the sub-scenarios, the biological measures were always more effective than the mechanical measures (Fig. 5).

Participatory Approach

According to PHEA, the main source of income in Danangou village was off-farm work such as trade, construction and transportation (86% of total farm income), while it was farm work in Leipingta (70% of total farm

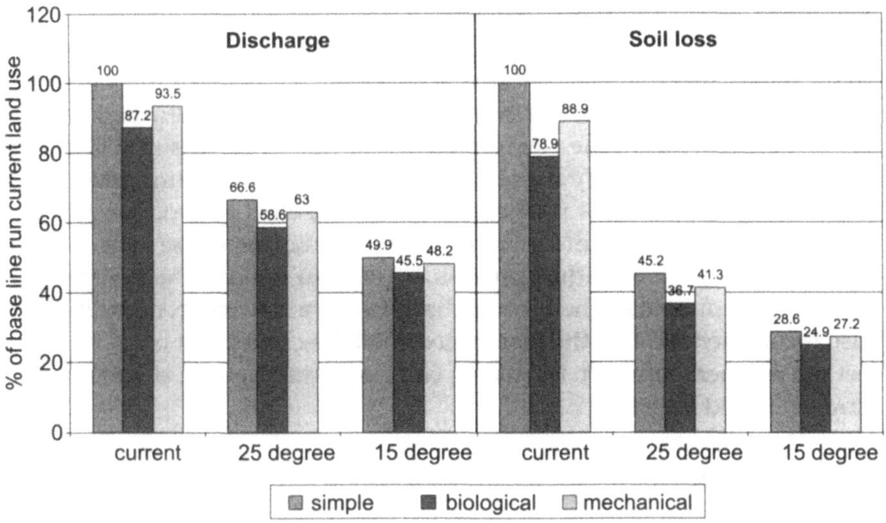

Fig. 5 Results of scenario simulations.

income). Farmers seemed to produce enough food for everyday life. However, with little or no savings, and a large variation in annual rainfall, they have no food security. The conditions of Danangou village were better than those of Leipingta village, which was due to the fact that Danangou is situated closer to Ansai town, giving farmers from Danangou better opportunities for off-farm activities. The balance between inflow and outflow of cash was positive for most of the interviewed families in Danangou village, while it was negative for Leipingta village. This means that most of the families in Danangou were able to save, while farmers in Leipingta did not have enough money to cover their everyday living expenditure.

The PHEA indicated that, with scenario 25°, income from farm production would be at the same general level as today for both villages, if farmers were to get support from the government for food and seedlings (as planned), as well as benefit from cropping intensification. However, with scenario 15°, income from farming in both villages would be reduced by almost 30% even though it was assumed that the villagers would get support from the government as well as benefits from cropping intensification. In the long term, however, the situation might be better if income from orchards and cash trees became available. With both scenario 25° and scenario 15°, farmers with off-farm activities seemed to manage better than farmers with only farm work both in the short and the long run. They are therefore more likely to convert sloping land into other land use types. This shows that the conversion work in connection with the central government policy on re-greening the Loess Plateau should focus first on the areas where off-farm activities are available or potentially available. If areas with fewer off-farm activities such as Leipingta village are to be involved in the conversion work, government support will be critical in sustaining the current living standards and farming activities of farmers, since farming is not only the source of food production, but also the source of employment for the villagers. Furthermore, the conversion of land use should be a gradual process. It is recommended that scenario 15° should start only after the farmers have benefited from scenario 25°. It is also recommended that the conversion work be carried out only if farmers get government support, including compensation for loss of land, health care, and seed breeding.

DISCUSSION

The EROCHINA project was successful because it combined measurements, land evaluation, modelling and PA methods. Although it was a research project and not a conservation project as such, there are lessons to be learned for monitoring and evaluation of projects.

First, such projects should involve the farmers at all stages in the project, including formulation of strategies and evaluation of effectiveness of such strategies. The EROCHINA project showed that by establishing a trusted partnership with the local farmers, both general trends and detailed factors, including gender aspects, could be investigated (full PA results presented in Hoang-Fagerström et al., 2003a,b). This partnership was established by researchers making repeated visits and involving the farmers during the whole research process. Such a good understanding is likely to improve adoption of the measures recommended. However, for optimal use in conservation projects, the project duration should be longer than the three years of the EROCHINA project.

Second, scientists should always realize that any conservation programme must be economically beneficial to the local farmers. Their livelihoods depend on their crop yield, so they simply cannot afford to lose yield because of conservation measures. This study indicated that once off-farm work becomes a significant source of income, conservation measures become more acceptable.

Third, measurement of erosion can be a valuable tool to investigate erosion in a more quantitative way. Such quantitative data are useful to compare with more qualitative data obtained from farmers. They can also help to pinpoint where conservation measures should be most effective. Finally, monitoring of erosion can give data about the actual effect of the strategies once implemented. It should be kept in mind that erosion is variable in time since it depends on variable rainfall characteristics. To obtain reliable data on the effects of conservation measures, long-term monitoring is needed at catchment scale. Plot-scale monitoring can have its uses too, but cannot incorporate spatial relationships that occur in the catchment. For some measures (like reforestation) the full effects will only become apparent after a number of years.

Erosion modelling can provide a good way to explore beforehand what the consequences of certain measures could be. Although such model predictions can only tell us what might happen, they are a means to visualize in a spatial way what the effect of conservation measures might be before these measures are actually implemented and can therefore be a valuable tool in land use planning. It should be realized that modelling can provide no certainties about what is going to happen but can only give indications. To be of practical use in conservation projects, relatively simple erosion models should be used.

Visualization and diagramming methods were especially important communication tools in this study, since even for Chinese researchers it was difficult to understand the local dialect. One important aspect noticed was that most of the existing visualization methods available in literature are mainly intended for gathering information from farmers; simple visualization methods for feeding back the research information to farmers are still lacking. The posters developed in this work are a contribution in this direction.

CONCLUSIONS

Crop farming on slopes of all degrees is the main land use in Danangou catchment. This type of land use has severe environmental consequences, in particular extensive soil erosion. The local people are highly aware of alternative conservation land uses suitable for different parts of the catchment. Simulations with the Limburg soil erosion model indicated that significant decreases in soil and water loss could be achieved by changes in land use, especially restricting cropland to gentler slopes. Other conservation interventions were predicted to be relatively less effective but could nevertheless result in some reduction of runoff and erosion. However, due to poor living standards and the comparative lack of off-farm employment in one village studied, any effort to improve land use toward more sustainable practices would seem to be difficult without support from the government. It is recommended that implementation of any such conservation measure be carried out by an active collaboration between farmers, extension workers and researchers. Measurement and modelling of erosion can give valuable data to be used in such projects.

ACKNOWLEDGEMENTS

The authors are grateful to the European Union (Contract Number: IC18-CT97-0158) for funding. A valuable part of the fieldwork was undertaken by colleagues from the Institute of Soil and Water Conservation (Yangling, Shaanxi), Beijing Normal University (Beijing) and the Research Centre for Eco-Environmental Sciences (Beijing), to whom the authors are indebted. Finally, the authors express their thanks to farmers in Danangou and Leipingta villages for their active participation in the project.

References

Chen, L., I. Messing, S. Zhang, B. Fu and S. Ledin. 2003. Land use evaluation and scenario analysis towards sustainable planning on the Loess Plateau in China – case study in a small catchment. Catena 54: 303-316.

Conway, G.R. and J.A. McCracken. 1990. Rapid rural appraisal and agroecosystem analysis. pp. 221-235. *In:* M.A. Altieri and S.B. Hecht. [eds.] Agroecology and Small Farm Development. CRC Press, Inc. Florida.

De Roo, A.P.J., C.G. Wesseling and C.J. Ritsema. 1996. LISEM: a single-event physically based hydrological and soil erosion model for drainage basins: I: Theory, input and output. Hydrol Processes 10: 1107-1117.

Douglas, I. 1989. Land degradation, soil conservation and the sediment load of the Yellow River, China: review and assessment. Land Degrad. Rehabil. 1: 141-151.

Francis, C., J. King, J. de Witt, J. Bushnell and L. Lucas. 1990. Participatory strategies for information exchange. Am. J. Alternat. Agr. 5: 153-160.

Hessel, R., V. Jetten,B. Liu, Y. Zhang and J. Stolte. 2003a. Calibration of the LISEM model for a small Loess Plateau catchment. Catena 54: 235-254.

Hessel, R., I. Messing, L. Chen, C. Ritsema and J. Stolte. 2003b. Soil erosion simulations of land use scenarios for a small Loess Plateau catchment. Catena 54: 289-302.

Hoang-Fagerström, M.H., I. Messing and Z.M. Wen. 2003a. A participatory approach for integrated conservation planning in a small catchment in Loess Plateau, China. Part I. Approach and Methods. Catena 54: 255-269.

Hoang-Fagerström, M.H., I. Messing, Z.M. Wen, K.O. Trouwborst, M.X. Xu, X.P. Zhang, C. Olsson and C. Andersson. 2003b. A participatory approach for integrated conservation planning in a small catchment in Loess Plateau, China. Part II. Analysis and Findings. Catena 54: 271-288.

Hoang-Fagerström, M.H., Tran Duc Toan, H. Sodarak, M. van Noordwijk and L. Joshi. 2004. How to combine scientific and local knowledge to develop sustainable land use practices in the uplands – A case study from Vietnam and Laos. Proceedings of the workshop "Poverty reduction and shifting cultivation stabilization in the uplands of Laos PDR: Technologies, approaches and methods for improving upland livelihoods". Luang Prabang, 27-30 January 2004, 16 pp.

Jetten, V. and A.P.J. De Roo. 2001. Spatial analysis of erosion conservation measures with LISEM. pp. 429-445. *In:* R. Harmon and W.W. Doe. [eds.] Landscape Erosion and Evolution Modeling. Kluwer Academic/Plenum, New York.

Jiang Deqi, Qi Leidi and Tan Jiesheng. 1981. Soil erosion and conservation in the Wuding River Valley, China. pp. 461-479. *In:* R.P.C. Morgan. [ed.] Soil Conservation: Problems and Prospects. Wiley, Chichester.

Liu, G., M. Xu and C. Ritsema. 2003. A study of soil surface characteristics in a small watershed in the hilly, gullied area on the Chinese Loess Plateau. Catena 54: 31-44.

Messing, I. and M.H. Hoang-Fagerström. 2001. Using farmers´ knowledge for defining criteria for land qualities in biophysical land evaluation. Land Degrad. Dev. 12: 541-553.

Messing, I., L. Chen and R. Hessel. 2003a. Soil conditions in a small catchment on the Loess Plateau in China. Catena 54: 45-58.

Messing, I., M.H. Hoang-Fagerström, L. Chen and B. Fu. 2003b. Criteria for land suitability evaluation in a small catchment on the Loess Plateau in China. Catena 54: 215-234.

MRDP. 1998. PRA Methodologies Used in Extension Work. Agricultural Publish House, Hanoi (In Vietnamese).

Pretty, J.N. 1995. Participatory learning for sustainable agriculture. World Dev. 23(8): 1247-1263.

Pye, K. 1987. Aeolian Dust and Dust Deposits. Academic Press, London.

Stolte, J., B. van Venrooij, G. Zhang, K.O. Trouwborst, G. Liu, C.J. Ritsema and R. Hessel. 2003. Land-use induced spatial heterogeneity of soil hydraulic properties on the Loess Plateau in China. Catena 54: 59-75.

Van den Elsen, E., R. Hessel, B. Liu, K.O. Trouwborst, J. Stolte, C.J. Ritsema and H. Blijenberg. 2003. Discharge and sediment measurements at the outlet of a watershed on the Loess Plateau of China. Catena 54: 147-160.

Wan, Z. and Z. Wang. 1994. Hyperconcentrated Flow. IAHR Monograph. Balkema, Rotterdam.

Wu, Y., K. Xie, Q. Zhang, Y. Zhang, Y. Xie, G. Zhang, W. Zhang and C.J. Ritsema. 2003. Crop characteristics and their temporal change on the Loess Plateau of China. Catena 54: 7-16.

Zhang, J., W.W. Huang and M.C. Shi. 1990. Huanghe (Yellow River) and its estuary: sediment origin, transport and deposition. J. Hydrol. 120: 203-223.

Part 4

Social, Economic and Institutional Aspects

19

Monitoring and Evaluation of a Village's Progress-driven Attitude

Aad Kessler[1] **and Hideo Ago**[2]

[1]Technical Coordinator JGRC Project, Sucre, Bolivia. Present address: Eykmanstraat 7, 6706 JT Wageningen, The Netherlands. E-mail: Aad.Kessler@wur.nl

[2]Sub-Director JGRC Project, Sucre, Bolivia. E-mail: hideo_ago@chushi.maff.go.jp

ABSTRACT

The intervention strategy for soil and water conservation implemented by a project of the Japan Green Resources Corporation in the inter-Andean valleys of Bolivia has motivated a majority of farmers to execute soil and water conservation practices. The key to the success of the applied strategy is the emphasis given to laying a solid foundation for sustainable rural development as an essential step before actually executing conservation activities. The most important aspect of this solid foundation is the existence of a progress-driven attitude in a village, which implies good internal organization and collaboration, the presence of village leaders, and awareness of the importance of natural resources for the village's future development. The monitoring and evaluation of the change that a village undergoes during the building of the solid foundation is essential to detect weak points in the villager's attitude. This is done by means of a set of easily measurable and practical criteria. When according to the criteria a village has achieved a progress-driven attitude, planning and execution of conservation and other activities can continue. This way, a project can largely avoid failures due to lack of motivation and preparation of a village.

INTRODUCTION

From 1999 to 2003, the project "Participatory Rural Development based on Soil and Water Conservation" (in this chapter referred to as "the project") executed research activities in three rural villages in the watershed of Rio Grande, situated in the Department of Chuquisaca in Bolivia. The project is co-funded by the Japan Green Resources Corporation (JGRC) and the Prefecture of Chuquisaca. Its principal objective is to validate soil and water conservation (SWC) practices and adequate intervention methodologies, which ideally should lead to more harmonious and realistic institutional interventions in rural areas, especially regarding SWC.

The project area involves three representative villages of the inter-Andean valleys of Chuquisaca, with altitudes between 2,500 and 3,200 m asl (Fig. 1). The climate is semi-arid with a yearly precipitation of 400-500 mm, concentrated in the months of November to March. In recent decades, human pressures on the scarce arable lands and overgrazing of rangeland have increased. Consequently, marginal lands on steep slopes have been taken into production, land degradation has augmented, and the quality of vegetation, soil and water has declined (JGRC, 2003). Poverty and a constant struggle for survival are a general phenomenon in Chuquisaca, which is one of the least developed departments in Bolivia, with a score of 0.49 on the Human Development Index (UNDP, 1997).

Although most farmers value present-day life in their village more than city life (Vargas, 1998), a growing number of farmers are forced to migrate elsewhere. Hence, rural households in the region are multiple and have significant non-agrarian components (Bebbington, 1999). Despite farmers' interest in improving soil fertility, the absence of an extension service has always limited technology transfer. As a consequence, traditional agricultural and conservation practices are still widely used but are not able to maintain soil fertility or control erosion (JGRC, 2002). Projects that aimed at rural development have not changed this situation and their rare conservation activities have largely failed because of erroneous intervention strategies.

According to van Niekerk (1997), the most likely future scenario for this region is an "impossible" situation, in which the peasantry continues to limp along, caught between migration and low productivity agriculture. We think that this vision is too pessimistic and that there are always possibilities that can improve the existing situation. The point is that we soil conservationists and our projects must pay (even) more attention to the roots of the solution, i.e. work with motivated and well-organized farmers with favourable attitudes towards their own

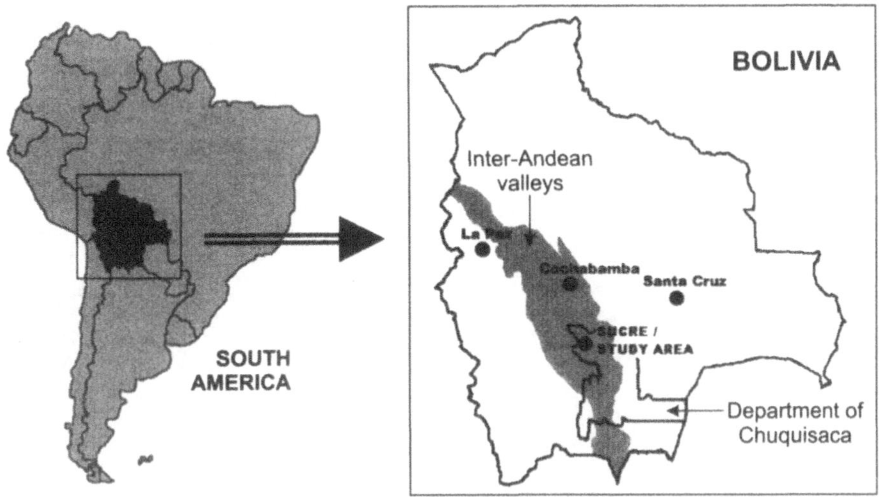

Fig. 1 Location of the study area in the inter-Andean valleys of Bolivia.

sustainable development. This approach and strategy are outlined in this paper, with major emphasis on how to monitor and evaluate such an attitude in rural villages. Figures 2-7 show the landscape of the study area and how SWC has been attempted.

A NEW INTERVENTION STRATEGY FOR SWC

Based on four years of research, the project validated a new intervention strategy for SWC within a framework of sustainable rural development, which is composed of two major phases (Fig. 8). In the first phase (laying a solid foundation for sustainable development) a conservation attitude at municipal level and a progress-driven attitude at village level are generated. The second phase (planning and sustainable execution based on integrated natural resources management) is necessarily executed after having achieved the required attitude (change), given that planning and execution can only be successful when done with organized villages and trained people.

Before proceeding to the essence of this paper, which is to present how a progress-driven attitude at village level is monitored and evaluated, both phases of the intervention strategy will be briefly explained.

Phase 1 Activities

Dealing with SWC issues is comparable to building a house: a solid foundation must be laid before the house can be built. However, projects

Fig. 2 Stone lines have been made on the contour.

Fig. 3 A-frames made from locally available material.

Fig. 4 Using the A-frame for construction of stone lines.

Fig. 5 Building stone lines during the SWC contest.

Fig. 6 The farmer and his stone lines, Sirichaca.

Fig. 7 Visiting farmers viewing bench terraces.

generally start almost directly with execution activities and overlook the importance of building a solid foundation. Hence, the activities that are executed in this first phase are generally not – or insufficiently – taken into account by projects.

A first set of activities is executed at municipal level to generate a conservation attitude among all persons involved. Given that municipalities are the (financial) counterparts of most rural development projects, the generation of a conservation attitude in the municipal authorities regarding the importance of natural resources and the need to plan and execute conservation activities is a big step towards more sustainable spending of the municipalities' budget. For example, in the state of Parana, Brazil municipalities have proved able to play an important role in SWC (Ago and Kessler, 1996).

A second set of activities is executed at village level and focuses on the generation of a progress-driven attitude at village level. During this phase, workshops are given about village organization and collaboration, with special emphasis on the functioning of the agrarian syndicate and a women's group. Topics related to natural resources conservation, which are of crucial importance in all workshops, point at reviving farmers' interest in, for example, the potential advantages of SWC. The Conservation Leaders form a very important group that receives special training. They are persons who are interested in SWC and are willing to acquire and subsequently transfer their knowledge about SWC to other villagers, leading the way to sustainable development in the village. The role of the extension worker, and especially the building of a relationship of trust with the villagers, is very important.

The financial costs for the execution of all activities in Phase 1 are minimal and principally refer to training material and the salary of the extensionist. However, the investment in time can be substantial and varies between one and two years. During this period it is recommended that a project not start with the planning and execution of other activities, although some small activities can be started to keep farmers motivated. Preferably, these activities take into account priority needs of the village.

Phase 2 Activities

Phase 2 consists of all planning and execution activities for sustainable rural development in a village. Planning at village and household level must ensure that all possible opportunities in a village are taken into account and that families are trained in formulating and expressing those activities at household level that can contribute to sustainable household systems. Execution of almost all activities is carried out through group

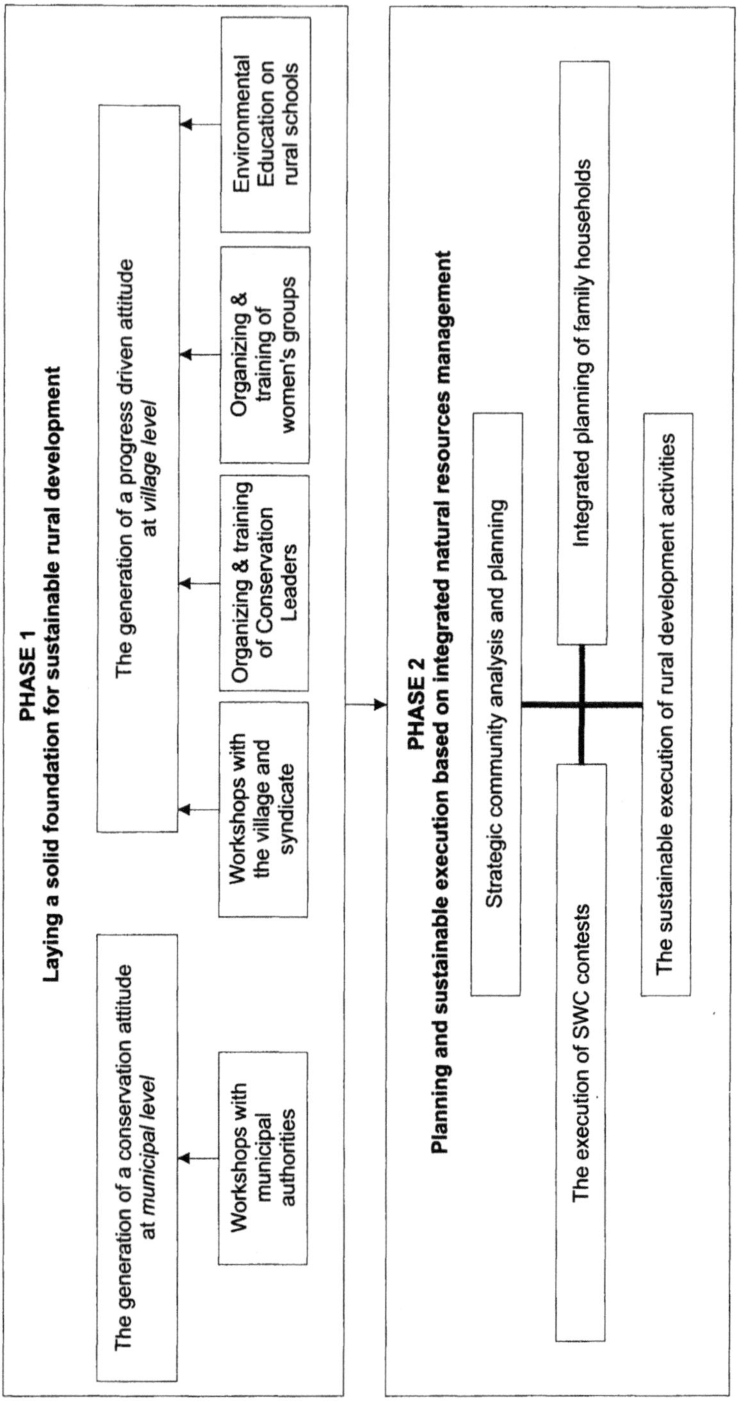

Fig. 8 The two phases of the intervention strategy.

work. Groups are organized according to specific objectives and work out Integrated Group Projects, in which all activities that contribute to the sustainable achievement of the objectives are incorporated. Conservation practices are executed by means of contests between organized groups and are based on farmer-to-farmer training. Because sustainability is the ultimate goal during execution, training and capacity building are constantly emphasized in this phase.

A VILLAGE'S PROGRESS-DRIVEN ATTITUDE

A village achieves a progress-driven attitude when it complies with the following five basic conditions. These conditions can also be used to describe the concept of a village's progress-driven attitude.

1. *Organization and leadership*. The agrarian syndicate, which is elected every year, is the leading and representative organization of a Bolivian rural village. Agrarian syndicates were established in all villages after the Agrarian Reforms in the 1950s as organizations that should lead village development. However, the agrarian syndicates have become, in many villages, organizations without power, leadership or vision, and often consist of persons for whom being a member is a kind of punishment. In the intervention strategy, a well-organized agrarian syndicate that actually leads village development is essential, because nothing will be achieved in a village that is led by persons who do not care about the village's future, and whose decisions are not respected by the villagers. Moreover, farmers' participation is more effective if projects work with strong local boards (Zoomers, 1998). Obviously, the training of Conservation Leaders can make a very positive contribution to the establishment of stronger agrarian syndicates.
2. *Responsible participation*. Village meetings and workshops are obligatory and generally not seen as very beneficial. The events always start later than planned and only the most influential villagers talk while the rest do not pay attention. Most people attend passively and often families send one of their children to these meetings as their representative, in order to avoid having to pay the penalty for not showing up. In this situation, only a few villagers make decisions and the rest of the villagers are not even consulted. In the proposed intervention strategy, an important condition is responsible participation, i.e. that at least one adult household head attend the village meetings and workshops and that all persons actively participate.

3. *Effective collaboration*. Village development implies working together to achieve set goals. However, in the rural area of Bolivia, the old tradition of reciprocal group work has almost been lost. Individualism and achievement of personal goals (money) have unfortunately become more important, although at the same time social networks have been kept intact. In the intervention strategy, collaboration among villagers, principally re-establishing group work, is an essential condition for sustainable village development. This means that all villagers respect the communal decisions and that different groups within the village are officially recognized. It is essential that a spirit of doing things together is created.
4. *Mutual trust*. Rural development in Bolivia needs the active intervention of development institutions because, for many of the basic needs, investments are desperately needed. In the past, most of the investments have not achieved the expected results, mainly because of inappropriate intervention strategies that were not directed towards obtaining sustainable results. As a consequence, in many villages development institutions are no longer trusted, which makes it almost impossible to achieve an effective collaboration. Villagers are mostly interested in tangible results and products (e.g. food or seeds), and will try to obtain the maximum short-term advantage from any institution that shows up in the village. In the intervention strategy, the establishment of a relationship of trust between the village and the project is essential. This means that the extensionist must identify with the village and must try to become a respected person and a friend of the villagers.
5. *Environmental awareness*. In order to continue with natural resources conservation activities, it is essential that the majority of the villagers become aware of the importance of soil, water and vegetation. Due to the absence of an extension service in Bolivia, most rural villages are cut off from any information source. As a consequence, natural resources management has not changed much over the years. Farmers still execute some traditional SWC practices, but their impact is not enough to conserve, for example, the productivity of soils. Although most farmers are aware of declining soil productivity and vegetation loss, causes are generally not correctly identified. In the intervention strategy, awareness about all aspects of natural resources conservation is one of the main factors that contribute to a positive attitude towards sustainable development, and an important condition before proceeding to Phase 2.

MONITORING AND EVALUATION

Monitoring and evaluation (M&E) of a village's progress-driven attitude at various times is crucial for determining whether a village and its inhabitants are sufficiently prepared for the planning and execution of conservation and other rural development activities. For this purpose, for each basic condition that has to be met in Phase 1, a set of evaluation criteria is formulated, which measures the respective progress of the village as well as the changes that have occurred. Given that total compliance with all evaluation criteria takes years, in Phase 1 only the most essential criteria must be totally complied with. The other criteria require "good progress" during Phase 1; their total compliance can still be achieved during the second phase. In Table 1 the required progress for Phase 1 for all evaluation criteria is presented. Only when this required progress has been achieved is it appropriate to start Phase 2 of the intervention strategy.

Monitoring and evaluation is a task of the extensionist, given that this is the person most involved in the community and able to give an exact opinion regarding all criteria. During the execution of Phase 1, evaluations should be done several times, in order to obtain a good overview of the villagers' attitude change and the respective progress measured for each criterion.

The first evaluation must be done in the first two months of the intervention. This is the zero-evaluation, in which the extensionist describes the initial status of the village regarding all criteria. This description is the starting point for measuring the progress made by the village.

The second evaluation is undertaken after having finished the first round of intensive awareness-raising workshops with the whole village. At this point, generally within six months after entrance in the village, some progress should be noticed, because during these workshops special attention is given to organizational aspects in the village and the role of the agrarian syndicate. Moreover, Conservation Leaders are being trained and a relationship of trust with the villagers should have been established.

The following evaluations should be done according to the opinion of the extensionist, for example, every three or six months. This way, a sequence of progress reports is obtained from a village concerning the solidity of the foundation for sustainable development, showing exactly where progress is made and which aspects should still be emphasized by the project.

Just before deciding to start with the second phase of the intervention strategy in a village, a final evaluation must be executed. The desired result of this evaluation is that a village has achieved the required

Table 1 Evaluation criteria of a village's progress-driven attitude and required progress in Phase 1.

Condition	*Evaluation criteria*	*Required progress in Phase 1*
1. Organization and leadership	The agrarian syndicate has responsible persons who know and fulfil their duties	good progress
	Decisions taken by the agrarian syndicate are complied with (communal works, village meetings, etc.)	good progress
	Members of the agrarian syndicate who don't comply with the requirements of their role are replaced	good progress
	Members of the agrarian syndicate are elected for their leading capacity (and not as a punishment)	good progress
2. Responsible participation	Village meetings are held each month on fixed dates	total compliance
	Villagers who participate in the village meetings are village members and responsible persons (adults)	total compliance
	Participants in village meetings and workshops show up punctually (the event starts at the fixed hour)	good progress
	There is active – and respected – participation of all the villagers in the meetings and workshops	good progress
3. Effective collaboration	There are trained Conservation Leaders who are willing to transfer their knowledge and train other farmers	total compliance
	There exists a consolidated women's group whose members actively participate in village development	good progress
	The village actively tries to solve daily communal problems in the village	good progress
	There is a good level of collaboration for executing communal and group activities	good progress
4. Mutual trust	The majority of the villagers participate in the meetings and workshops of the project	total compliance
	There is friendly behaviour towards the project staff on behalf of the great majority of the villagers	total compliance
	The project is considered important and priority is given to the extensionist during village meetings	good progress
	The villagers share confidential information with the extensionist and/or other project staff	good progress

(Table 1 Contd.)

(Table 1 Contd.)

5. Environmental awareness	There are favourable opinions regarding natural resources conservation, people are interested	total compliance
	A large majority of the villagers has participated in the project workshops about natural resources management	total compliance
	There is willingness to implement regulations for natural resources management	good progress
	School children are aware of the importance of natural resources conservation in their village	good progress

progress for Phase 1 in all evaluation criteria. If this is not the case, the project should decide whether it is worth the effort to repeat certain activities of Phase 1 and wait for the village to lay the required foundation. The alternative is the withdrawal of the project from the village, given that without the required progress in generating a progress-driven attitude in a village the second phase cannot start.

EXPERIENCES OF THE PROJECT

Progress in the three villages of the project has been very different. In the following, a summary of the experiences is presented in order of most positive to least.

- Kaynakas is a remote village that has never received aid from development institutions. When the project was looking for a third intervention village, the leaders of the agrarian syndicate of Kaynakas requested that the project be implemented in their village. In other words, right from the start the villagers of Kaynakas were very interested. As a result, the execution of activities in Phase 1 was very easy: meetings started at the fixed hour, the whole village assisted, the project staff was treated very well. The only constraint was that women did not dare to speak in the village meetings, but this changed completely after some workshops with the newly organized women's group. Less than a year after the start of the project, the required progress in Phase 1 was met for all criteria, and the village was ready for the second phase. It is now evident that the villagers of Kaynakas have achieved tremendous progress in their development: SWC practices have been executed everywhere, vegetable gardens produce regularly, skill workshops function daily, active village leaders can negotiate with the municipality about new investments, etc.

- In Tomoroco, upon entrance of the project, the agrarian syndicate did not function well, communal works were no longer executed, and individualism among villagers reigned. After almost two years of intervention, this situation changed favourably, and a solid foundation for sustainable development was laid in the village. Although most of the criteria still required strengthening during the second phase, the project decided to go ahead with planning and execution. Nowadays, the agrarian syndicate is led by Conservation Leaders who are chosen for their leading ability and knowledge. In the village there exist more than 10 organized groups, mostly for the management and use of drinking water and irrigation systems, but also for the daily execution of skilled activities such as carpentry (for young men) and a sewing workshop for women. Of all families, 80% have participated in the SWC contest and implemented SWC practices on their most important fields. Migration also has been drastically reduced, now that the project has provided new opportunities in the village.
- Sirichaca is situated not far from Sucre. When the project started, the migration rate was about 70-80%, the majority of them male villagers, leaving the village almost without workforce. Due to migration, Sirichaca was a totally unorganized village, with members of the agrarian syndicate living most of the year outside the village. Children were sent to the village meetings in order to avoid paying penalties. Despite the efforts of the project, this situation did not change and according to the criteria even after two years a progress-driven attitude had not been generated. Nevertheless, the project started some planning and execution activities, such as the improvement of wells for drinking water, workshops in skills for men and women, and the SWC contest. The results were very disappointing. Although many families participated in the SWC contest, afterwards the executed practices were not maintained and in some cases they were even removed by farmers who returned from migration. The organized groups for the skill workshops disintegrated as soon as the training ended and the improvement of the wells did not even begin because the groups could not agree on how to work together. Also, migration did not diminish, as it is still the best income-generating alternative in Sirichaca.

These three different experiences show that much of the success of generating a progress-driven attitude in a village depends on how the villagers behave. Some factors, such as the absence of leaders during large periods of the year, high migration rates, very poor agricultural conditions, or the lack of village initiatives, might indicate serious constraints for laying a solid foundation for sustainable development. This is the case in Sirichaca, where the project finally decided to withdraw owing to lack of progress. It should be understood that in this kind of village a progress-driven attitude might never be generated, and that therefore the execution of a SWC project, or any development project, will never be successful. Therefore, before a project starts an intervention, a short diagnosis should be executed in all potential intervention villages to select those villages in which organizational and physical conditions are more favourable. This way, a project can greatly improve the success and sustainability of its intervention.

CONCLUSIONS

Laying a solid foundation for sustainable development is indispensable when sustainable results are to be obtained in the second phase: planning and execution. At village level, this means that a project during the first phase of its intervention (for almost two years) must focus exclusively on the generation of a progress-driven attitude among the villagers. As long as a village does not meet the required progress in each of the criteria formulated for the basic conditions of a progress-driven attitude, a project should not start with the execution of development activities. If this is not taken into account, the risk of failure and non-sustainable execution of development activities is very high. These probable failures will only lead to frustrated stakeholders and project staff, and to the loss of money that could have been spent much better in other villages with more favourable conditions.

The task of any rural development project should be to include in its objectives and strategy activities for generating a progress-driven attitude at village level. Motivation and the genuine participation of farmers are crucial. Obviously this implies that the desired tangible results of a project cannot be achieved immediately, given that generally two years should be invested in laying a solid foundation for sustainable development. However, the final results of rural development projects, especially those focused on SWC and natural resources management, will be much more sustainable and failures can be avoided. Moreover, the cost of Phase 1 activities – not taking into account general project costs and staff – are very low. Hence, it is principally an investment in time and patience that a project must make.

With the lessons learned from the experiences of the project and the tools presented briefly in this chapter, laying a solid foundation for development in rural villages is now within reach of any project. More effective spending of donor aid through government and non-government organizations is possible if actual intervention strategies are modified. In the case of Bolivia, much work is still to be done at municipal level, where most decisions are made regarding where to invest and how to spend government and donor-aid budgets. The required foundation is solid only when support is obtained from municipalities for conducting follow-up activities.

Based on the experiences of the JGRC project, in the Department of Chuquisaca a new SWC project has been formulated. In this project, which will start in 2005 with the collaboration of Sucre University, special emphasis is given to the training of agronomy students in SWC and participatory techniques. These students, who for the most part are sons of farmers themselves, are the future extension workers in the region and the promoters of a progress-driven attitude in their home villages. In this sense, one of our farmers gives the best description of what is needed to reverse a future scenario of an "impossible" situation: "In order to make progress, we should walk together."

References

Ago, H. and C.A. Kessler. 1996. El manejo integrado y la planificación participativa para enfrentar la degradación de tierras, FAO, Santiago, Chile.

Bebbington, A. 1999. Capitals and Capabilities. A Framework for Analysing Peasant Viability, Rural Livelihoods and Poverty in the Andes. Policies That Work. Working Paper Series. Sustainable Agriculture Programme, International Institute for Environment and Development, London.

JGRC. 2002. Estudio de Sistematización: Prácticas tradicionales de conservación de suelos y aguas en Chuquisaca. Serie "Estudios e Investigación". Documento 1.

JGRC. 2003. Guía General: Estrategia de Intervención para el Desarrollo Rural Sostenible, basado en la conservación de suelos y aguas. Serie "Guías y Manuales". Documento 1.

UNDP. 1997. Índices de Desarrollo Humano y otros indicadores sociales en 311 municipios de Bolivia. 137 pp.

van Niekerk, N. 1997. La cooperación internacional y las políticas públicas: el caso de las zonas andinas de altura de Bolivia. Seminario sobre Estrategias Campesinas, Sucre, Bolivia.

Vargas, M. 1998. La migración temporal en la dinámica de la unidad doméstica campesina. *In:* Estrategias Campesinas en el Surandino Boliviano. KIT/CEDLA/CID.

World Bank. 2000. Bolivia: Poverty Diagnostic 2000, Green Cover Draft.

Zoomers, A. 1998. En busca de políticas apropiadas de desarrollo rural en la región andina: explorando los límites entre estrategias campesinas e intervenciones externas. *In:* Estrategias Campesinas en el Surandino Boliviano. KIT/CEDLA/CID.

20

Environmental Assessment of Conservation Initiatives within the Darby Creek Watershed in Ohio, USA

Ted L. Napier

Professor of Environmental Policy, Department of Human and Community Resource Development; The School of Natural Resources; and the Graduate Programme in Environmental Science of the Ohio State University; 2120 Fyffe Road, Columbus, Ohio 43210, USA.

E-mail: napier.2@osu.edu

ABSTRACT

Research was initiated within the Darby Creek watershed in central Ohio to assess the effectiveness of voluntary soil and water conservation programmes implemented using information, education, technical assistance and partial economic subsidies to motivate land owner-operators to adopt and to use conservation production systems. Studies using cross-sectional research design, time series methodology, and multiple group comparisons were implemented in 1991, 1994 and 1999. Study findings revealed that participation in conservation programmes implemented within the Darby Creek watershed was not significantly related to adoption of soil and water conservation production systems at the farm level. Factors commonly argued to be significantly related to adoption of soil and water conservation practices were shown not to be predictive of adoption behaviours at the farm level in any of the watersheds examined. Findings produced from each of the assessment studies consistently demonstrated the failure of the voluntary approach to encourage adoption of conservation production practices within the Darby Creek watershed and the other watersheds examined. Alternative public policy approaches are discussed.

INTRODUCTION

Research findings reported in this chapter are based on data collected from agriculturalists operating farms within the Darby Creek watershed located in central Ohio, from land owner-operators in watersheds in three Midwestern states, and from farmers in the Upper Scioto River watershed in central Ohio. The findings from all the watershed studies are compared with the findings for the Darby Creek watershed.

The first study was conducted within the Darby Creek watershed in central Ohio. The watershed is composed of two major tributaries of the Scioto River watershed: the Big Darby Creek watershed and the Little Darby Creek watershed. The Darby Creek watershed is adjacent to the city of Columbus, which is the capital and the largest city in Ohio. Columbus suburbs have been invading the Darby Creek watershed for decades and in a few years it is highly probable the entire watershed will become a part of nearby towns and cities. Figure 1 shows the Darby creek watershed in central Ohio.

Water quality in both the Big Darby Creek and the Little Darby Creek is considered to be very good by contemporary standards and both of the streams are used extensively for recreational fishing, waterfowl hunting, and canoeing. Contamination of water resources within both streams is

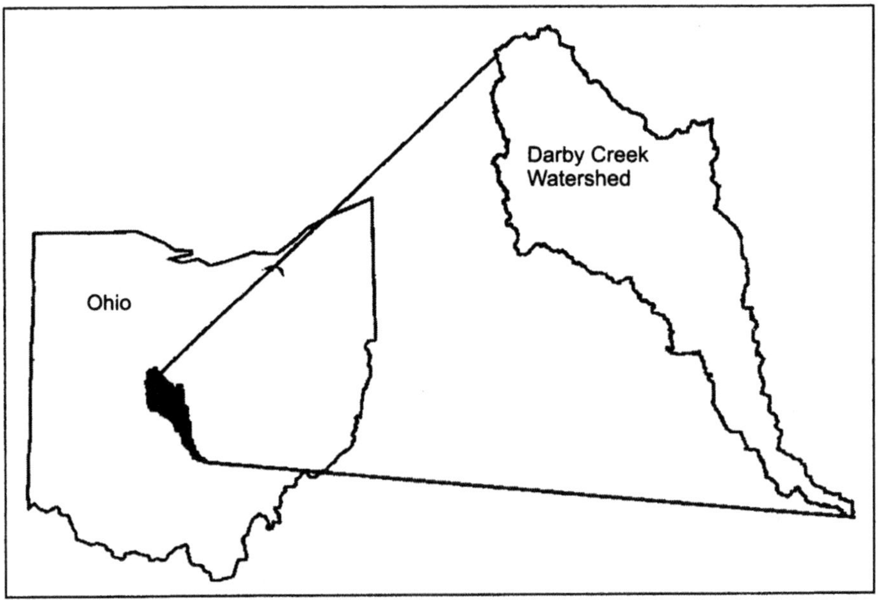

Fig. 1 Darby Creek watershed, central Ohio.

primarily due to fertilizer and pesticide runoff from agricultural land. Soil displacement due to erosion is relatively inconsequential owing to the flat topography of the watershed (Napier and Johnson, 1998). Land within the watershed is well suited for production agriculture. The soils are deep, fertile and well drained. The major agricultural crops produced are corn and soybeans, with wheat being introduced occasionally into the crop rotation system. Swine and cattle contribute a small percentage to farm income. There is animal production, but the vast majority of animals raised for market are produced by Amish farmers. Amish and Mennonite farmers constitute about 15% of the total population within the watershed and they are well-respected members of the communities in which they live (Sommers and Napier, 1993).

Technology-intensive production systems are the standard system of agricultural production within the watershed, except for Amish farmers. While Amish farmers continue to employ horse-drawn equipment for farming purposes and apply very small quantities of farm chemicals (Sommers and Napier, 1993), most farmers within the watershed, including Mennonite farmers, use large-scale agricultural technologies and apply large quantities of farm chemicals. Farmers who use technology-intensive production systems also achieve high levels of output per unit input, which results in low levels of residual chemicals that contribute to water pollution. Assessment of nutrient application rates per unit output within the study region strongly indicates that non-Amish farmers have been applying nutrients in an environmentally responsible manner and probably do not contribute disproportionately to chemical contamination of water resources (Napier and Sommers, 1996; Napier and Tucker, 2001a).

During the summer and fall of 1991, several conservation organizations decided that environmental problems within the Darby Creek basin were of sufficient magnitude to justify the commitment of extensive human and economic resources to address non-point pollution issues within the watershed. Under the leadership of the Ohio Cooperative Extension Service (OCES), a conservation programme was organized with the express purpose of motivating land owner-operators within the watershed to adopt and to use soil and water conservation (SWC) production systems. The primary goal of the project was to improve water quality within the Darby Creek watershed via use of education, information, technical assistance, and small economic subsidies (Napier and Johnson 1998). Such an approach is the standard means of implementing SWC production systems in the United States (Napier, 1990).

Conservation programmes employing information, education, technical assistance, and economic subsidy components were initiated within the watershed in 1991 to encourage adoption of conservation practices at the farm level. Information programmes were launched to inform land owner-operators about environmental problems created by use of inappropriate agricultural production systems. Educational programmes were provided to make potential adopters aware of existing conservation technologies and techniques that could be employed to address environmental issues identified within the watershed. A number of public relations activities were implemented to sensitize people within the region about the benefits associated with protecting water quality in the watershed. Canoe trips were organized to provide key political actors within the region and influential local residents the opportunity to observe in close proximity the beauty of the Darby Creek watershed. A number of special activities were organized within the watershed to demonstrate how local people could benefit economically from SWC efforts. Regional printed and electronic media were employed to document conservation efforts underway within the watershed in hopes of building public support for them among central Ohio residents. Educational programmes for children were developed and implemented to create an awareness of environmental problems within the watershed in hopes youngsters would be able to influence conservation behaviours of their parents. Small economic subsidies were offered to motivate farmers within the watershed to adopt no-till production technologies. The subsidies were designed to reduce the capital investment required for land owner-operators to adopt no-till equipment. Technical assistance was made available by state and federal agencies to ensure that conservation production systems would be implemented in an appropriate manner.

Huge quantities of economic and human resources were allocated to conservation efforts within the Darby Creek watershed. Information obtained from colleagues within the Natural Resources Conservation Service/United States Department of Agriculture (NRCS/USDA) revealed that more than one million dollars per year had been allocated for salaries to people working within the watershed. On a yearly basis, five NRCS/USDA agents were being assigned to work there. One and often two OCES agents were assigned to the watershed and one person was committed to watershed conservation efforts by the Nature Conservancy. Three full-time employees of the Ohio Department of Natural Resources, one-quarter time person from the Ohio Environmental Protection Agency, five graduate students funded by grants from Americorp, and a host of volunteers were actively working in the watershed. In addition to the field agents and volunteers, several hundred thousand dollars of funding per year were made available for conservation efforts within the watershed. The funding was used for cost-sharing and other economic subsidies. All of these economic and human resources allocated to the Darby Creek watershed were focused on approximately 1,170 partials of agricultural land and a total geographic area of approximately 150,000 ha. Had the economic resources allocated to conservation efforts within the Darby Creek been made available to purchase property rights to farmland that was contributing to water pollution, practically all of the highly erodible cropland within the watershed could have been purchased by the state for permanent retirement from production agriculture.

IMPACT ASSESSMENT OF CONSERVATION PROGRAMMES WITHIN THE DARBY CREEK WATERSHED

Data were collected in 1991 from 166 land owner-operators within the Darby Creek watershed to establish a data base for future comparisons of the conservation behavioural changes observed there. Data were collected using a systematic random sampling procedure demonstrated to be appropriate for drawing samples from dispersed populations (Napier, 1971). The sampling procedure consisted of the random selection of the first household each day data were collected within the watershed. After the initial household was chosen at random, every other occupied residence was selected and an adult member of each household was asked to complete a structured questionnaire (see Appendices 1 through 6 for examples of how study variables were assessed). Once a person consented to complete a questionnaire, the field data collector and the study participant agreed on a time agreeable to both for the completed questionnaire to be retrieved (usually two days later). Approximately 90% of all persons contacted by data collectors completed questionnaires that were included in the statistical modelling. The distribution of the sample drawn was recorded on detailed maps and monitored by the principal investigator throughout the data collection phase of the project. Examination of the maps at the conclusion of the data collection phase demonstrated the sample was proportionally distributed over the study area. Given the large sample size, the high response rate, and the wide distribution within the study area, it is argued the sample is representative of residents within the watershed.

The findings from the 1991 study indicated that respondents were primarily grain producers who tended to employ conventional agricultural production systems. Study participants reported they sometimes used fall tillage and fall application of fertilizer and seldom used no-till. Deep ploughing was used more often than other tillage systems and winter application of manure was sometimes used.

During the winter of 1991, conservation programmes were implemented within the Darby Creek watershed that emphasized education, information, technical assistance, and small economic subsidies, similar to those traditionally employed by the NRCS/USDA (Napier 1990). Numerous conservation programmes were implemented within the watershed that stressed raising awareness of pollution issues and the need to change existing production systems.

To assess the impact of the conservation programme implemented within the watershed, data were collected from 245 farmers in 1994 using the identical sampling technique employed in 1991 and the same measurement devices (Napier and Johnson, 1998). The response rate in 1994 was approximately 86%, which indicates little sampling error from refusals of residents to participate in the study. Examination of maps used to plot approximate location of study respondents revealed the sample was proportionally distributed throughout the watershed. Given the similarities of the methodologies employed in both studies, it was deemed appropriate to compare the findings for the two time periods. It was hypothesized that three years of conservation efforts within the watershed should have produced significant changes in conservation behaviours and that farmers would exhibit more extensive adoption of SWC production practices than reported in 1991.

Study findings revealed that significant changes in farm production systems had occurred over time. Unfortunately, all of the observed changes were not consistent with research expectations and some of the changes noted were not desirable from the perspective of SWC. Use of conservation practices such as no-till and conservation tillage increased between 1991 and 1994; however, environmentally degrading production practices such as use of fall tillage, fall application of fertilizer, and deep ploughing also increased in frequency. Study findings also indicated that the distribution of production practices over time became more polarized. Apparently conservation efforts undertaken within the watershed convinced farmers either to adopt or to reject conservation production practices and to cease experimentation with innovative conservation systems. Such an outcome is highly undesirable in terms of long-term conservation efforts because lack of experimentation with new production systems will retard more extensive adoption of conservation innovations. The conclusion drawn from the time series study was that SWC practices had not been uniformly adopted within the watershed. It was also observed that it may be more problematic in the future to motivate farmers to adopt SWC production systems because of polarization of conservation behaviours among land owner-operators within the watershed.

A conservation adoption index was calculated from information collected about production practices in use within the watershed in 1994. The response values attached to the frequency of use of each of the farm production practices assessed in the studies were summed to form a composite index of adoption behaviours for each respondent. No differential weighting values were employed during the computation of the conservation index scores at this time period. Cross-sectional analysis using correlation and regression analyses revealed that none of the variables included in the statistical modelling was useful for predicting adoption of conservation production practices at the farm level.

Factors such as access to conservation information, participation in fertilizer education programmes, participation in conservation clubs, use of extension agents, participation in local conservation programmes, access to government financial support, and access to government technical assistance were shown not to be useful for predicting adoption of conservation production systems at the farm level. Study findings strongly suggested that voluntary conservation initiatives, such as those implemented within the Darby Creek watershed, had failed to produce positive outcomes in terms of increased adoption of conservation practices. It was also concluded that economic and human resources allocated to develop and to implement the conservation programmes within the watershed had not been effectively used to motivate land owner-operators to adopt conservation production systems at the farm level (Napier and Johnson, 1998).

While the time series evaluation study reported here raised serious questions about the effectiveness of conservation programmes being implemented within the watershed, the findings were basically ignored by project proponents. Conservation efforts implemented in 1991 were continued even though research had demonstrated the efforts had made little impact on conservation behaviours of farmers within the watershed. Various organizations that were involved with the conservation programmes from the beginning were unwilling or unable to change the direction of existing conservation efforts.

MULTIPLE GROUP COMPARISON OF DARBY CREEK WATERSHED SWC PROGRAMMES

While the longitudinal study findings raised serious questions about the utility of the voluntary education, information, technical assistance, and subsidy approach employed in the Darby Creek watershed, researchers monitoring conservation behaviours within the watershed were reluctant to conclude that conservation programming had no positive impact on conservation adoption behaviours. It was reasoned that use of a conservation index composed of summed frequencies of use of all production practices without any differential weighting of the various types of conservation practices may have introduced measurement error into the statistical modelling and could have masked possible positive outcomes. To control for the possibility that measurement error could have been operative, a study was organized to develop a better composite indicator of adoption behaviours and to replicate the Darby Creek watershed study. It was also reasoned that multiple group comparisons would provide a better means of comparing what was occurring within the study watershed than using time series data collected within the Darby Creek watershed.

A study was organized in 1998 to collect data from 1,011 land owner-operators within three watersheds in three Midwestern states to assess use of conservation production systems (Napier, 2000; Napier and Tucker, 2001a,b; Napier, et al., 2000; Tucker and Napier, 2001; Robinson and Napier, 2002). Systematic random samples of land owner-operators within two of the three study watersheds were drawn using the same sample selection procedures employed in the two previous Darby Creek studies. Respondents were drawn from the Lower Minnesota River watershed located in southeastern Minnesota and from the Maquoketa River watershed in northeastern Iowa using the systematic sampling approach described above. A systematic random sample was not drawn in the Darby Creek watershed in 1998 because so few farmers remained in the area at the time of the data collection that it was impossible to systematically sample the group and have a sufficiently large number of respondents needed for statistical modelling. Therefore, anyone identified as being engaged in production agriculture was included in the sampling in the Darby Creek watershed. Since 1991, when the conservation programme was first initiated within the Darby Creek watershed, the number of farmers had been reduced by almost 1,000 because of transition of farm land to non-farm uses and the tremendous increase in the average number of hectares farmed. Data were collected in all three watersheds during the fall of 1998 and winter of 1999. A total of 551 land owner-operators were surveyed in Minnesota, 355 in the Iowa watershed, and 105 in the Ohio watershed. The locations of the Iowa and Minnesota watersheds are presented in Fig. 2.

The primary criterion for selection of watersheds for investigation was the extent of involvement in conservation programming by watershed farmers. Farmers within the Darby Creek watershed had

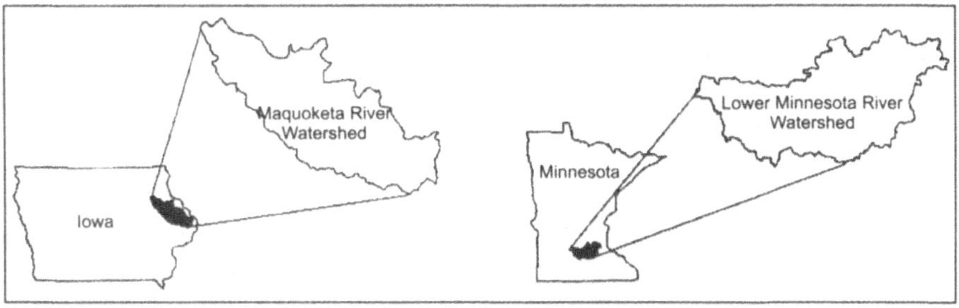

Fig. 2 Location of the Iowa and Minnesota watersheds.

received substantially more conservation programming than farmers in the other two watersheds. Land owner-operators within the Iowa watershed had received more than land owner-operators within the Minnesota watershed. If conservation programme participation is important as it is proclaimed to be for motivating land owner-operators to adopt and to use conservation production systems, then it was expected that farmers within the Darby Creek watershed would report significantly more adoption of conservation production practices than farmers within the other two watersheds. It was also expected that Iowa farmers would report more adoption of conservation production systems than Minnesota farmers.

The topography of the Minnesota and the Iowa watersheds differed considerably from that of the Darby Creek watershed. Land within the Iowa watershed ranged from slightly rolling to steep slopes. The Minnesota watershed ranged from flat land near the river to steep slopes to a higher flat plain. Soil within both the Iowa and the Minnesota watersheds was deep and fertile. In both watersheds erosion was a significant environmental issue and pollution from farm chemicals was prevalent in the waterways. Farmers within the Iowa watershed produced both feed grains and animals for market. Minnesota farmers produced feed grains and dairy products.

Data were collected from study participants in all three watersheds using standardized measurement devices. Adoption behaviour was assessed using the same measurement instruments that had been used in the two previous Darby Creek watershed studies. Factors assessed in the study are as follows: farm production systems presently used on the farm at the time of the data collection, access to information systems, access to government funding for implementing SWC programmes, access to technical assistance to implement SWC programmes at the farm level, characteristics of the farm enterprise, characteristics of the primary farm operator, perception of risk associated with using farm chemicals, perception of return to investment in SWC, and anticipated changes in land uses due to inter-generational transfer of farmsteads.

Study findings revealed that the average number of acres being cultivated in the Darby Creek watershed was about twice the number being farmed in the other two watersheds. While average acreage owned was similar in all three watersheds, rented acreage in the Darby Creek watershed was more than twice that in the other two watersheds. This is due primarily to absentee ownership of land resources in the Darby Creek watershed. While many land owners had retired from farming, they retained ownership of the land resources to capitalize on the rapidly increasing land values associated with suburban sprawl. Agricultural land owners rented their land to individuals still active in production agriculture to avoid taxes assessed on the basis of potential use. Tax rates on agricultural land are a small fraction of the tax rates on residential property.

Access to technical assistance was similar for farmers in the Iowa and the Darby Creek watersheds but was very low in the Minnesota watershed. The same finding was noted for access to government economic support for implementation of conservation programmes at the farm level. The average age of primary farm operators was similar for all three watersheds, as were educational achievement level and years farming. Gross farm income was highest within the Darby Creek watershed and second highest in the Iowa watershed (Napier, 2000; Napier et al., 2000).

Findings for farm production practices employed at the time of the study revealed significant differences among the farmers in the three watersheds. Fall tillage was used by more than 80% of the farmers in the Minnesota watershed every year and approximately one-third of the farmers in the Iowa and the Darby Creek watersheds. No-till was used by more than 50% percent of the farmers in the Darby Creek watershed every year, about 12% of the farmers in the Iowa watershed, and less than 3% of the farmers in the Minnesota watershed. Fall application of fertilizer was used by about one-third of the farmers every year in the Darby Creek watershed, about 14% of the farmers in the Iowa watershed, and about 24% of the farmers in the Minnesota watershed. Winter application of manure was used yearly by about two-thirds of the Iowa farmers, by about 46% of the Minnesota farmers, and by about 31% of the Darby Creek farmers. Deep ploughing was used by more than 53% of the Minnesota farmers on a yearly basis, about 22% of the Darby Creek farmers, and less than 6% of the Iowa farmers. Chisel ploughing was used on a yearly basis by more than 51% of the Minnesota farmers, less than 32% of the Iowa farmers, and 29% of the Darby Creek farmers.

A composite conservation adoption index was computed from the responses to the various agricultural practices assessed in the study and was examined using regression modelling (see Table 2 footnote for computation of this index). Data for each state was examined independent of the others to eliminate the possibility of an averaging effect over the study watersheds. Study findings revealed that little of the variance in composite conservation index scores was explained by a number of factors often suggested as being strong motivators of adoption of SWC production systems at the farm level (Napier, 2000; Napier and Tucker, 2001b).

The findings from the multiple watershed study again brought into serious question the utility of using the voluntary information, education, technical assistance, and subsidy approach to motivate land owner-operators to adopt and to use SWC production systems at the farm level. While farmers within the Darby Creek watershed had received much more human and economic assistance to implement SWC programmes than farmers in the other two watersheds, study respondents in the Darby Creek watershed did not consistently report more extensive use of conservation production systems. It was concluded that the voluntary conservation approach commonly used in the US was not very effective for motivating land owner-operators to adopt and to use conservation production systems in any of the study watersheds (Napier, 2000).

One of the most important conclusions drawn from the multiple-state study was that use of SWC production systems varied significantly among the study watersheds. No-till had been adopted extensively within the Darby Creek watershed but not in either of the other two watersheds. Chisel ploughing had been adopted by Minnesota farmers even though conservation programmes had not been emphasized within the watershed. The common assertion that conservation programmes produce more extensive adoption of conservation production systems was not consistently supported by study findings.

Another important observation was the lack of whole farm planning within any of the three study watersheds. While respondents in each of the watersheds had adopted some conservation production practices, they also continued to use production practices that are considered to degrade soil and water resources. Practices such as mould-board ploughing, fall application of fertilizer, and winter application of manure could negate the positive impacts of conservation tillage or no-till production practices. It was concluded from the study findings that assessment of conservation performance at the farm level requires consideration of all production systems in use and not whether or not a specific farming practice has been adopted. To conclude that a land owner-operator is a good steward of the land because he/she has adopted no-till production system is not valid. The same can be said of farmers who adopt and use other forms of conservation tillage systems such as chisel ploughing with one-third ground cover at planting time. The total farm production system must be considered before any conclusion can be drawn about conservation behaviours at the farm level.

COMPARISON OF TWO WATERSHEDS IN CENTRAL OHIO

While the findings from the multiple watershed project contributed extensively to the assessment of the effectiveness of the conservation programmes implemented within the Darby Creek watershed, concern was expressed that variability among the three watersheds could have obscured positive adoption behaviours among land owner-operators within the Darby Creek watershed. To control for possible error introduced by comparing conservation behaviours in watersheds located so distant from each other, a study was organized to collect data from a watershed located close to the Darby Creek watershed. In this way, variability introduced by distance, climate, farm specialization, socio-demographic characteristics of the primary farmer operator, and numerous other variables could be controlled.

Farmers operating agricultural land within the Upper Scioto River watershed were chosen as the comparison group. Data were collected from 113 land owner-operators living within three counties located north of Columbus through which the Scioto River passes. Data were collected during the late winter of 1999 and early summer of 1999. The Upper Scioto River watershed is adjacent to the Darby Creek watershed. Farmers there exhibit many of the same characteristics as those operating farms within the Darby Creek watershed. In both watersheds farmers tend to specialize in grain production, farms tend to be large, farmers have been operating their own farms for many years, technology-intensive production systems are the norm, and the physical characteristics (soil type, climate, topography) are basically the same. The same questionnaire used to collect data within the Minnesota, Iowa and Darby Creek watersheds was employed to collect data within the Upper Scioto River watershed. Systematic random sampling was used to select study participants within the Upper Scioto River watershed, as had been done in the three other watersheds.

Study findings revealed that farmers within the Darby Creek and the Upper Scioto River watersheds were similar on many of the factors chosen for comparison purposes. Characteristics of respondents for the two study groups are presented in Table 1.

Table 1 Characteristics of respondents in the Darby Creek watershed (N = 105) and the Upper Scioto River watershed (N = 113).

Characteristic	*Darby Creek*	*Upper Scioto*
	Age in years	
Mean	48.6	51.3
SD	11.9	13.0
	Education in years	
Mean	12.7	12.6
SD	2.1	2.4
	Years farming	
Mean	23.8	30.0
SD	13.4	14.3
	Hectares usually cultivated	
Mean	334.3	303.9
SD	362.6	319.6
	Percent grain farmer	
Mean	68.6	88.1
SD	40.1	24.2
	Gross farm income (%)	
< $60,000	21.9	24.8
60,000-119,999	18.1	21.3
120,000-179,999	12.4	27.4
180,000-239,999	8.7	5.4
240,000-299,999	4.8	8.0
300,000-359,999	2.9	0.9
> $359,999	16.2	8.0
Missing data	15.2	4.4
	Debt-to-asset ratio (%)	
0-10	32.4	23.9
11-20	12.4	19.5
21-30	9.5	17.7
31-40	7.6	15.0
41-50	4.8	8.0
51-60	6.7	5.3
61-70	2.9	0.0
71-80	1.9	0.0
81-90	1.0	0.9
91-100	0.0	0.0
Missing data	21.0	9.7
	Received technical assistance (%)	
Yes	27.6	21.2
No	72.4	78.8
	Received government funding for conservation (%)	
Yes	21.0	8.8
No	79.0	91.2

Findings presented in Table 1 show that in both watersheds the average age of the primary farm operator was almost identical, average education was basically the same, the average number of hectares usually farmed was very large, gross farm income was high, and debt-to-asset ratios were low (Napier and Bridges, 2002).

Respondents in the Upper Scioto River watershed reported low participation in government economic programmes to encourage adoption of conservation production systems and they also indicated they received less technical assistance. These findings were expected, since the Upper Scioto River watershed had not been targeted to receive the conservation programmes that land owner-operators within the Darby Creek watershed had been provided. If conservation programmes implemented within the Darby Creek watershed were achieving their objectives, it was expected that adoption of SWC production systems would be significantly higher among the Darby Creek respondents.

Findings for production practices employed at the time data were collected are presented in Table 2 and show that variability existed between the two watersheds in terms of conservation behaviours reported by study respondents. Study findings for agricultural production practices are presented in Table 2

Fig. 1

Findings presented in Table 2 demonstrate that Darby Creek farmers did not consistently report greater use of conservation practices (Napier and Bridges, 2002). Examination of specific production practices in use at the time of the data collection revealed that fall tillage was employed yearly or every other year by a larger percentage of farmers in the Darby Creek watershed than in the Upper Scioto River watershed. Similarly, deep ploughing yearly or every other year was much higher within the Darby Creek watershed than in the Upper Scioto River watershed. These production practices are most often defined as contributing to degradation of soil and water resources. No-till was used yearly or every other year by a larger percentage of farmers in the Darby Creek watershed than in the Upper Scioto River watershed. No-till is considered to be a conservation production practice, if implemented appropriately. Chisel ploughing was used more extensively yearly or every other year in the Darby Creek watershed than in the Upper Scioto River watershed. There was little difference between the two groups of farmers in terms of winter application of manure and for use of banded application of fertilizer. Soil testing was reported to be conducted yearly or every other year by about 40% of the respondents in each watershed. Crop rotation was reported to occur yearly or every other year by a larger percentage of Darby Creek respondents.

Table 2 Frequency of use of agricultural production practices in the Darby Creek (N = 105) and Upper Scioto River (N = 113) watersheds (in percentages)

	Never use	*Once every 5 yrs*	*Once every 4 yrs*	*Once every 3 yrs*	*Every other yr*	*Every yr*	*Missing data*
				Fall tillage*			
Darby	18.1	6.7	3.8	13.3	21.0	34.3	2.9
Upper Scioto	9.7	8.0	7.1	23.0	22.1	29.2	0.9
				Fall application of fertilizer*			
Darby	23.8	2.9	2.9	13.3	14.3	35.2	7.6
Upper Scioto	19.5	8.0	8.0	17.7	15.0	31.0	0.9
				Soil testing+			
Darby	8.6	7.6	3.8	32.4	22.9	20.0	4.8
Upper Scioto	8.0	4.4	8.8	33.6	13.3	25.7	6.2
				No-till+			
Darby	19.0	1.9	1.0	5.7	14.3	50.5	7.6
Upper Scioto	17.7	5.3	12.4	9.7	11.5	38.1	5.3
				Chisel ploughing with one-third ground surface covered with residue at planting+			
Darby	24.8	6.7	3.8	12.4	19.0	28.6	4.8
Upper Scioto	34.5	8.0	8.8	14.2	8.8	25.7	0.0
				Ridge tillage+			
Darby	98.1	1.9	0.0	0.0	0.0	0.0	0.0
Upper Scioto	85.0	6.2	0.9	0.9	0.0	1.8	5.3
				Deep (mould-board) ploughing*			
Darby	46.7	11.4	1.0	2.9	7.6	21.9	8.6
Upper Scioto	55.8	15.9	7.1	7.1	0.9	8.8	4.4
				Winter application of manure*			
Darby	53.3	3.8	2.9	1.9	1.0	30.5	6.7
Upper Scioto	44.2	6.2	3.5	8.8	10.6	19.5	7.1
				Banded (in furrow) application of fertilizer+			
Darby	39.0	7.6	2.9	1.0	2.9	40.0	6.7
Upper Scioto	40.7	2.7	6.2	12.4	7.1	31.0	0.0
				Side-dressing of fertilizer during growing season+			
Darby	34.3	3.8	2.9	7.6	6.7	39.0	5.7
Upper Scioto	27.4	3.5	8.0	14.2	8.8	35.4	2.7
				Banded application of herbicides+			
Darby	67.6	1.0	1.0	3.8	1.0	16.2	9.5
Upper Scioto	68.1	2.7	7.1	2.7	3.5	9.7	6.2

(Table 2 Contd.)

(Table 2 Contd.)

			Mechanical weed control+				
Darby	44.8	4.8	13.3	7.6	2.9	26.7	0.0
Upper Scioto	58.4	4.4	3.5	10.6	6.2	15.9	0.9
			Use of nitrification inhibitor+				
Darby	57.1	1.9	2.9	4.8	5.7	16.2	11.4
Upper Scioto	62.8	6.2	2.7	4.4	6.2	10.6	7.1
			Crop rotation+				
Darby	1.9	0.0	1.9	5.7	11.4	71.4	7.6
Upper Scioto	2.7	0.0	1.8	12.4	18.6	54.0	10.6
			Contour planting+				
Darby	81.0	0.0	1.0	0.0	1.9	6.7	9.5
Upper Scioto	77.0	5.3	3.5	1.8	1.8	1.8	8.8
			Buffer strips+				
Darby	61.9	2.9	1.9	0.0	1.0	20.0	12.4
Upper Scioto	70.8	7.1	3.5	2.7	1.8	10.6	3.5
			Integrated pest management+				
Darby	69.5	13.3	3.8	1.9	0.0	11.4	0.0
Upper Scioto	71.7	4.4	3.5	3.5	2.7	10.6	3.5
			Precision farming+				
Darby	85.7	1.9	1.0	0.0	1.0	10.5	0.0
Upper Scioto	76.1	0.9	3.5	3.5	6.2	7.1	2.7

+ Responses weighted 0 through 5 with *never use* receiving a value of 0 and *every year* a value of 5.

* Responses weighted 5 through 0 with *never use* receiving a value of 5 and *every year* a value of 0.

Fall tillage, deep ploughing, and winter application of fertilizer were defined as being the worst types of farm production practices in terms of environmental degradation. No-till and conservation tillage (chisel ploughing with one-third ground cover at planting time) were defined as being the most environmentally benign production systems. Weighting values were multiplied by 2 to given differential weight to adoption of these practices. A panel of agricultural experts was used to determine what variables should be weighted differentially and the value used to differentially weight them. The weighting values for all of the 18 practices assessed in the study were summed to form a composite Conservation Production Index (CPI) score for each respondent. The CPI values were designated as the dependent variable for statistical modelling.

The study findings focused on adoption and use of specific agricultural production practices demonstrated that farmers in both watersheds had adopted some SWC production systems. The findings also revealed that study respondents in both watersheds were also using some degrading production practices.

A conservation adoption index was computed using all of the production practices examined in the study as it had been constructed for the Minnesota, Iowa and Darby Creek comparison (Napier and Bridges, 2002). The index scores for the Darby Creek and Upper Scioto River watersheds were compared using one-way analysis of variance to determine whether significant differences existed between the two study groups. Analysis of variance findings demonstrated that the two groups were not significantly different at the 0.05 level in terms of the adoption index scores. This finding again brings into question the legitimacy of employing the education, information, technical assistance, and subsidy approach to motivate land owner-operators to adopt and to use SWC production systems.

Study findings strongly indicate that extensive investment in human and economic resources within the Darby Creek watershed to motivate land owner-operators to adopt and to use SWC production systems at the farm level was not successful in accomplishing that objective. When compared with study groups that had received much less attention from conservation agencies, land owner-operators within the Darby Creek did not consistently report significantly greater adoption of conservation production practices.

POLICY IMPLICATIONS OF MONITORING STUDIES WITHIN THE DARBY CREEK WATERSHED

Evidence secured from all of the studies undertaken within the Darby Creek watershed to assess the impacts of conservation programming on farmer adoption behaviours consistently revealed that voluntary conservation programmes failed to accomplish the stated goal of motivating land owner-operators to adopt and to use conservation production systems at the farm level. Millions of dollars of public economic resources and extensive human resources have been allocated to efforts within the Darby Creek watershed that have not resulted in demonstrably greater use of conservation production systems. Had the resources been used differently, it is highly likely that the results would have been much more positive in terms of adoption of conservation practices within the watershed.

No consideration was given within the Darby Creek conservation initiative to permanently retiring environmentally sensitive cropland. Limited economic resources could have been used to purchase cropping rights to sensitive agricultural land and the land would have been protected from degradation. Farm land could have been purchased using the economic resources employed to fund the programmes shown to have been ineffective and conservation corridors established that would have protected water quality for decades. Economic resources could have been used to subsidize farmers to reduce fertilizer application rates, which would have reduced nutrient contamination of the waterways within the watershed.

Another option that could have been implemented within the Darby Creek watershed is the targeting of specific environmental problems and the formation of public policies to address them. For example, public policies to protect streambanks could have been established. Conservation funds could have been focused on streambank protection and riparian corridors established along waterways. Such targeting would have been very useful, if cropping rights and development rights had been secured to prevent land use changes in the future.

Economic resources used to provide information, technical assistance, and partial economic subsidies to local farmers could have been used to hire conservation professionals to monitor farming operations within the Darby Creek watershed. Such an approach would have acted as an incentive for land owner-operators to adopt and to use production systems that protect environmental quality. Formation of specific public policies focused on protection of water quality within the watershed combined with on-farm monitoring would have been the most effective means of assuring compliance with environmental objectives within the watershed. However, such an approach would have been the least acceptable method of achieving environmental goals among local farmers. Land owner-operators strongly support economic subsidy approaches and strongly oppose command and control approaches to achieve conservation objectives.

Impact assessment findings produced within the Darby Creek watershed in Ohio clearly indicate that traditional approaches for implementing SWC are probably no longer useful for motivating farmers to adopt and to use conservation production systems at the farm level. This assumes, of course, that traditional approaches were at one time useful for that purpose. The findings generated within the other watersheds also strongly suggest that alternative policy approaches should be considered.

One of the approaches that should be considered in the future is the use of public conservation policies combined with command and control mechanisms for enforcement. Such an approach will require significant changes in attitudes among landowner-operators and will require courage on the part of public policymakers to create and implement such policies. Given the present orientation among agriculturalists and public policymakers within the United States, such conservation policies and programmes will not be developed and implemented in the near future.

References

Napier, T.L. 1971. The impact of water resource development upon local rural communities: Adjustment factors to rapid change. Doctoral Dissertation. The Ohio State University.

Napier, T.L. 1990. The evolution of US soil conservation policy: From voluntary adoption to coercion. pp. 627-644. *In:* J. Boardman, I.D.L. Foster, and J.A. Dearing. [eds.] Soil Erosion on Agricultural Land. John Wiley and Sons, New York.

Napier, T.L. 2000. Use of soil and water protection practices among farmers in the North Central region of the United States. J. Am. Water Resour. Assoc. 36(4): 723-735.

Napier, T.L. and T. Bridges. 2002. Adoption of conservation production systems in two Ohio watersheds: A comparative study. J. Soil Water Conservat. 57(4): 229-235.

Napier, T.L. and E.J. Johnson. 1998. Impacts of voluntary conservation initiatives in the Darby Creek Watershed of Ohio. J. Soil Water Conservat. 53(1): 78-84.

Napier, T.L. and D.G. Sommers. 1996. Farm production systems of Mennonite and non-Mennonite land owner-operators in Ohio. J. Soil Water Conservat. 51(1): 71-76.

Napier, T.L. and M. Tucker. 2001a. Use of soil and water protection practices among farmers in three Midwest watersheds. J. Environ. Manag. 27(2): 269-279.

Napier, T.L. and M. Tucker. 2001b. Factors affecting nutrient application rates within three Midwestern watersheds. J. Soil Water Conservat. 56(3): 220-228.

Napier, T.L., M. Tucker and S. McCarter. 2000. Adoption of conservation production systems in three Midwest watersheds. J. Soil Water Conservat. 55(2): 123-134.

Robinson, J.R. and T.L. Napier. 2002. Adoption of nutrient management techniques to reduce hypoxia in the Gulf of Mexico. Agr. Syst. 72: 197-213.

Sommers, D.G. and T.L. Napier. 1993. Comparison of Amish and Non-Amish Agriculturalists: A Diffusion-Farm Structure Perspective. Rural Sociol. 58(1): 130-145.

Tucker, M. and T.L. Napier. 2001. Determinants of perceived agricultural chemical risk in three watersheds in the Midwestern United States. J. Rural Stud. 17: 219-233.

Appendix 1. Measurement instrument used to assess perceived level of risk posed by farm chemicals.

Factor assessed	*Possible responses*								
	No risk		*Little risk*		*Moderate risk*			*Serious risk*	
Water quality	0	1	2	3	4	5	6	7	8
Food safety	0	1	2	3	4	5	6	7	8
Food quality	0	1	2	3	4	5	6	7	8
Health of applicator	0	1	2	3	4	5	6	7	8
Health of farm animals	0	1	2	3	4	5	6	7	8
Wildlife	0	1	2	3	4	5	6	7	8
Destruction of beneficial plants	0	1	2	3	4	5	6	7	8
Destruction of beneficial insects	0	1	2	3	4	5	6	7	8
Human health	0	1	2	3	4	5	6	7	8
Air quality	0	1	2	3	4	5	6	7	8

A composite index of perceived risk associated with farm chemicals was computed by summing the values circled by respondents.

Appendix 2. Measurement instrument used to assess farm specialization.

Corn ____	Oats____	Swine____	Hay____	Other crops____
Soybeans____	Dairy____	Poultry____	Fruit____	
Wheat____	Beef____	Sheep____	Vegetables____	

Respondents were asked to indicate the percentage of gross farm income received from each farm product listed during the last 3 years. Only farm income was assessed. Grain specialization as a variable was computed by summing the percentages for corn, soybeans, wheat, and oats. Animal specialization was computed by summing the percentages for dairy, beef, swine, poultry, and sheep. Other specialization was computed by summing the percentages for hay, fruit, vegetables, and other crops.

Appendix 3. Measurement instrument used to assess perceived impacts of adoption of SWC production systems at the farm level.

Possible responses

Large decrease	Moderate decrease	Slight decrease	No change	Slight increase	Moderate increase	Large increase
1	2	3	4	5	6	7
How will production COSTS change if farm operated in a manner to conserve soil and water resources?						
1	2	3	4	5	6	7
How will farm OUTPUT change if farm operated in a manner to conserve soil and water resources?						
1	2	3	4	5	6	7

Respondents were asked to circle a number along the continuum from large increase to large decrease that best represented their perceptions about the impact of operating their farm in a conservation-oriented manner. The values selected by respondents were used in statistical modelling.

Appendix 4. Measurement instrument used to assess expected financial return to investments in SWC.

Highly unlikely		Somewhat unlikely		Slightly unlikely		Slightly likely		Somewhat likely		Highly likely
0	1	2	3	4	5	6	7	8	9	10

Respondents were asked to indicate the likelihood that they would receive a return to their investments in SWC. They were asked to circle a number along a continuum that best represented their expectations. The value circled was used in the statistical modelling.

Appendix 5. Measurement instrument used to assess sources of information about SWC.

Hired consultant___	Ohio Department of Natural Resources___	Financial institutions___
EPA___	Soil conservation districts___	Friend___
NRCS___	Extension Service___	Family member___
FSA____	University programmes____	USGS___
Conservation club___	Nature Conservancy___	Farm implement dealer___
Farm cooperative___	Neighbor___	Agri-chemical dealer___
Mass media___	Grain elevator operator___	

Respondents were asked to indicate the frequency of use of various sources of information by noting the number of times they had used each information source during the previous year.

Factor analysis was used to develop several variables that measured use of different types of information systems. Factor analysis revealed that several of the sources clustered together to form unique combinations of information systems. Identified factor clusters were used to compute composite measures of different information systems.

Appendix 6. Measurement devices used to assess characteristics of the primary farm operator.

Age at time of data collection ____ age in years

Education at time of data collection ____ years of formal education

Years primary farm operator has been operating his/her own farm ____ years

Debt to asset ratio (respondents were instructed how to compute the percentage) _____ percent

Acres usually cultivated each year ______ number of acres

Acres owned ______ number of acres

Acres rented ______ number of acres

Days usually worked off-farm for wages each year _____ number of days

Received government technical assistance to implement soil and water conservation programmes at the farm level _____ yes ____ no

Received government economic assistance to implement soil and water conservation programmes at the farm level _____ yes _____ no

The data entered by respondents were used for statistical modelling.

21

Monitoring and Evaluation of Watershed Initiatives: Lessons from Landcare in Australia

Allan Curtis

Professor, Integrated Environmental Management and Director of the Institute for Land, Water and Society, Charles Sturt University, P.O. Box 789, Albury, NSW 2640, Australia.
E-mail: acurtis@csu.edu.au

ABSTRACT

Landcare has been operating as a national programme in Australia for 16 years. With over 4,000 local groups involving in excess of 120,000 participants across different state jurisdictions, Landcare provides a powerful example of the potential of local watershed organizations. At the same time, there have been important criticisms of Landcare, including that insufficient attention has been given to articulating the roles and responsibilities of different natural resource management actors, and that critical issues affecting the long-term capacity of Landcare have not been addressed. In this chapter I draw on my experience over almost 20 years as a researcher and as a Landcare participant to identify some key lessons from the Australian experience with monitoring and evaluation of Landcare. In doing this I attempt to identify some of the topics that evaluations of watershed groups should examine and reflect on the opportunity for monitoring and evaluation to influence policy and management.

INTRODUCTION

Landcare has been operating as a national programme in Australia for 16 years. With over 4,000 local groups involving in excess of 120,000 participants across different state jurisdictions, Landcare provides a powerful example of how to establish local watershed organizations (Curtis et al., 2002). At the same time, there have been important criticisms of Landcare, including that it is a case of "too little too late", that government is shifting responsibility to communities, and that Landcare has failed to address the structural issues leading to land and water degradation (Lawrence, 2000). At the heart of these critiques are fundamental questions for evaluators and programme managers about the outcomes that can be expected from watershed group activity (Campbell, 1992; Chamala, 1995; Kenney, 1995). The Landcare experience with these evaluation issues appears to offer important lessons for an international text exploring the theory and practice of monitoring and evaluation for watershed initiatives.

After providing some background on Landcare in Australia, the chapter will be structured around a discussion of what I believe are the key lessons from Landcare for monitoring and evaluation of watershed initiatives. Although there will be some reflections on the achievements and limitations of Landcare, readers are referred to other literature for a more comprehensive discussion of those topics.

Much of the following discussion is based on my research and that of research partners, including work to evaluate the Landcare programme while a university researcher. I will also be drawing on my experience, first as a Landcare participant and a Ministerial appointee to regional natural resource management (NRM) committees and second as a researcher working in the Commonwealth's Department of Agriculture, Fisheries and Forestry-Australia, the lead agency for the National Landcare Programme (NLP).

BACKGROUND

Landcare as National Programme

Landcare can be viewed as part of a lengthy process by which Australians adapted emerging theories of rural development to an Australian context (Curtis, 1998). Landcare groups emerged first in 1986 in the state of Victoria, where a small vanguard of soil conservationists, extension agents and farmers were attracted by the core elements of rural development theory that emphasized (1) self-help supported by change agents, (2) human resource development rather than technology transfer, (3) public participation, and (4) cooperative efforts at the local community scale (Curtis 1998). Early experience with Landcare, first in Victoria and then the state of Western Australia, confirmed overseas evidence that participation through local organizations could accomplish broad-based rural development (Chambers, 1983; Esman and Uphoff, 1984).

Recognizing the potential of Landcare groups as a potent force for improved NRM, in 1988 the federal government committed 360 million Australian dollars for the Decade of Landcare programme. The programme initially had limited government funding available for coordination and project work, principally for education and demonstration activities to be undertaken by groups in collaboration with state agency advisors.

Establishment of the five-year, $1.25 billion Natural Heritage Trust (NHT) in 1997 significantly altered the course of Landcare. The NHT employs cost-sharing principles that identify the community and private benefits from specific work funded by government on private land. The NLP was one of the programmes funded under the NHT and the Trust's extension. In 1999-2000, the NHT funded 870 NLP projects worth $71 million of the $200 million spent through the NHT. In 2003, the NLP again became a stand-alone programme, with an annual budget of $38 million. These funds supported on-ground work by community groups, the National Landcare Facilitator, a peak Landcare advisory body, and some innovative industry partnerships. The Australian Government committed similar levels of funding for the period up to 2007.

Landcare Group *Modus Operandi* and Activity

Most Landcare groups have developed in rural areas, membership being voluntary and open to any local person. Groups frequently operate at small watershed or community scales and are encouraged to view their activities holistically, using a systems approach. They have no legislative backing and are only informally linked to local government and regional planning bodies. While the focus of group activity is usually on privately owned or leased land managed by group members, groups also work on roadsides, reserves and other public lands.

Given that Landcare group *modus operandi* is not prescribed, there is a great variety in the activities of groups. Campbell's (1994) text provides a number of informative case studies. The summary that follows is necessarily a highly generalized picture. Groups are involved in a variety of rural development activities facilitating learning and action:

- Meetings held to discuss issues, identify priorities, liaise with agency staff, prepare funding submissions and debate resource management issues;

- Workshops conducted to develop property and catchment plans and enhance management and planning skills;
- Field days, farm walks and demonstration sites to identify best practices;
- Education and promotional activities such as tours, conferences, workshops, newsletters and field guides to facilitate dialogue and information exchange;
- On-ground actions such as tree planting and seed collection, building salinity and erosion control structures, pest plant and animal control, and erecting fencing to manage stock and feral animal access to habitats.

The benefits of Landcare membership for landholders were seen as the following:

- Sharing problems and ideas;
- Working more effectively to address common problems;
- Learning about land management;
- Planning at the property and catchment scale so that resource management is based upon a shared understanding of important physical, social and economic processes operating within and beyond the farm gate;
- Accessing financial and technical assistance from government; and
- Having greater opportunities for social interaction (Campbell, 1994; Curtis and De Lacy, 1995).

Local Landcare Groups and Regional Planning

Although the federal government has greater financial resources, NRM authority rests primarily with the six Australian states and the Northern Territory and Australian Capital Territory. By 1992, most Australian states had established regional Catchment Management Committees (CMC). The committees are now important bodies, largely responsible for developing and implementing the regional catchment strategies that guide the expenditure of state and federal NRM funds. In many instances the community representatives on these regional groups are ministerial appointees and usually include representatives from Landcare groups, local government, and other NRM stakeholders. Landcare groups have become an important delivery mechanism for the regional CMC but they are not formally linked to them (Curtis and Lockwood, 2000; Ewing, 2000).

LESSON FOR MONITORING AND EVALUATION

Articulate the Programme Logic

Evaluators attempt to assess the value of programmes in order to improve their efficacy in ameliorating social problems (Cook and Shadish, 1986). Programme evaluation is an important but challenging undertaking and there is a plethora of expert opinions about how it should be accomplished (Patton, 1990). As a researcher approaching the task of evaluating Landcare in the early 1990s, the programme provided particular challenges. There were the now familiar issues of what could reasonably be expected of a relatively small number of volunteers operating with limited resources for short periods of time to address difficult, long-term issues where there is considerable uncertainty about how to proceed. There were other issues also likely to bedevil evaluators of watershed initiatives, including the large numbers of Landcare stakeholders, some of whom held opposing expectations of the programme, considerable variation in programme implementation across the Australian states and between local watersheds, and, in the beginning, little programme documentation about programme objectives.

There is a substantial body of literature that identifies the unravelling of programme logic or underlying theory as the critical first step in programme evaluation (Rossi and Freeman, 1985; Chen, 1990; Prosavac and Carey, 1992). Making the programme logic explicit is seen as the first step in identifying intermediate objectives that can be employed to assess programme effectiveness. Evaluators can turn to a number of sources in their efforts to unravel programme logic: they can approach staff, clients, and other stakeholders for their views; they can review literature on the programme under scrutiny or similar programmes; they can examine programme documentation; and they can observe the programme in action (Rossi and Freeman, 1985; Chen, 1990). To a large extent I accomplished this task through the scrutiny of programme documents, the examination of relevant literature, including on rural development, agricultural extension, and volunteerism, personal experience of Landcare groups and direct contact with Landcare stakeholders, and analysis of data collected through the first surveys of Landcare groups (Curtis et al., 1993).

Drawing on this information I was able to articulate a programme logic that has been widely accepted (Cary and Webb, 2001). From a federal government perspective, Landcare was a catalytic rural development programme intended to engage a large proportion of the rural population and produce more informed, skilled, and adaptive private resource managers. In turn, these managers would adopt a stronger stewardship ethic and increase their adoption of recommended practices that would assist the move to more sustainable agriculture and enhance biodiversity conservation (Curtis and De Lacy, 1996a).

Drawing on this programme logic, and working with a number of colleagues over more than a decade, I have undertaken research focused on programme monitoring and evaluation, including studies exploring the following:

- Motivations for participation and the extent Landcare participation has moved beyond the expert farmer group (Curtis and De Lacy, 1996b; Curtis and Van Nouhuys, 1999);
- Nature and impact of learning through Landcare group participation (Curtis and De Lacy, 1995; Millar and Curtis, 1997);
- Contribution of Landcare groups and networks of groups, including in building social capital (networks, trust and norms) (Curtis et al., 1995; Sobels et al., 2001);
- Impact of Landcare participation on awareness, concern, and knowledge of land and water degradation issues (Curtis and De Lacy, 1996a; Curtis et al., 2000; Curtis and Cooke, 2005);
- Effect of Landcare participation on landholder confidence in the efficacy of recommended practices (Curtis and Byron, 2002; Curtis and Robertson, 2003);
- Link between Landcare participation and landholder stewardship and between stewardship and adoption of recommended practices (Curtis and De Lacy, 1995, 1998);
- Impact of Landcare participation on landholder adoption of practices recommended to enhance the sustainability of agriculture and biodiversity (Curtis and De Lacy, 1996a; Curtis et al., 2001, 2005);
- Extent Landcare activity leads to action beyond the property, including shaping CMC processes and addressing structural issues such as gender and property relationships that are driving land and water degradation (Curtis et al., 1997; Curtis, 2000; Sobels et al., 2001);
- Influence of Landcare activity on non-Landcare participants (Curtis and De Lacy, 1996a); and
- Burnout in Landcare members, leaders and coordinators (Byron and Curtis, 2000, 2002; Byron et al., 2001).

This research involved a variety of approaches at different scales, including the following:

- Mail surveys completed by group leaders that were mailed to a random selection of groups in five of six Australian states to gather information about group membership, leadership, activities, and organizational characteristics expected to shape group outcomes. These surveys were repeated in one state (Curtis et al., 1993; Curtis, 2000; Curtis and Cooke, 2005).
- Mail surveys completed by random selections of private landholders in seven CMC regions to assess Landcare impact on key programme objectives and the extent and impact of burnout (Byron et al., 2001; Curtis and De Lacy, 1996b; Curtis et al., 2001; Curtis and Robertson, 2003).
- The use of a Geographic Information System (GIS) to compare landholder and expert assessments of the occurrence of saline-affected areas to explore Landcare contribution to landholder awareness of salinity (Curtis et al., 2005).
- Semi-structured interviews and focus groups with women Landcare participants in one CMC region (Curtis et al., 1997).
- Participant observation and semi-structured interviews to explore learning through Landcare and industry-sponsored single-issue groups (Millar and Curtis, 1997).
- Participant observation, document studies and semi-structured interviews as part of case studies exploring the impact of Landcare networks (across two states) (Sobels et al., 2001).
- Participant observation in one CMC and a mail survey to all representatives on CMC in one state to explore the contribution of Landcare participants to CMC decision-making processes (Curtis et al., 1995).

Australian researchers approaching the evaluation of Landcare have agreed that it is unreasonable to assess group effectiveness by measuring changes in the biophysical condition of watersheds. Instead, the focus has been on exploring Landcare contributions to learning and action (Carr, 1993; Ewing, 1995; Lockie, 1995; Campbell, 1997). Unfortunately, all stakeholders have not shared this view of the Landcare programme logic. Landcare has achieved very high levels of public recognition and it seems that the wider Australian public, and many Landcare members, thought that Landcare was a large budget programme investing in on-ground work that would ameliorate environmental degradation. One of the lessons is to avoid over-selling programmes to the public and to clearly explain to participants what the programme is expected to achieve. Landcare programme managers have also struggled to respond to a real or perceived need for evidence of programme impact on biophysical conditions at the watershed scale. One approach to managing these unrealistic expectations is to ensure that key stakeholders, including the Commonwealth Treasury, understand and sign off on the underlying programme logic and an accompanying evaluation strategy.

Distinguish Roles and Responsibilities of Landholders, Watershed Groups and Regional Planning Bodies

The Australian experience suggests that local watershed organizations need to be embedded within a supportive institutional framework that identifies realistic roles for landholders, watershed groups, and regional planning bodies such as the CMC (Curtis et al., 2002). All Australian jurisdictions have now accepted the need to have both local Landcare groups and regional CMC. This process has largely occurred by trial and error and has generated considerable tension, particularly among Landcare participants who have been concerned that government is expecting too much of individuals and groups and that the establishment of the CMC was undermining Landcare (Ewing, 2000). The process of defining the roles of individuals, groups and regional bodies needs to be negotiated, should be commenced as early as possible, and should be considered as part of programme development. Monitoring and evaluation should then focus on the extent that roles are clearly defined, realistic, and appropriate in the minds of different stakeholders.

From my observations the Australian experience suggests there are four important roles for regional organizations such as CMC: (1) to aggregate and express regional needs, (2) to establish priorities for allocation of government resources, (3) to provide accountability for expenditure of public funds, and (4) to link and support independent local groups. We also need to establish realistic expectations of what local watershed groups can accomplish. It seems that the most important roles for local groups are therefore (1) to mobilize participation, (2) to initiate and support learning, (3) to pull down resources to support local efforts, and (4) to undertake on-ground work to the extent that resources are available. Of course, it must also be understood that levels of group activity will vary from group to group and over time. In an era of two-income families and considerable off-property work, it is unreasonable to expect landholders in developed economies to take leading roles in managing large watershed projects. They simply do not have the time. On the other hand, we can expect individual landholders (1) to participate in watershed group activities, (2) establish community priorities, and (3) undertake work on their properties or those of others as time and other resources permit.

Be Purposeful: Set Out to Learn

Landcare programme evaluation has largely been approached through short-term consultancies to address specific requests for information, often to justify ongoing programme funding. One exception would be the investment by the Commonwealth in additional questions on Landcare in the national farm surveys conducted by the Australian Bureau of Resource Economics (Mues et al., 1994, 1998). To my knowledge, programme managers at state and Commonwealth levels have not articulated a comprehensive research agenda and there has been limited investment in research to underpin programme improvement. Most research exploring the social dimensions of Landcare has been undertaken by a small group of academics, mostly through doctoral studies or small consultancies (Carr, 1993; Curtis and De Lacy, 1995; Ewing, 1995; Lockie, 1995; Byron and Curtis, 2000; Sobels et al., 2001; Curtis and Cooke, 2005). As a result, only a small proportion of the social research community in Australia has contributed to the tasks of supporting programme improvement or providing evidence to substantiate programme impact.

A related issue has been the excessive use of scarce programme and volunteer resources to develop and assess Landcare programme funding applications as opposed to setting out to learn from programme implementation. Regional, state and Commonwealth panels have assessed project applications submitted by Landcare groups, usually for small sums of money, in a process that typically takes up to nine months. Until very recently, only limited resources have been expended in monitoring project implementation, identifying project impacts, and exploring the wider lessons for programme management.

Set Out to Sustain, Not Just Establish Watershed Groups

The experience with Landcare provides considerable insight into the topics that should be the focus of a monitoring and evaluation programme that aims to assist programme improvement and to contribute to the difficult task of sustaining watershed initiatives.

Most Landcare groups appear to have established supportive group cultures in that there is a strong social connexion or bond in the group, resources are shared fairly, and people are willing to compromise to reach decisions acceptable to most members. Each of these attributes was linked to significantly higher levels of group activity (Curtis and Cooke, 2005).

As successful as Landcare has been, it is my view that inadequate attention to a number of critical topics has limited Landcare group impact and threatens to undermine the long-term viability of the programme. Indeed, in Victoria, where we have the only longitudinal survey data, there is now evidence of a decline in the health of the Landcare movement (Curtis and Cooke, 2005). For example, there are now fewer groups than in 1998, these groups include a smaller proportion of the landholders in their localities, groups are less active in engaging the wider community, and there appears to be an unsustainably high level of member attrition (Curtis and Cooke, 2005).

There is also evidence of burnout among Landare members and group coordinators and links between burnout and the programme management issues identified below (Byron and Curtis, 2000, 2002). Burnout has been described as a process in which continued exposure to stressful situations leads to a syndrome characterized by emotional exhaustion, depersonalization and reduced personal accomplishment (Maslach and Jackson, 1981). Over time, increased levels of burnout could be expected to lead to reduced personal engagement, group activity and outcomes (Byron and Curtis, 2000; Byron et al., 2001).

There has been little evidence of a sophisticated approach to the management of Landcare. In most groups there has been little discussion of leadership succession, and with high administrative workloads for group leaders, most groups have difficulty attracting leadership candidates. A substantial minority of groups have not set annual priorities, and among those that have, there has often been no documented outcome of these processes. Only minimal attention has been given to the important tasks of volunteer member recruitment and retention. A small proportion of groups employ their own coordinators and most of these work part-time and across a number of groups (Curtis, 2000; Curtis and Cooke, 2005). A related issue is that paid coordinators are mostly young, inexperienced, poorly trained, and inadequately supported as they attempt to balance conflicting role interpretations by groups and agencies (Byron and Curtis, 2002). We now have a substantial body of evidence linking these management issues to higher burnout, reduced levels of group activity, and lower scores on a range of qualitative measures of group outcomes (Curtis et al., 2000; Byron and Curtis, 2002; Curtis and Cooke, 2005).

The extensive literature on volunteers provides a useful source of information for those setting out to anticipate and manage these management issues (Brudney, 1990). For example, understanding the forces motivating volunteers, including the importance of factors such as personal gain or improvement, the accomplishment of specific tasks, and social interaction, would provide a sound framework for monitoring and evaluating the reasons for changes in participation and participant satisfaction with group activities.

Action to ameliorate these management issues needs to be taken at a variety of scales, across individual groups, networks of local groups, regional CMC, a state, and the nation. Effective monitoring should also enable group leaders, coordinators and programme managers to take timely and cost-effective interventions that would prevent these issues from becoming endemic problems. Different research approaches and levels of resources would be required depending on the scale, topic and perceived importance of particular issues. For example, the volunteer literature highlights the importance of effective leadership for the attainment of group outcomes. With other colleagues I have undertaken research monitoring Landcare group leadership that has included studies exploring gender relationships (Curtis et al., 1997), administrative workloads (Curtis, 2000), the extent of leadership stability or turnover (Curtis, 2000; Curtis and Cooke, 2005), perceptions of leadership effectiveness (Byron and Curtis, 2000), and burnout among group leaders (Byron et al., 2001). Again, these studies involved a mix of methods including surveys completed by group leaders, semi-structured interviews and focus groups with women Landcare participants in one CMC region, structured face-to-face interviews with leaders from every group across two CMC regions, a diary kept by Landcare leaders for all groups in two CMC regions to quantify time spent on administrative activities, and mail surveys to random samples of Landcare members and leaders in three CMC regions (one in each of three states).

For some of these issues our findings have at times been counter-intuitive. For example, when examining turnover in Landcare leadership tenure we expected to discover evidence to support anecdotal reports that many leaders were staying in office too long. Our early research did establish that there had been little change in leadership personnel in some groups. However, it was far more common for group leaders to turn over rapidly, with the mean length of office for incumbent chairpersons in Victoria at 2.3 years in 1998 (Curtis et al., 2000). A more recent survey showed that there had been a significant increase in the mean length of leadership tenure (to 4.5 years for chairpersons). Our analysis of this survey data showed that shorter leadership tenure was significantly linked to higher levels of group activity (Curtis and Cooke, 2005). These findings confirm the value of monitoring key leadership parameters.

Strive to Inform Management and Policy

The obvious question is, did this research activity shape management and policy decisions and what has been learned about influencing the actors? Before responding, I think it is important to remember that the researcher is only one actor attempting to influence the direction of management and policy. It is also very difficult to identify the influence of research, even when management or policy changes appear to mirror research findings and when change occurs soon after the research has been published or presented.

Despite a large body of research findings that has been widely accepted, including by programme managers, there has not been what I would describe as a coherent and determined attempt to address the ongoing Landcare group management issues identified earlier. This inertia appears to have resulted from a combination of factors, including, the lack of volunteer group management expertise among most agency staff, rivalries between state and national agencies responsible for Landcare, and the influence of a broader thrust towards reduced government involvement in social and economic life.

I do not think that the limited uptake of monitoring and evaluation findings reflects a lack of confidence in the research findings, or a failure to effectively communicate research findings. For example, we have communicated our research findings through reports, including succinct executive summaries, and through invited presentations to programme managers. Nor do I think this situation represents a lack of commitment in government to evidence-based policy development. Perhaps part of the explanation can be gleaned from the fact that our research has been used extensively to underpin state and national agency evaluations of Landcare, and to provide evidence to substantiate bids for ongoing or renewed programme funding. My recent experience working as a researcher in the Department of Agriculture, Fisheries and Forestry-Australia (DAFF), suggests there is an overwhelming focus on addressing the immediate concerns of Ministers and their advisors, and very limited time or enthusiasm for agency programme managers to carefully plan refinements and build the stakeholder support required to implement change. Frequent movements of staff between divisions and programmes also appears to constrain the capacity of state and national agencies to effectively manage programmes.

The latest survey of Victorian Landcare groups (Curtis and Cooke, 2005) was commissioned by DAFF as part of a three-year, $2 million programme of NLP monitoring and evaluation. I was one of the architects of this programme while working in DAFF. I am optimistic but uncertain about the extent that this work will shape future policy and programme management. So my answer is that monitoring and evaluation research has had some impact at the national level. I am more confident that our research has influenced the work of regional agency staff and Landcare coordinators working at that scale, particularly in those regions where we have had collaborative projects, mostly at the initiative of the local staff or CMC.

Other fairly obvious lessons include the importance of attempting to develop longitudinal studies, the value of combining research approaches, and the need to be persistent. Our surveys of Landcare groups in the state of Victoria over 15 years is one of the few longitudinal studies of Landcare or watershed group anywhere and has provided important understanding of trends in leadership tenure reported earlier and of changes in participant's concerns about the adequacy of agency and government support. These data sources have also enabled us to undertake analyses that have provided robust information about trends in the level of group activity and of the factors influencing changes in activity related to community education and on-ground works, including changes in government funding, access to agency staff and group coordinators, and levels of landholder participation in groups.

The mail surveys completed by group leaders also enabled us to identify potential management issues, including concerns about burnout, the conflicting role expectations placed on group coordinators, and the assignment of gender roles in groups. These issues were then explored more fully using participant observation, semi-structured interviews and focus groups, often through regional case studies. At the same time, there have been a number of instances when the survey process was either poorly supported or inadequately funded. On each occasion we worked with our partners in the state agency to move ahead, confident in our assessment of the value of the survey data.

CONCLUSIONS

Reflecting on my research and experience with monitoring and evaluating Landcare, I have suggested that the key findings are as follows:

Work with key stakeholders to articulate the programme logic and develop an evaluation framework. Effective monitoring and evaluation is unlikely to occur unless there is an agreed evaluation framework. This is not a simple task for watershed initiatives that have multiple stakeholders, operate across a number of jurisdictions, and are attempting to address difficult long-term issues where the way forward is uncertain. Programmes evolve and the task of developing the evaluation framework also needs to be thought of as an iterative process. If this important task is to be effectively accomplished, programme managers need to have adequate resources and they need to be competent and determined. One of the lessons from Landcare is that it is also important to continue to explain the programme logic to stakeholders, including watershed group participants and those funding programmes to prevent misunderstandings about programme objectives.

Local watershed organizations need to be embedded within a supportive institutional framework that identifies realistic roles for landholders, watershed groups and regional planning bodies such as the CMC. An important lesson from the Australian experience is that effective community-based NRM requires both local watershed groups and regional planning processes and that each must have clearly defined roles and responsibilities. Again, this is a task that needs to be negotiated as part of programme development, remembering that roles will evolve over time. An important lesson from Landcare is that there must be realistic expectations of what private landholders can accomplish as part of their voluntary contributions to watershed groups. Monitoring and evaluation should focus on the extent that the roles of individuals, groups and regional bodies are clearly defined and appropriate from the perspective of different stakeholders.

Be purposeful: set out to learn. With a sound evaluation framework, programme managers can set out to gather the data that will enable them to refine programme delivery and to assess programme outcomes. The Landcare experience highlights the importance of avoiding excessive emphasis on assessment of project applications from watershed groups at the expense of investments that will assess programme outcomes, including those that were not anticipated, and identify key lessons that will enhance programme delivery. Monitoring and evaluation is likely to be more purposeful if programme managers establish partnerships with researchers.

Set out to sustain, not just establish, watershed groups. The Landcare experience suggests a number of important volunteer programme management issues that should be the focus of monitoring and evaluation that aims to sustain watershed initiatives. These topics include leadership succession, priority setting, member recruitment and retention, decision-making processes within groups, and the administrative workloads of group leaders. Research investigating burnout is one way of exploring many of these issues.

Strive to inform management and policy. While the research by my colleagues and me has contributed to programme evaluation and has been used to substantiate bids for programme funding, there appears to have been little action to address the ongoing group management issues identified by our research. A number of factors appear to have contributed to this inertia on the part of programme managers and government. First, there has been a general lack of expertise about volunteer group management among most agency staff. Second, the number of state

Fig. 1 Eroded site prior to replacement of annual weeds by perennial pasture sown by Landcare.

Fig. 2 School students growing native trees for Landcare revegetation projects.

Fig. 3 Mobile display showcasing best-practice farming.

Fig. 4 Trialing perennial pasture establishment to prevent erosion and increase profitability.

Fig. 5 Landcare sponsored catchment tour to learn about the work of other groups.

Fig. 6 Demonstrating local relevance of best-practice: a partnership beween Landcare and government agencies.

Fig. 7 Evidence of landscape-scale charge as a result of Landcare.

jurisdictions and strong rivalries between different jurisdictions has constrained the capacity to take action. Third, programme managers are generally consumed by the immediate concerns of government and have little time to invest in the complex negotiations required to develop and implement monitoring and evaluation. Fourth, the frequent turnover of staff within the lead agencies appears to have limited both staff knowledge of the programme and commitment and capacity to implement programme improvements. Finally, there has been the pervasive influence of the thrust to reduced government involvement in most aspects of society. These are formidable challenges and require a determined response from government. Nevertheless, monitoring and evaluation can make a difference. Those undertaking monitoring and evaluation must engage programme managers and policy decision-makers and they must be highly responsive when opportunities for influence are presented.

References

Brudney, J.L. 1990. Fostering Volunteer Programmes in the Public Sector: Planning, Initiating and Managing Voluntary Activities. Jossey-Bass, San Francisco.

Byron, I. and A. Curtis. 2000. Landcare in Australia: burned out and browned off. Local Environ. 6: 313-328.

Byron, I. and A. Curtis. 2002. Maintaining volunteer commitment to local watershed initiatives. Environ. Manag. 30(1): 59-67.

Byron, I., A. Curtis and M. Lockwood. 2001. Exploring burnout in Australia's Landcare programme: a case study in the Shepparton region. Soc. Nat. Resour. 14: 903-912.

Campbell, A. 1994. Landcare: Communities Shaping the Land and the Future. Allen & Unwin, Sydney, Australia.

Campbell, A. 1997. Facilitating Landcare: Conceptual and Practical Dilemmas. pp. 143-153. *In:* S. Lockie and F. Vanclay. [eds.] Critical Landcare. Centre for Rural Social Research, Charles Sturt University, Wagga Wagga.

Cary, J. and T. Webb. 2001. Landcare in Australia: community participation and land management. J. Soil Water Conservat. 56(4): 274-278.

Carr, A. 1993. Community involvement in Landcare: The case of Downside, Report No. 93/2, Centre for Resource and Environmental Studies, Australian National University, Canberra.

Chamala, S. 1995. Group effectiveness: from group extension methods to participative community Landcare groups. pp. 5-43. *In:* S. Chamala and K. Keith. [eds.] Participative Approaches to Landcare: Policies and Programmes. Academic Press, Brisbane.

Chambers, R. 1983. Rural Development: Putting the Last First. Longman, New York.

Chen, H.T. 1990. Theory-driven Evaluations. Sage, Newbury Park, California.

Cook, T.D. and W.R. Shadish. 1986. Programme evaluation: the worldly science. Annu. Rev. Psychol. 37: 193-232.

Curtis, A. 1998. The agency/community partnership in Landcare: lessons for state-sponsored citizen resource management. Environ. Manag. 22(4): 563-574.

Curtis, A. 2000. Landcare: approaching the limits of volunteer action. Austral. J. Environ. Manag. 6: 26-34.

Curtis, A. and I. Byron. 2002. Understanding the Social Drivers of Catchment Management in the Wimmera Region. Johnstone Centre, Charles Sturt University, Albury, Australia.

Curtis, A. and P. Cooke. 2005. Landcare in Victoria: Approaching Twenty Years. Institute for Land, Water and Society, Charles Sturt University, Albury, Australia.

Curtis, A. and T. De Lacy. 1995 Landcare Evaluation in Australia: towards an effective partnership between agencies, community groups and researchers. J. Soil Water Conservat. 50(1): 15-20.

Curtis, A. and T. De Lacy. 1996a. Landcare in Australia: does it make a difference. J. Environ. Manag. 46: 119-137.

Curtis, A. and T. De Lacy. 1996b. Landcare in Australia: beyond the expert farmer. Agr. Human Values 13(1): 20-31.

Curtis, A. and T. De Lacy. 1998. Landcare, stewardship and sustainable agriculture in Australia. Environ. Values 7: 59-78.

Curtis, A. and M. Lockwood. 2000. Landcare and catchment management in Australia: lessons for state-sponsored community participation. Soc. Nat. Resour. 13: 61-73.

Curtis, A. and A. Robertson. 2003. Understanding landholder management of river frontages: the Goulburn Broken. Ecol. Manag. Restor. 4(1): 45-54.

Curtis, A. and M. Van Nouhuys, M. 1999. Landcare participation in Australia: the volunteer perspective. Sustain. Dev. 7(2): 98-111.

Curtis, A., J. Birckhead and T. De Lacy. 1995. Community participation in landcare policy in Australia: The Victorian experience with regional landcare plans. Soc. Nat. Resour. 8: 415-430.

Curtis, A., I. Byron and J. MacKay. 2005. Integrating socio-economic and biophysical data to underpin collaborative watershed management. J. Environ. Manag. 41(3): 549-563.

Curtis, A., P. Davidson and T. De Lacy. 1997. The participation and experience of women in landcare. J. Sustain. Agr. 10(2/3): 37-55.

Curtis, A., T. De Lacy and N. Klomp. 1993. Landcare in Victoria: are we gaining ground? Austral. J. Soil Water Conservat. 6(2): 20-27.

Curtis, A., M. Lockwood and J. MacKay. 2001. Exploring landholder willingness and capacity to manage dryland salinity in the Goulburn Broken Catchment. Austral. J. Environ. Manag. 8: 20-31.

Curtis, A., B. Shindler and A. Wright. 2002. Sustaining local watershed initiatives: lessons from Landcare and Watershed Councils. J. Am. Water Resour. Assoc. 38(5): 1207-1216.

Curtis, A., M. Van Nouhuys, W. Robinson and J. MacKay. 2000. Exploring Landcare effectiveness using organisational theory. Austral. Geogr. 33(2):159-170.

Esman, M.J. and N.T. Uphoff. 1984. Local Organisations: Intermediaries in Rural Development. Cornell University Press, New York.

Ewing, S. 1995. Small is beautiful: the place of the case study in Landcare evaluation. Rural Soc. 5(2/3): 38-43.

Ewing, S. 2000. The place of Landcare in catchment management structures. pp. 113-118. *In:* Proceedings of the International Landcare 2000 Conference: Changing Landscapes – Shaping Futures, 2-5 March 2000, Melbourne, Australia.

Kenney, D.S. 1995. Are community-based watershed groups really effective?: confronting the thorny issue of measuring success. Chron. Community 3(2) 33-38.

Lawrence, G. 2000. Global perspectives on rural communities - trends and patterns. pp. 144-151. *In:* Proceedings of the International Landcare 2000 Conference: Changing Landscapes – Shaping Futures, 2-5 March 2000, Melbourne, Australia.

Lockie, S. 1995. Rural Gender Relations and Landcare. pp. 71-82. *In:* S. Lockie and F. Vanclay. [eds.] Critical Landcare, Key Papers Series, No. 5. Centre for Rural Social Research, Charles Sturt University, Wagga Wagga.

Maslach, C. and S.E. Jackson. 1981. Maslach Burnout Inventory Manual. Consulting Psychologists Press, Palo Alto, California.

Millar, J. and A. Curtis. 1997. Moving farmer knowledge beyond the farm gate: An Australian study of farmer knowledge in group learning. Eur. J. Agr. Educ. Ext. 42: 133-142.

Mues, C., L. Chapman and R. Van Hilst. 1998. Landcare: Promoting Improved Land Management Practices on Australian Farms, A survey of landcare and land management related programmes. Australian Bureau of Agricultural and Resource Economics, Canberra.

Mues, C., H. Roper and J. Ockerby. 1994. Survey of Landcare and land management practices: 1992-93. Report No. 94/6. Australian Bureau of Agricultural and Resource Economics, Canberra.

Patton, M.Q. 1990. Qualitative Evaluation Methods. Sage, Newbury Park, California.

Prosavac, E.J. and R.G. Carey. 1992. Programme Evaluation: Methods and Case Studies. Prentice Hall, New Jersey.

Rossi, P.H. and H.E. Freeman. 1985. Evaluation: A Systematic Approach. Sage, Beverly Hills, California.

Sobels, J., A. Curtis and S. Lockie. 2001. The role of Landcare networks in rural Australia: exploring the contribution of social capital. J. Rural Stud. 17(3): 265-276.

22

Social Cost-Benefit Analysis as an Evaluation Tool with Case Studies of Small-scale Irrigation and Micro-hydroelectric Generation (Microhydel) Projects in North West Pakistan

John Cameron

Reader in the School of Development Studies/Overseas Development Group, University of East Anglia, Norwich, UK. E-mail: john.cameron@uea.ac.uk

ABSTRACT

Soil- and water-based development projects are complex activities that involve many technical linkages, affect many people's lives, and are very difficult to value. Market prices may be highly distorted by institutional factors and markets may not even exist for some factors. Social cost-benefit analysis (SCBA) was developed in the 1960s to meet many of these problems and allow decision-makers to allocate scarce investment resources in a globally, socially optimal pattern. It fell out of favour in the late 1970s when it over-reached itself in terms of claiming to offer technical closure to political questions. In the 1990s, SCBA has revived and taken to heart the challenges of valuing the physical environment. In this revived form, it can assist in evaluating soil- and water-based development activities at any stage in the project cycle. The chapter describes SCBA principles and practice and offers examples of small water-based projects in northern Pakistan.

INTRODUCTION

This chapter is aimed at managers and monitoring and evaluation (M&E) professionals involved in activities using soil and water. It assumes some broad familiarity with the discipline of economics, but no more than a broad familiarity. If the techniques described here are to be applied in the field, then a professionally trained economist will be needed to advise the M&E team, but this economist need not, and arguably should not, be the team leader. The chapter also assumes a familiarity with a spreadsheet software package, but no more than assumed normal today for a M&E professional.

The case is made that social cost-benefit analysis (SCBA) can help meet some of the current strategic challenges faced by M&E users and practitioners working on water and soil projects. Many of these challenges were present early in the 1990s (Cameron, 1993) but have become greater, rather than less, especially for soil and water activities that are especially demanding on M&E systems because of their complexity in direct and indirect impact. The challenges are the following:

- Combining information on physical and socio-economic systems. Monitoring and evaluation of soil and water activities is expected to cover changes in the physical environment (e.g. water contamination and soil degradation) and changes in the human condition (e.g. the five dimensions of livelihoods, see Cameron, 1999). The range of potential variables is enormous and a systematic framework is needed to keep the analysis manageable.
- Modeling causality and linking concepts and variables. Both physical and human processes are complex in themselves and even more complex in combination. Monitoring and evaluation also has responsibility for showing causality from the planned activities of a developmental agency to impact on human and environmental well-being (logical framework analysis attempts to play this role but the causality linkages are unclear in complex projects).
- Identifying observable and measurable indicators. Logical frameworks have been invaluable for M&E systems in forcing stakeholders to agree on objectively verifiable indicators. But evaluation frameworks need to maintain pressure on stakeholders to specify observable indicators if the impacts of alternative activities are to be objectively compared.
- Coping with data gaps and inaccuracy. Identifying observable indicators does not guarantee they can be measured accurately. In addition, there may be gaps as baseline data may never have been collected and there may be ethical or political reasons for not

identifying or observing control groups. Evaluation frameworks need to provide explicit room for incorporating such concerns.

- Weighting indicators into composite indices. Evaluative comparisons also frequently require aggregation of indicators into a single number index. This requires weighting of indicators, de facto a form of relative valuing or pricing.
- Incorporating time/uncertainty to achieve sustainability goals. Sustainability is a keyword in wider development thinking as well as in M&E. It involves explicit consideration of longer-term processes. Evaluations clearly cannot be delayed indefinitely to assess impact and sustainability and hence evaluation frameworks need to be able to incorporate long-term processes and the associated inevitable uncertainty.

The argument here is that an updated SCBA can help M&E meet all these challenges (Heal, 1997). This will be demonstrated through a case study after a brief overview of the history of SCBA and its changing relationship to M&E.

THE RISE, FALL AND RISE OF SBCA AND ITS CHANGING RELATIONSHIP TO M&E

In the 1990s, SCBA re-emerged as a development economics tool, due to growing concern with the natural environment and to a lesser extent distributional concerns, both very relevant to analysis of M&E of water and soil activities.

The technique was originally developed in the 1960s in answer to continuing demands on the State to build basic infrastructure, growing confidence in a mixed economy with associated widespread market prices, innovations in electronic data processing capacity, and shortage of investible savings and international purchasing power.

In the late 1960s, Little-Mirlees and UNIDO developed SCBA techniques that gave answers to a number of technical questions in pricing costs and benefits. This gave economists the apparent power to comparatively appraise any developmental activities against an international standard in terms of their net benefits to the global human condition. This framework included the following:

- chains linking developmental activities to final outcomes,
- a numeraire to give an international standard for comparison,
- relative valuation of activities in terms of socially appropriate shadow prices, and
- valuation of time through discounting.

An appraisal or evaluation decision then could be made by ranking activities using net present values (NPV) or benefit/cost ratios or internal rates of return. The framework also gave systematic insights into choice of techniques and assigning distributional weights.

The basic SCBA model builds on standard commercial, financial CBA. Financial CBA is what a commercial enterprise would use to appraise or evaluate an investment activity. It accepts market prices (including interest on borrowing), pays the taxes it cannot avoid (or evade), and welcomes any subsidies. If it can displace or externalize costs on to other economic agents (other producers, consumers, government, neighbours, the human species), it will. The end result is simply financial profitability for the enterprise as a single institution. Neo-classical economists would claim this is necessary and sufficient for appraising activities and free markets will deliver the best of all possible economic worlds as part of a wider neo-liberal developmental agenda.

But SCBA claims the right for an analyst to modify the prices used in the commercial accounts. This modification was claimed to be valid if and when competitive market forces are not operating as assumed by neo-classical economics and/or the distribution of wealth is not considered just. Social cost-benefit analysis claimed to capture market "failures" such as the following:

- Externalities. People's lives may be affected by activities in ways that do not enter into the commercial accounts and may not be directly marketed at all. They may experience monetary or non-monetary costs and benefits as a result - air and water pollution are obvious examples.
- Public goods. An activity may allow people to get benefits by paying for them individually. They can free-ride once the activity is underway as the producer cannot be sure how much benefit they are individually receiving; their demand is non-separable from others - flood control often has something of this characteristic.
- Imperfect competition. An imperfect market will allow some economic agents to use power to become price setters in their own interest. Monopoly producers can set prices above the social optimum, monopsonistic purchasers can set prices below the social optimum - control of a spring was used as an early example of the effect of monopoly on pricing.
- Civil society institutional conventions. Social conventions may not permit the full operation of market forces; some goods and services will not be allowed to be bought and sold, for others prices may only operate within a socially restricted range - common property

or pooled rights over forest and/or water usually has such characteristics.

- Government regulatory and fiscal actions. Governments intervene to affect many markets through regulations, taxes and subsidies at regional, national and international levels - varying taxes and subsidies on agricultural products have implications for imputed values of soil and water.

In SCBA, all these forms of market failure could justify modifying observed prices to so-called shadow prices. A shadow price may be higher or lower than observed prices depending on the specific nature of the market failure (see Cameron 2001 for a set of training notes on the economics of SCBA). Social cost-benefit analysis involves evaluating an activity from the public perspective; at its most ambitious this is a global perspective. But SCBA always involves judgements on accuracy of data and its interpretation into shadow prices and risks and uncertainty surrounding the future. Therefore, sensitivity tests are always needed. These can undermine the claims of SCBA to rank developmental activities on purely technical criteria but allow healthy deliberation over variables that may crucially influence project performance. The need for sensitivity tests was seen as a weakness in the 1970s and contributed to the perception of SCBA as "smuggling" political judgements into technical evaluations. This explicit concern with data inaccuracy and conceptual interpretation can now be seen as a strength rather than a weakness.

Most of the original attention in SCBA focused on government interventions affecting markets through regulations, taxes and subsidies at regional, national and international levels. The neo-liberal structural adjustment policies of the past 20 years have substantially reduced governments' interventions in economic activity and removed regulations and reduced variations in taxation and subsidy rates. Generally, confidence in open market forces was high among leading developmental donors and therefore SCBA was on the decline just at the time interest in M&E was rising in the late 1970s and early 1980s.

This chronology might suggest that M&E and SCBA are alternative or even rival approaches to assessing developmental activities. There are three reasons that could be advanced for such a view:

- M&E, especially as developed by non-government organizations and similar agencies, had a strong element of anti-economism.
- SCBA must quantify variables, while M&E has been strongly influenced by participatory approaches that stress qualitative data.
- SCBA developed as an appraisal technique for developmental blueprints, while M&E is concerned with later stages of the project

cycle or blueprint-denying process approaches to activity design and implementation.

The case made here is that none of these reasons disqualify SCBA as a M&E tool. Used sensitively, SCBA can be a good servant to M&E, especially in soil and water activities. As developmental concerns in the 1990s shifted towards environmental stresses plus income gains for target groups (notably the most excluded of the excluded, especially women), SCBA is re-emerging as a useful tool.

SCBA AND CURRENT ENVIRONMENTAL AND DISTRIBUTIONAL CONCERNS IN SOIL AND WATER CONSERVATION PROJECTS

Social cost-benefit analysis is being increasingly used as a technique for including environmental factors in projects. Most funding agencies wish to incorporate environmental concerns in soil and water conservation projects alongside socioeconomic factors. Social cost-benefit analysis can help in this aim.

- Where negative environmental effects of a project are likely, then SCBA can put values on the damage.
- If the damage is reversible, then costs for reinstating the natural environment of the project to the pre-project condition could be included in the SCBA.
- If the project generates disposable waste, then costs for the environmentally responsible and healthy disposal of that waste could be included in the SCBA.
- If the damage is irreversible, then costs of a compensating environmental improvement, not necessarily in the project area, could be included in the SCBA. Even if not implemented such inclusion will lower the rate of return for such damaging activities.

The early 1970s witnessed the first statements in the development debate on environmental issues. The Club of Rome reports used exponential relationships to suggest that human population growth, use of non-renewable resources, and generation of pollution in combination would result in a global sustainability crisis in the 21st century. But the reports did not engage with the role of valuation and markets in rationing resources.

In the early 1980s, neo-classical economics was in dominance, claiming that market forces would work directly and indirectly to prevent environmental crises. In the shorter term, markets would ration non-renewable and difficult-to-renew resources by price rises. In the longer term, profits-induced technological change would prevent environmental melt-down. But such reasoning did not engage with the realities that some goods cannot be replaced and some physical environmental processes are irreversible.

This market-driven techno-optimism was dented by the discovery of a growing hole in the earth's ozone layer - and to a lesser extent claims that human activity was warming up the atmosphere and setting chaotic climatic change processes in motion. The identification of global environmental concerns was rapidly followed by numerous studies of local situations in which the natural environment was under pressure and natural wealth was being degraded or lost by complex combinations of human activities and "natural" processes. Soil and water considerations often featured in these studies. It was hard to point to anywhere in the global ecosystem where natural wealth appeared to be increasing.

Social cost-benefit analysis can include environmental factors in activity assessment by combining physical and value estimates. It seeks observable prices to give a foundation to its value analysis. Almost every local environment has been fundamentally modified by human activity to produce some products of clear marketable value, but much of what we now value about the physical environment does not have a clear market price, though there may be clues to what could be an appropriate shadow price. Environmental economics has become increasingly ingenious in devising pricing techniques from indirect judgements and behaviour, e.g. contingent valuation (Jorgensen et al., 2001).

- Products may already be collected by local people at a relatively sustainable rate, albeit on a common property or pooled resource basis, and there may be a market price for these products. The discounted market value of these products, if collected on a sustainable or renewable basis, can form the basis of imputing values to the physical environment. However, if the concern is the whole ecosystem and its biodiversity, then this will be a low estimate of the value.
- What people are willing to spend on tourism may give market values to relatively untouched environments. But environmental concerns are not reducible to the leisure and aesthetic choices of more affluent people.
- In higher income economies, people have been asked directly and indirectly what they would be willing to pay to preserve the more natural environments in their countries or pay to live close to them. Such methods frequently give relatively high values to the remaining areas of natural beauty.

- Finally, there is the possibility of considering what people would pay for private property rights over the natural resources, which would presumably include values of possible patents on the life forms in that ecosystem.

None of these methods give a complete sense of value of the environment for SCBA purposes, but they may give a starting point for further analysis leading to acceptable shadow prices.

Concerns with pollution may result in identifying productive activities that will be damaged by a pollution-causing project and using the NPV loss of those activities as the opportunity cost of the pollution. It may be possible to identify the costs that people are already paying to avoid the hazards. In the richer economies, property prices may give a guide. The cost of technology to stop the pollution at source may be included in an SCBA, even if it is unlikely that this technology will be adopted, making it less likely that a polluting project will be selected.

Where negative environmental effects of a project are occurring, then there are techniques for making explicit calculations of social costs of the damage:

- If the damage is reversible, then costs for reinstating the natural environment of the project to the pre-project condition should be included in the SCBA even if this reinstatement is unlikely to occur.
- If the project generates waste or involves resettlement, then costs for the environmentally responsible and healthy disposal of that waste or resettlement should be included in the SCBA.
- If the damage to landscape quality is irreversible, then costs of a compensating environmental improvement, not necessarily in the project area, should be included in the SCBA. Even if not implemented, such inclusion will lower the rate of return for such damaging activities.

The precautionary principle argues that if the project has great environmental uncertainties, then, given the complexity of ecosystems, it should be postponed until we possess sufficient knowledge of the ecosystem to act with reasonable certainty. Social cost-benefit analysts may be in the front line of identifying the need for exercise of the precautionary principle as they must scope risks and uncertainties as part of the SCBA exercise.

High discount rates tend to work against environmental responsibility: for example, reinstatement costs at the end of a 20-year project discounted at 12% are worth only a tenth of what they would have been at the start of the project, and a 40-year project reduces their value to a hundredth. But SCBA analysts have a responsibility to build environmental costs into evaluations even if they may not ever be paid. The SCBA analysts' task is to assess the full social benefits and costs of an activity, not to prejudge the effect of choosing a discount rate.

In the middle of the 1970s, many SCBA analysts thought they could introduce distributional concerns into SCBA by identifying the costs and benefits associated with particular social groups and allocating them differential weights. The weights were always a matter of judgement, and in that sense political. This was seen as a weakness at that time and contributed to the side-lining of SCBA.

In the 1990s, governments were making political statements about the importance of reducing poverty and increasing gender equality. An SCBA analysis could support this strategy by applying weights to costs and benefits accruing to women and people judged to be in poverty. It is now much more acceptable not to have technical closure in project decision-making and to present decision makers with a range of choices, including differing distributional weightings for groups of people with differing socio-economic characteristics.

Thus SCBA is capable of meeting not only the technical challenges in evaluating soil and water activities, but also the environmental and political needs of today. In the following sections, a general description of the SCBA process is followed by examples demonstrating how these challenges and needs may be met in practice.

THE SCBA PROCESS

An SCBA process can begin by using a modified logical framework to scope the numerous linkages between activities and final impacts using a brainstorming approach with people who have had direct experiences of similar activities in the locations where the developmental agency is operating. The brainstorming also can use a stakeholders' model to identify groups of people likely to be affected both positively and negatively by the activities. Risk analysis can be used to identify variables that would be relevant to sensitivity tests.

With the results of these brainstorming activities to hand, a list of all physical inputs, outputs and effects giving rise to costs and benefits can be made in a single column on an Excel spreadsheet. A time-scale in terms of years is then put on to the columns of the spreadsheet starting from the year of construction. The analysis also requires a choice of a final year:

- either when flows of costs and benefits have steadied and discounting reduces present values to insignificant levels.
- or when a scheme is believed to require such heavy capital expenditure that a substantial new investment activity would be needed beyond normal operation and maintenance activities.

In that final year, decisions have to be made about the socially acceptable end-of-project environmental status, possibly including complete reinstatement or environmentally compensating activity elsewhere.

Estimates of physical quantities of inputs and outputs are then made and fed into the chronology. Market prices for unit quantities of all inputs and outputs are then identified wherever possible from primary and secondary sources. Each of these market prices is then scrutinized and discussed to decide whether it should be modified to a socially more appropriate shadow price. The scrutiny focuses on institutional factors that are affecting the observed prices and then modifies these observed prices in the direction of removing those institutional effects to reveal a shadow price. In some cases, such as taxes and subsidies, the effects can be relatively easily quantified. In others, such as foreign exchange rates, standard formulae often exist at national level. Where there are believed to be private monopolies controlling peak season transactions, then quantification can be made by finding low season transactions outside the monopoly relationship and assessing where a market clearing price might lie, taking scale of activity into account.

In some cases, missing prices for some inputs and outputs can be derived from related inputs and outputs that had observable prices. Land is an example: even though land is not readily bought and sold in many rural societies, changes in land use can be given an induced price by observing changes in net value of the land's produce.

Greater challenges arise where physical changes induced by the developmental agency's activities have no observable prices and/or the linkages are very indirect. Greenhouse effects of carbon absorption changes and effects on water availability on the Indus plain are examples. Best estimates of values for such changes can be made using secondary material where available or allocating a notional value added. Such estimates would be prime candidates for sensitivity tests.

Once a complete set of shadow prices representing social values and real scarcities is obtained, then these are applied to the quantities of inputs and outputs and a spreadsheet expressed solely in shadow price values is produced. Where particular costs and benefits were seen as accruing heavily to vulnerable target stakeholder groups, then weightings could be applied to reflect distributional concerns. All assumptions about the time pattern of physical activities, pricing of resource use and physical effects, and distributional weightings should be fully documented.

This spreadsheet is then augmented by three rows to calculate total costs and benefits and net costs and benefits in each year. A standard discounting formula is then applied to net costs and benefits in each year to take account of the effects of time. Either NPVs can be calculated for a target discount rate or an internal rate of return reducing the sum of discounted net costs and benefits to zero can be calculated.

This most likely estimate scenario can then be adjusted to best case and worst case scenarios by inspecting the risk analysis and the documentation on assumptions made. All risks and assumptions that tended to increase benefits and decrease costs are quantified and used for a best case scenario. To construct a worst case scenario, all risks and assumptions that would increase costs and decrease benefits are quantified.

All three scenarios with their associated documentation are then offered to decision-makers for consideration in the final evaluation process. Presenting at least three scenarios is a clear signal to the decision-makers that they are being presented with not merely a technical calculation, but an indicative exercise in which they must exercise judgement.

CASE STUDY

The Aga Khan Foundation has been working on rural development in the northern areas of Pakistan for decades. The terrain can be classified as a mountain desert in which water comes primarily from glacial melts. The water flows rapidly into the Indus River system, taking soil with it.

Settlements are small and tightly clustered, concentrated primarily around locations where some control of water is possible. Part of the Aga Khan Foundation's efforts have been to increase that water control at local level through small-scale irrigation and microhydel works and these efforts had been documented, but without formal SCBA (Iqbal, 1994; Akram, 1999; Kenward, 1999).

The UK Department for International Development (DFID) has been providing some financial support to the Foundation for its work in the region. In 2000, DFID wished to evaluate the impact of its contributions. Net Present Values at a minimum social real rate of return of 12% (at the time a global rule-of-thumb rate for DFID evaluations) were requested by DFID as one indicator in that evaluation process.

At the time of the evaluation exercise, the schemes had been constructed and were in operation. But processes producing benefits were still in train and maintenance and repair costs would need to be met in the future. The author of this chapter provided training and advisory support to the Foundation M&E staff in developing the SCBA for individual, local projects. One irrigation project and one microhydel SCBA are described here as together they bring out a wide range of issues. The process of deriving the SCBA was as described in the previous section of this chapter, though, as with every real SCBA exercise, data proved incomplete.

The capital works for the two projects had been made several years before the evaluation and information was only available in financial terms at the then current prices. It was decided to normalize all prices on the year 2000 and this required calculations of appropriate price deflators (de facto inflators in this case) reflecting the materials used in the capital works. Thus the modeling simulated all the projects as if they had started in 2000, but the choice of end year was different in the two projects.

The irrigation project was allowed to run for 40 years, as the capital works were seen as near permanent (they were gravity fed). After 40 years, at 12% per annum discounting had reduced the net value multiplier to 0.01 reducing the NPV at year zero to one-hundredth of the year 40 value. As the project was seen as enhancing biodiversity in the local environment, there was no need to include compensating environmental activities in the SCBA. There was an interesting discussion on whether there would be a negative effect on water availability for cultivation on the plains and what might be the positive carbon sink effect of the additional trees. The fact that a negative and positive effect could both be identified led to a collective conclusion that neither would be included, as quantification would be both difficult and contentious.

The microhydel project was given a life of 20 years, which was judged to be the physical life of the equipment even with regular maintenance. The end-of-project status decision was to restore the environment to its pre-project status and sell the equipment for its scrap value. In reality, it would be hoped that the equipment would be replaced, but the group wished to account for some environmental impact before the discounting process rendered the impact insignificant.

The comparative effect of both end-of-project decisions meant that the microhydel project was made relatively unattractive compared with the irrigation project in evaluating environmental impact.

The irrigation project's benefits were increased by the extension of low external input agroforestry. Time profiles were developed for fruit and timber physical production. Shadow prices were introduced for agricultural labour (a Pakistan national figure of 0.75 of the going labour market wage rate). Also, wheat was given the international border price. While it was judged that fruit was sold in a fair market, the timber market was seen as having a monopoly element and timber was given a shadow price of 0.81 of the observed open market price.

The microhydel project's physical main benefits were seen as time saved, and it was noted that much of the time saved would be women's. Women's time was seen as valued whether or not it was used for strictly defined economic activity. Increased handicraft production in the evenings was also included. The group discussed whether additional benefits from children able to do homework in the evenings and increase their human capital should be included. But such effects were seen as socio-economically ambiguous and not included. Disposal of potential environmental hazards in the forms of spent bulbs and tube lights were included.

The complete "best bet" estimate SCBA are included at the end of this chapter. The NPV outcomes were very positive for both projects at a 12% discount rate. Using benefit/cost ratios as comparative indicators that are scale neutral, the irrigation project has a superior performance at just over 1.6 compared with almost 1.3 for the microhydel project.

Sensitivity tests for the irrigation project made the following assumptions:

- Best case scenario: the price of timber increases significantly as a result of the increasing population and the cost of producing fodder has been overestimated.
- Worst case scenario: the system is massively damaged due to an exceptional flood in year 18 of the project and the cost of fodder has been underestimated.

  In neither case is the benefit/cost ratio significantly affected.

  Sensitivity tests for the microhydel project made the following assumptions:

- Best case scenario: more time is saved in domestic activities and handicrafts make a greater profit.
- Worst case scenario: the system requires periodic exceptional maintenance activities due to technical and climatic reasons and handicrafts income is lower than in the "best bet" case.

The worst case scenario made little difference to the benefit/cost ratio, but the assumptions of the best case scenario raised the ratio to close to 2.1.

The evidence from the SCBA suggests that small water control projects, either for irrigation or microhydel, are certainly justifiable for international official development assistance. Of course, decision-makers should look at the sensitivity tests and ask for re-calculations if they feel the worst case scenarios are still too optimistic in terms of all the uncertainties and judgements involved.

CONCLUSION

Social cost benefit analysis is not presented here as a panacea for evaluating water and soil development activities. It is a servant of evaluation, not a master. The difference between SCBA today and in its previous heyday in the 1970s is that it has learned humility. In its best practice, SCBA today is not afraid to be explicit and transparent about assumptions and judgements involved. It faces decision-makers with choices they can make with good deliberative reason and does not seek technical closure. The journey of Nobel Prize Economics winner Amartya Sen, from an early developer of SCBA ideas (Sen, 1962) to new millennium development philosopher (Sen, 2001), has relevance here. Sen has always perceived economics as a way of clarifying evaluation of resource allocation for better-informed deliberations, not as closing debates, and this chapter is offered in that spirit.

References

Akram, M.U. 1999. Evaluation of Lift Irrigation Projects. AKRSP, Gilgit.

Cameron, J. 1993. The challenges for M&E in the 1990s. Proj. Apprais. 8(2) 91-96.

Cameron, J. 1999. Trivial Pursuits? - Reconciling sustainable rural development and the global economic institutions. IIED/DFID, London.

Cameron, J. 2001. Notes on CBA and worked examples of rural development projects on Excel Spreadsheets, www.uea.ac.uk/dev/publink/cameron/cba.shtml

Heal, G. 1997. Valuing Our Future: Cost-Benefit Analysis and Sustainability, ODS Discussion Paper, UNDP, New York.

Iqbal, J. 1994. A Study of Thing Sahan Ayun Hydel Project. AKRSP, Gilgit.

Jorgensen, B.S., MA. Wilson and T.A. Heberlein. 2001. Fairness in the contingent valuation of environmental public goods: attitude toward paying for environmental improvements at two levels of scope. Ecol. Econ. 36(1): 133-148.

Kenward, S. 1999. A Socio-Economic Impact Assessment of Hanuchal's Irrigation Channel. AKRSP, Gilgit.

Sen, A. 1962. Choice of Techniques: An Aspect of the Theory of Planned Economic Development. Blackwell, Oxford.

Sen, A. 2001. Development as Freedom. Oxford University Press, Oxford.

23

Institutionalization of Monitoring and Evaluation in Natural Resource Management: Experiences and Issues

Jan Willem Nibbering

Natural resource management specialist who worked as consultant advising the NWFP Forest Department; presently with Ministry of Foreign Affairs, P.O. Box 20061, 2500 EB, The Hague, The Netherlands. E-mail: jw.nibbering@minbuza.nl

ABSTRACT

A process of institutionalizing monitoring and evaluation in the natural resource management sector of the North West Frontier Province of Pakistan has been undertaken. In response to forest and land degradation and increasing poverty in the province, a new participatory and integrated management approach, which involves both rural communities and the provincial Forest Department, was introduced and relies heavily on monitoring and evaluation. Monitoring and evaluation at the village level will primarily be the task of newly established Village Development Committees, with support from the Forest Department and other agencies. Appropriate performance, progress and effect indicators have been developed at the village level. Monitoring and evaluation also needs to be integrated into the operational and strategic planning cycles of the Forest Department. At these planning levels mechanisms for participation need to be created. To move from the old to the new situation, change is needed in attitudes, notably of Forest Department staff, for whom monitoring used to mean "staff control" only. It also requires major efforts in capacity building. Donors and the highest layers of government should fully endorse the process.

INTRODUCTION

In the North West Frontier Province (NWFP) of Pakistan, the forestry sector is undergoing a process of reform. In order to halt the degradation of natural resources and reduce poverty in the province a participatory and integrated approach has been adopted. The mandate of the NWFP Forest Department has been broadened so as to include not only commercial forestry and protection of natural forests, but also range management, watershed protection, and the conservation of biodiversity, while at the same time combating poverty of local communities. In order to be better equipped for its new tasks, the department has adjusted its organizational structure and modified its planning methodology. Monitoring and evaluation (M&E) is an essential part of the new approach and is therefore being institutionalized as a joint responsibility of the department, local communities and NGOs.

The implications of institutionalizing M&E in the sector, i.e. making it a permanent activity of existing or new institutions, will be explored. It involves new roles for different stakeholders. Suitable indicators have to be formulated and an effective system of data collection, aggregation, processing and evaluation needs to be set up. New objectives such as socioeconomic improvement and soil and water management also need to be integrated into M&E. This chapter aims to explore how M&E in the NWFP forestry sector has been conceived and operationalized and what issues have been confronted in this process.

First, we will briefly describe land use and the worsening state of natural resources in the province. Then we will present the new planning approach that has been developed in response to these conditions and clarify how M&E has been organized as part of this approach. Finally, we will discuss major challenges in getting the M&E system in place. These are about ensuring acceptance, defining responsibilities, and building capacity.

NWFP, POVERTY AND LAND DEGRADATION

The North West Frontier Province (Fig. 1) has a total area of 101,760 km^2 and a population of about 19 million, of which more than 80% live in rural areas. The southern and central parts of the province consist of low-lying basins divided by rugged hills. In the north the area is mountainous, culminating in the high and steep ranges with peaks of over 7,000 m asl (Fig. 2). Summers are hot and winters are mild except in the north, which has a Himalayan temperature regime. The climate is generally semi-arid with less than 500 mm of rainfall, but not in the northeast, where the subcontinental monsoon creates a humid environment with 1,000 mm of rainfall or more.

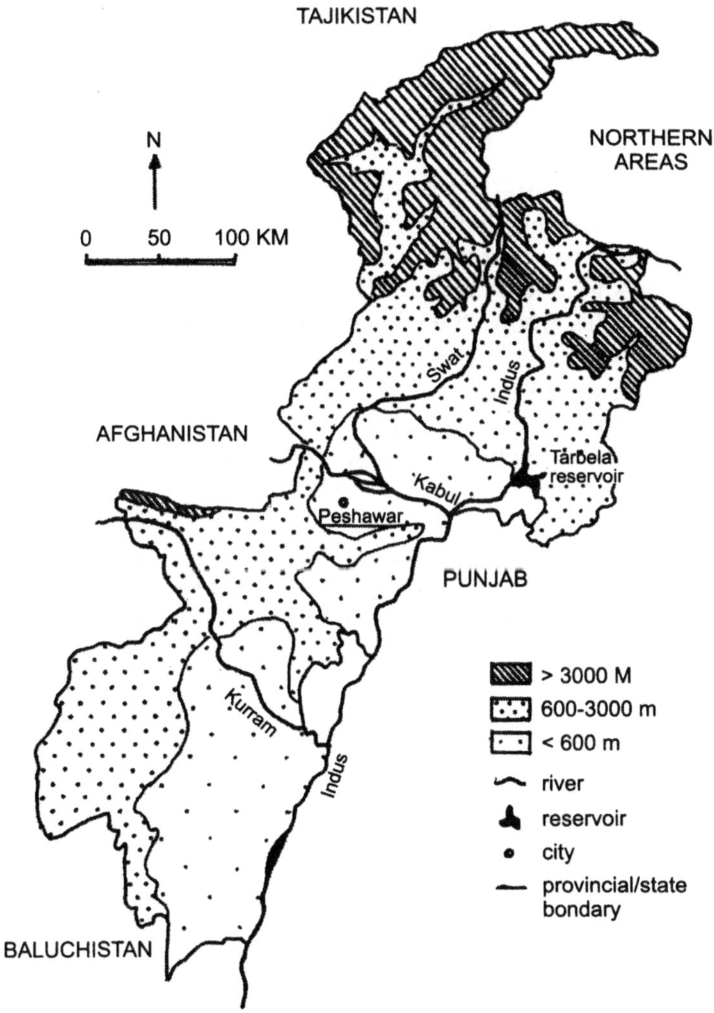

Fig. 1 The North West Frontier Province (adapted from Atlas of Pakistan, 1985).

Dense coniferous forests ceding to alpine grasslands and bare rock at higher altitudes cover the northern mountains. At elevations of 1,000 to 2,500 m, the forest has been greatly reduced and the land is used for grazing and farming at the valley bottoms and on small terraces on the hillsides. Here, and in the adjoining northern areas, originate big rivers such as the Indus and Swat, which serve as important sources of the country's water supply, which is used for irrigation, drinking water, and the generation of hydropower. The central basins are largely irrigated, while subtropical vegetation grows in the dry hill areas surrounding them. In the arid south, forests are scarce and most land is used as rangeland. About 17% of the total land area in the province is wooded, 48% is rangeland and 15% is in agricultural use (Forestry Sector Master Plan, 1992).

More than 80% of the forested area is officially under the control of the Forest Department, but land-owning clans in local communities often have substantial user and revenue rights. Farmland is privately owned. Most holdings are small and largely operated by tenants (Fig. 3). Non-cultivated hillsides around the villages are often under communal use for grazing, fuelwood collection and the like.

Because of rapid population growth and the absence of agricultural development and alternative employment, poverty has grown over the last 50 years. Consequently, the pressures on land and forest resources for agriculture and animal husbandry, timber and firewood production have increased dramatically (Fig. 4). Local customs, which played a key role in environmental conservation and management, have often broken down, while actual control over forest resources ended up in a few hands. Commercial felling, cutting of trees for fuelwood, clearance for

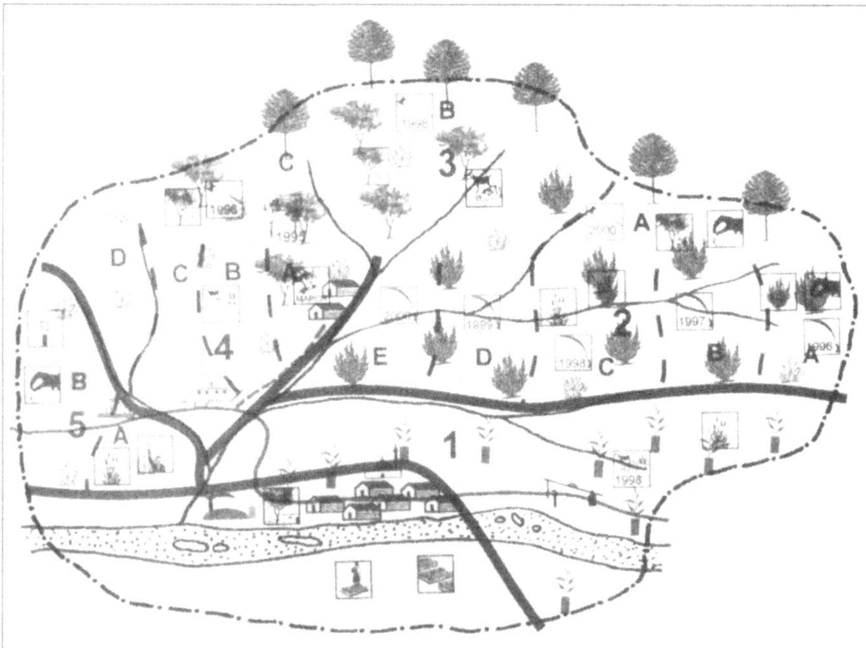

Fig. 2 Example of a village activity map (photo by J.M. Samyn).

Fig. 3 River valley in northern NWFP (photo by J.M. Samyn).

Fig. 4 Small holders living on the hillsides, NWFP (photo by J.M. Samyn).

cultivation, and uncontrolled grazing led to unprecedented forest and land degradation. As a result, soil erosion accelerated, the productivity of the land declined, and hydrological conditions in the catchments worsened, which has exacerbated flood problems in lower areas. As a result of excessive siltation, the economic lifespan of the economically very important Tarbela reservoir in the Indus basin has been considerably reduced (van Dijk and Hussein, 1994).

THE NEW MANAGEMENT APPROACH

In earlier times, the Forest Department was predominately preoccupied with "scientific" forestry geared to commercial timber production on state-owned forest lands. Its main task was to prevent damage to the forest by keeping local communities out. Initially, this system worked well for its purpose. The wooded hills and mountains were still scarcely populated and the highly centralized management structure in the department fitted well with the overall system of the day. The department managed the forest under forest working plans, which were essentially oriented towards harvesting timber.

Owing to its style of working, the department increasingly faced difficulties in developing harmony with local communities that depended on natural resources for their livelihood, while the people considered the department to be responsible for depriving them of their traditional right to use their resources. It became evident that new management forms were needed involving the local communities. The growing universal tendency for more democracy reinforced this perception (Ahmed and Mahmood, 1998).

In the 1980s, donor-assisted projects were started in various parts of the province and developed new participatory management approaches. They also broadened the concept of forest management by including rangeland and wasteland management as well as tree planting on farmland. All forms of natural resource management were dealt with in an integrated manner considering their ecological, economic and social interactions. Village Land Use Planning (VLUP) was developed on communal and private land and Joint Forest Management (JFM) on government land. They enable village communities to manage the natural resources within or near their village domain in a participatory manner. In VLUP a representative Village Development Committee (VDC) is created in which the interests of the various stakeholder groups (land owners, tenants, pastoralists, or different clans) are reconciled and agreements are reached on use and management of the natural resources of the village. It provides a mechanism and procedures through which community needs can be identified and suitable investments and management measures negotiated and worked out. For JFM a special committee is established with representatives from the VDC and the Forest Department.

Through VLUP, a multi-purpose village plan is prepared and implemented. The main focuses are the non-cultivated hillsides, but interventions elsewhere in the village are also possible. The plan provides guidelines for the implementation of such interventions as reforestation, natural regeneration, forest and range management, agroforestry, and gully control, as well as income-generating activities such as beekeeping, collection of medicinal herbs, and sericulture. The interventions are based on the understanding of the different livelihood strategies developed by different groups of people. In VLUP the community goes through a process of awareness raising, analysis of land use, livelihood analysis, social organization, preparation of the village plan, and finally implementation and monitoring of the interventions selected under the plan. The main device for participatory planning is the "activity map". It shows in subsequent stages the initial state and use of the village area and the locations and timing of the agreed interventions as well as their completion (Fig. 5). Village plans are prepared for a period of five years and revised annually.

The hillsides (white) are divided in units and blocks, showing their initial state and use and planned and implemented interventions by means of standardized sticker-symbols of different colours.

It was decided to mainstream the participatory planning methodology developed by previous projects into the working procedures of the Forest Department. With the use of the new methodology, the department is expected to help the communities to organize themselves, where possible, with the assistance of NGOs, to assist VDCs in preparing village plans, to give technical advice on interventions, and to provide funds on a cost-sharing basis. Terms of partnership are drawn up and signed by the VDC and the department. These constitute the legal basis on which the department will provide its assistance.

In order for planners to deal strategically with VLUP and to render it operational it has to be integrated into a broader planning system. This system is very different from the working plans that are prepared at a central planning office where management objectives and measures are worked out into detailed prescriptions. In the new set-up, the provincial forest policy provides the main directions for natural resource management and planning in the province. The department then prepares so-called strategic plans for 28 resource management units (RMUs). These plans contain an assessment of the natural resources and – this is new – the socioeconomic conditions in the RMUs, and they set priorities for natural resource management. They are drafted so as to allow enough flexibility to incorporate wishes and concerns of the local population and enable a more incremental planning process at the grassroots level. In this way they lay the basis for management options, which are no longer geared solely towards commercial timber production, but also towards the multiple needs of the population, watershed protection, and biodiversity conservation. They have a validity of 25 years and are adjusted every five years.

The strategic plans are worked out in operational plans for a total of 119 resource management sub-units (RMS). These plans prioritize villages for VLUP and JFM, outline institutional and organizational arrangements for implementation, such as the involvement of NGOs, estimate the scale of interventions, and present a budget. A time schedule for activities is also given. Operational plans are drawn up for a period of seven years and are revised annually (Nibbering and Samyn, 2002).

Fig. 5 Father and son carrying brushwood (*Dodonaea viscosa*) from hillsides for fuel. *Source:* Social Forestry Project Malakand–Dir.

To support the new approach, the department has reorganized itself. It has assumed a matrix structure with four specialized units supporting the RMU and RMS managers. These units deal respectively with human resource development, community development and extension, research and development, and planning and monitoring.

MONITORING AND EVALUATION

Monitoring and evaluation (M&E) is to become an integral and permanent part of planning and management in the new approach as a device to improve achievements and to be able to adjust to changing conditions and objectives. Due to the participatory nature of the approach it requires the involvement of all the actors in the process, appropriate institutional arrangements, and meaningful indicators.

Monitoring and Evaluation at the Village Level

The main focus of M&E in the new management approach is at the village level, because here management is actually brought to the field. Village organizations play a central role in natural resource management. Therefore, there must be appropriate and stable institutional arrangements for M&E at this level. Basically the VDC should monitor and evaluate the progress and the effects[2] of the interventions it undertakes. This will enable the VDC to make adjustments in implementation and further planning. Also, the Forest Department needs information on progress and effects to be able to assess its support to VDCs and village plans.

There is also a need to monitor and evaluate the institutional framework and how the various actors, the VDC, the Forest Department, other external agencies, and the village community as a whole perform within this framework. This is necessary to ensure that the participatory planning process is followed and that interventions undertaken will be socially and financially sustainable, and to improve the VDC's decision-making and managing capacity. It should be undertaken jointly by the VDC, the department, and possibly an NGO that may have been involved in the organization of the village community. The village community also has a role to play in monitoring and evaluating the performance of its own VDC and may do this through village meetings. The VDC and its partners

---

[2] In this chapter we use *effect monitoring* to indicate monitoring of the more or less immediate results and benefits of interventions, whereas *impact monitoring* has been reserved for long-term and wider environmental and socioeconomic effects of (combined) interventions, such as sedimentation in reservoirs or investments made by VDCs with funds generated through proper management of the natural resources of the village.

need to observe clear principles in the way they operate. These revolve around democracy, commitment, transparency, discipline, conflict resolution, gender equity, and social learning. Indicators related to these principles are presented in Table 1.

For monitoring the progress of the natural resource management interventions that the village community has decided to undertake, the village plan with its activity map will indicate what needs monitoring, when and where, and what indicators could be used.

Progress monitoring should take place at key stages in the implementation process and needs to be conducted by the VDC together with the department and, if applicable, other agencies as well, in order to ensure a common understanding of the state of affairs and general agreement on what action to take. The terms of partnership drawn up

Table 1 Examples of village organization indicators.

Major responsibility with:	*Indicators*
VDC	- VDC meetings are regularly held and attendance is adequate. - Members of VDC hold consultations with all stakeholder groups in the village when taking important decisions. - The VDC resolves conflicts particularly on natural resources. - Transparent accounting and proper record keeping is practised. - Women's organizations in the village are effectively consulted in decision-making. - The VDC participates effectively in support sessions, training courses and excursions and the content is put to use. - The VDC organizes and reports on M&E visits and sessions and proposes plan revisions where necessary.
Village community	- Village meetings are well attended. - Rules and regulations established by the VDC are being followed. - All social groups including women are able to express their needs and ideas. - Physical and financial contributions are made by the villagers. - The village plan is revised periodically.
External agencies	- Terms of partnership with the Forest Department and other agencies are drawn up. - Agencies regularly meet with VDC and the associated women's organization to participate in M&E visits and sessions and to discuss progress and problems.

between the VDC and the department are an important institutional instrument to regulate joint progress monitoring. They should specify the stages at which joint monitoring is required and enable monitoring of the mutual obligations of all the parties. A special monitoring committee may be established with representatives of each party. The findings of this committee are recorded in a report signed by all members.

Progress indicators for field interventions are given in Table 2. Some of them not only reflect progress but also show whether financial or work obligations have been met. Indirectly, they also reflect the capacity of the VDC and its partners in organizing the work and observing agreed rules and regulations.

Table 2 Indicators for monitoring the progress of field interventions.

Intervention	*Indicators*
Reforestation, planting of shrubs and grass on communal land	Indicators - Planting or sowing (according to agreed specifications). - Survival of plants. - Maintenance and guarding by VDC. - Financial records.
Range management, natural regeneration on communal land	Demonstration plots - Fences kept intact, no trespassing incidences in the plots. - Treatments followed according to plan. Large-scale application - Experienced herdsmen engaged from poor households. - Grazing plan followed. - No trespassing of animals, humans (for grass cutting) or both. - Financial records. - Livestock owners' contributions collected and deposited in the VDC joint account.
Agroforestry on farmland Infrastructure schemes (e.g. road, drinking water supply, drop structures)	- VDC records for plant distribution according to demand list. - Number of planted and surviving trees (through random checking of farms). - Design and specifications of estimate followed. - Contribution of VDC collected and deposited in joint account. - Labour by VDC provided. - VDC financial record keeping.

Finally, a VDC should undertake effect monitoring in order to see whether the interventions bring about the desired results. It will help diagnose problems, take remedial actions, or find alternative solutions. Again, effect monitoring will be most fruitful if it can be carried out jointly with the department. It will enable the Forest Department and other agencies to give advice on monitoring methods, to review interventions, and to evaluate and improve their own support and planning strategies. The same monitoring committee as above could be involved. Examples of indicators are given in Table 3.

The VDC will normally be motivated to monitor socioeconomic effects, in terms of supply of products needed by the population, for instance fuel wood, because these represent direct benefits for the community and the VDC needs this information to ensure a fair distribution of benefits among the villagers. But the VDC is less likely to rigorously quantify biophysical effects, for which it would also have insufficient capacity. Qualitative methods would be more appropriate for monitoring biophysical effects, for instance, based on visual comparison of treated areas with untreated areas. This is particularly relevant in

Table 3 Indicators for effect monitoring.

Objective	*Indicators*	*Methods*
Increase in vegetation cover	Tree and shrub density. Species composition and growth stage.	Observing and counting trees, species and growth stage.
Reduction of soil erosion, better infiltration of water into the soil	Better ground cover on erosion spots. Size of gullies.	Estimating ground cover; comparing gully size with marking pins.
Self-sufficiency in seedling production	Extent to which demand for seedlings is met by village nurseries.	Seedling supply records.
Increase in fuel wood production	Number of fuel tree species.	Counting and measuring fuel wood trees.
Increase in fodder production	Density of fodder trees and grass cover on communal hillside.	Counting trees, species and growth stage; measuring grass cover.
Increase in income of poor households and women	Income from income-generating activities, e.g. fruit trees, beekeeping, sericulture.	Interviews with beneficiaries. Records kept by beneficiaries.
Improvement in village infrastructure	Performance of infrastructure in operation; number of beneficiaries reached.	Site and work inspection; interviews with beneficiaries.

demonstration plots. Benchmarking could be undertaken with the aid of photographs of the original state of the resource concerned. Also, the local effects of interventions such as gully control, checking siltation in water ponds, protection of springs, and prevention of landslides could be monitored in a qualitative manner. Quantitative monitoring of biophysical effects would have to be conducted at higher planning levels with a greater role for the Forest Department and other agencies.

Simple records on indicators, financial reports and comments need to be maintained by the VDC for all the three types of monitoring. Simple standard formats have been designed for this, which allow VDCs enough flexibility to deal with local conditions.

Monitoring and Evaluation in Operational and Strategic Planning

Monitoring and evaluation also needs to be institutionalized at higher planning levels. The Forest Department must be able to monitor progress and effects of operational and strategic plans and have enough information to evaluate them and to review the provincial forest policy from time to time. Table 4 gives an overview of organizations (to be) involved at the various planning levels and their main responsibilities in M&E.

At the operational planning level the progress of VLUP, VDC performance and interventions in the villages of the RMS are monitored. The RMS manager and his team of foresters perform this task. At the strategic planning level the RMU manager is primarily responsible for M&E. Both are assisted by teams of specialists from the specialized units. Progress monitoring is more important for evaluating operational plans to allow adjustments in budgets, manpower and organizational requirements. Effect and impact monitoring is more important for evaluating strategic plans in order to be able to conduct ex-post cost/benefit analyses and review strategies. It is also necessary to monitor autonomous developments at this level as these may reinforce or weaken the effects of interventions, or they may alter planning objectives.

At the operational planning level, it is important that other agencies and NGOs involved in the implementation of operational plans participate, whereas the evaluation of strategic plans should also involve representatives of different stakeholder groups in the RMU, including other agencies with an interest in land use and poverty alleviation, as well as representatives of regional VDC organizations. Legitimate institutional mechanisms for this are still lacking. The newly established district governments in the province could provide for this. RMU boundaries

Table 4 Organizations involved in M&E at various planning levels.

Planning level	*Organization*	*Responsibilities in or with regard to M&E*
Village level	VDCs	M&E of their performance and development, progress and effects of interventions; involving entire village community.
	Forest Department (forester, RMS manager)	M&E of performance and development VDC, progress and effects of interventions, jointly with VDCs; internal reporting to RMS level.
	Local NGOs	M&E of performance and development of VDC, jointly with VDC and Forest Department.
Operational level (RMS)	Forest Department (RMS/ RMU manager)	M&E of mainly progress of operational plans (progress of VLUP in selected villages); internal reporting to RMU level.
	Other agencies	M&E of progress of their contributions to operational plans and of coordination with Forest Department.
	Local NGOs	M&E of progress of social organization in selected villages and of coordination with Forest Department.
	Clusters of VDCs in RMS	M&E of progress of VLUP in selected villages.
Strategic level (RMU)	Forest Department (RMU manager, conservator)	M&E of effects and impact of strategic plans: VLUP, JFM and natural resource management interventions in RMU; sharing of data with district-level organizations; internal reporting to provincial level.
	District government and agencies (agriculture, livestock, etc.)	Providing institutional mechanisms for evaluation; evaluation of effects in light of district and sector priorities.
	WAPDA, NGOs (IUCN, WWF, etc.), Pakistan Forestry Institute and other academic institutes	Monitoring of watershed management parameters, biodiversity; impact on poverty alleviation.
	VDC associations in RMU/ regional forest round tables	Evaluation of progress and effects with regard to priorities of interest groups; benefit distribution.
Provincial level	Forest commission, forest round table	Evaluation of progress and effects in light of provincial forest policy, distribution of benefits and performance of Forest Department.

(Table 4 Contd.)

(Table 4 Contd.)

	Provincial government and agencies	Evaluation of effects in light of provincial and sectoral priorities.
	Forest Department (chief conservator, directors of specialized units)	Providing M&E results to above provincial entities; contribution to provincial policy review.
Federal level	Federal government, Ministry for Environment, Inspector General of Forests	Monitoring of forest cover, impact on watershed conditions, with reference to national resource management strategies and international agreements.

generally follow watersheds, but adjustments have been made to achieve as much overlap as possible with districts in order to facilitate coordination.

Much of the monitoring data required for the operational and strategic planning levels have to be derived from aggregated data from the village level. To enable Forest Department staff to provide monitoring information in a regular and uniform fashion to higher planning levels, standard forms are used. The RMS manager compiles the data from VDC records and joint monitoring visits for his own use and submits a summary version of the monitoring results of his sub-unit to the RMU manager. The latter will in turn assemble the monitoring data for all the RMSs in his RMU and submit these to the Planning & Monitoring (P&M) Unit. The P&M unit compiles the data of all RMUs at the provincial level. It is expected that with time all data will be entered by the RMS manager into a computerized, GIS-based management information system, which is run by the P&M unit. Officers of this unit will also carry out spot checks on reported progress from time to time.

For effect monitoring at the strategic planning level other sources of information are also used. The evaluation of biophysical and especially downstream effects of interventions - which is most relevant at the strategic planning level - requires other sources and cooperation with other agencies. Village development committees do not have the capacity to make quantitative measurements of downstream effects and measurement of effects on water flow and sedimentation is meaningful only over larger geographic units. Moreover, the main beneficiaries of watershed management live downstream. Downstream effects may be monitored in selected areas with major roles for the Forest Department, the Agriculture Department, and the national Water and Power Development Agency (WAPDA), which has facilities for monitoring sedimentation and water discharge in river systems. Also, satellite imagery may be used to monitor impact of natural resource management interventions on land use and land cover, while sample surveys on impact on livelihoods could be outsourced to academic institutes or NGOs specialized in the matter.

At the provincial level, Forest Round Tables have been set up, in which various interest groups are represented and a Forest Commission will be instituted as well. They will monitor the performance of the Forest Department and other actors in the sector and assist in policy formulation and reviews. The department is expected to provide M&E data to these entities, but they may also undertake inquiries of their own.

Indirectly, monitoring results may provide information on deficiencies in skills and capacities of department staff, which can be used by the human resource development unit for setting up training programmes and enhancing their policies.

MAKING THE CHANGE

So far we have been concerned largely with the hardware of M&E: institutional requirements, indicators and information flows. Now we will shift our attention to the changes that are needed to get from the old situation to the new one, concentrating on attitudes and capacity. This is not to suggest that the M&E system as presented above has been the product of a blueprint design. Rather, it has been the outcome of an interactive process involving senior and junior department staff, local and international consultants, and a number of village communities in different settings. These actors have influenced the process in various ways, as will become clear below. This process has not been completed yet. While the performance of VDCs and the progress of the implementation of village plans are routinely monitored, effect monitoring is still very much under development. So is M&E at the operational and strategic planning levels.

It goes too far to suggest that collective M&E has been a totally new concept for village communities. In mountain villages where traditional structures still existed, councils of elders often stipulated the periodic resting of portions of communal rangeland as soon as it showed signs of degradation, in order to allow it time to recover. Village guards were assigned to prevent trespassing and violators had to pay fines. Even in those villages the systematic and participatory manner in which M&E has to be conducted has been new. Nevertheless, it generally met with ready acceptance as planning under VLUP is tailor-made and monitoring easily fell in the same pattern, also thanks to such participatory devices as the activity map. In some more stratified communities, however, as the first benefits (e.g. timber, fuel, fodder) derived from successful natural resource management came within reach, M&E proved to be a test of the willingness of land owners to allow tenants and pastoralists to participate and have their initially agreed, rightful share in the benefits. But generally, M&E sessions have proven to strengthen village organization because they allowed the villagers to see the merits of being organized and learn from mistakes. Evaluation sessions are usually eagerly attended (Fig. 6).

In contrast, institutionalizing M&E in a line department such as the Forest Department has been a major challenge. Here monitoring has traditionally meant "staff control" only, i.e. a means to check on subordinates. Therefore, it is essential that the purpose of M&E is well understood by both senior and junior staff. Senior staff should deal with M&E in a proper way so as to overcome the resistance of junior staff to M&E. Clearly, this matter surpasses the theme of M&E. A new work culture needs to be developed in the department with more emphasis on teamwork and less on commands. Secondly, for the Forest Department to perform its new role effectively, it can no longer limit itself to monitoring physical targets such as the number of acres planted. The department should equally foster an interest in monitoring the performance of the VDC and monitoring effects. This interest should come from the realization that the department is no longer an implementing agency but a facilitating one. The best way to learn to appreciate M&E for both the

Fig. 6 Joint M&E session on gully control, NWFP. Source: Social Forestry Project Malakand-Dir.

department and the village communities has been to discover its benefits by having its outcome fed back into subsequent actions. The increasing demand for showing results expressed by the federal and provincial governments - with better watershed management foremost in mind - also forces the department to take M&E seriously.

While M&E for village communities and the department may have posed problems of its own, to conduct M&E jointly has been another challenge for both. The Forest Department used to carry out its inspections entirely by itself and in the past relations between the department and village communities were characterized by suspicion and lack of respect. Now they must act together. While village communities must overcome their lack of confidence in the department, foresters need to step down to the villagers and acquire social skills to communicate effectively with them. In joint monitoring visits, quality control often has proven to be a bone of contention. In the eyes of the foresters, the work of the villagers was sub-standard, because the former were eager to point out that the latter were not fit for the job and forest staff did not perceive their role as that of advisors. More enlightened senior staff sometimes had to step in and mediate in such disputes. Joint evaluation has also been problematic, with villagers expecting the department to tell them what to do, like most other government agencies continue to do.

In NWFP the reforms were spearheaded by a donor-aided project through which the funds for large-scale field interventions were channelled as well. While the project was assisting the department to institutionalize M&E, the donor and the provincial government expected the project to report instantly on the results of the field interventions that had to be started at the same time. As a result, a parallel monitoring structure emerged within the project, which weakened the institutionalization of M&E within the department and led to a struggle of competence between the project manager and the P&M director. If M&E is to be institutionalized, responsibility for it should be placed directly where it belongs institutionally. It also means that donors and governments should wholeheartedly support the institutional reforms they initiated and not slip back to the short-term philosophy of a tree-planting project. They should withstand internal pressures of realizing rapid disbursements and showing to the public quick, tangible successes in the field.

Donor coordination is of crucial importance when it comes to institutionalizing M&E. Other forestry projects in different parts of the province supported by other donors should be willing to subscribe to the institutional reforms in the Forest Department and refrain from establishing different monitoring systems. It is the task of the department to point this out to other donors, thereby proving that it really "owns" the monitoring system that it has developed.

A special entity in the department was needed to manage a complex M&E system. Therefore, a P&M unit was formed to run a management information system in the department, to supervise the M&E procedures, and to improve both of these further. The P&M unit was moulded out of the existing Forest Management Centre, which had been in charge of the preparation of forest working plans, with monitoring explicitly being added to its responsibilities. It took some time to convince department staff that P&M officers were not supposed to do all the monitoring work but only to advise field staff on monitoring and conduct spot checks. If different officers were going down to the villages for different steps of the same process, this would make participatory natural resource management at the village level virtually impossible. The unit would also not have the staff available to do so. Similarly, staff of the community development unit only advises field staff on monitoring VDC performance and does not assume this task in the field itself.

Monitoring and evaluation requires new skills for all parties involved. It requires a full understanding of the purpose, scope, and mechanisms involved. It requires expertise in monitoring new topics such as poverty reduction, watershed management, and biodiversity conservation. Special introductory training programmes have been set up to help the department, NGOs and village communities in acquiring these skills. Much of this has been on-the-job training. Furthermore, organizational capacity had to be created to submit, process, store and retrieve data for evaluation and further planning. The computerization of information systems will make these tasks easier. This is a major effort in a department that has so far been overwhelmingly computer illiterate. Monitoring and evaluation also requires a capacity to coordinate monitoring efforts with other agencies such as WAPDA, another novelty for a department that used to work in isolation. It also finds it difficult to be confronted with data produced by others, such as data on siltation of reservoirs or satellite imagery showing high rates of forest and land degradation. In a situation where governance is more about blaming one another for what went wrong than about trying to solve the problem together, such a reaction is understandable. For similar reasons, the installation of a supervisory body in the shape of a forest commission has met much resistance in the department and has been delayed.

Monitoring requires manpower and takes time. Nevertheless, time is scarce for all and has opportunity costs. The time and manpower required for M&E should therefore be kept at manageable levels. The number of indicators and the number of monitoring records should be limited to what is essential for proper management. A series of field trials with monitoring, involving both department staff and VDCs, have led to a better selection of indicators and a reduction of the number of records. There are also other ways in which time can be saved. Villagers can save time if they can choose trustworthy representatives to join monitoring visits instead of the entire community making the effort, although at times this can be very useful. Forest staff may combine monitoring visits with other tasks they have to perform. More time may also become available for them for monitoring when the need for policing diminishes as village communities take greater interest in managing and protecting the natural resources around them. Therefore, a possible transition phase, where the old style of management has not been quite abandoned and the new style of management already introduced, should be kept as short as possible to avoid an increase in the burdens of field officers.

Evaluation requires analytical capacity and, apart from knowledge of monitoring indicators, a deeper insight into the management situation. This will be a major challenge at the strategic planning level, the more so because broader management concerns such as biodiversity and watershed management and other concerns such as security and the settlement of refugees from neighbouring Afghanistan come more to the forefront at this level. On these themes good monitoring data are often lacking, which may lead to endless and paralysing controversies, such as the one in 2002 on the supposedly adverse effects of large-scale planting of eucalyptus trees on hydrological conditions in the Malakand region. For the same reason, it is also difficult to value costs and benefits of biodiversity protection and watershed management, which is necessary to guide strategic planning in a better way. The Forest Department does not have the capacity to obtain these data by itself and needs to cooperate with other organizations, academic institutes and NGOs for the purpose.

CONCLUSION

For the institutionalization of M&E in natural resource management one has to tackle many obstacles that may not exist in a project situation. In a project sufficient funding and qualified people may be readily available; one can escape local bureaucracies, and there is usually a strong expectation from donors and governments to have a system in place through which progress is reported rapidly. Effects and impact of interventions and the performance of the institutional framework are hardly an object of M&E, however. In contrast, the institutionalization of M&E in a government department and the sector overall requires several conditions:

- Beneficiaries and regular department staff are convinced of its merits and will thus be persuaded to take up M&E.
- Appropriate supportive institutional structures at all relevant levels (department, village communities) are created that favour participation and collaboration.
- M&E is integrated into a planning hierarchy, with appropriate indicators that may be different for different stakeholders and at different planning levels.
- Manpower constraints are taken into account.
- Styles in departmental human resources management are changed and much effort goes into capacity building of all the actors, comprising (new) technical skills as well as social and managerial skills, also in the field of M&E.
- Appropriate data management technologies are developed.
- National and provincial governments as well as donors give the institutionalization process their full support and do not have it derailed by their own short-term interests.

But these are all temporary hurdles. In the end, the overriding advantage of institutionally embedded M&E is to have a permanent, stable and uncontested M&E system in place, which serves all stakeholders, has more scope for coordination with other organizations, and, with growing experience, can be improved over time. This will be beneficial for M&E of natural resource management and watershed management in particular, where complex and long-term efforts as well as effects require long series of sound and complementary monitoring data and comprehensive evaluations.

References

Ahmed, J. and J. Mahmood. 1998. Changing Perspectives on Forest Policy. Policy that Works for Forests and People No. 1. IUCN Pakistan and International Institute for Environment and Development, Islamabad and London.

Atlas of Pakistan. 1985. 1st edition. Survey of Pakistan, Rawalpindi.

Forestry Sector Master Plan. 1992. Islamic Republic of Pakistan and Asian Development Bank, Islamabad.

Nibbering, J.W. and J.M. Samyn. 2002. Integrated participatory forest management in a densely populated mountain region, North West Frontier Province, Pakistan. pp. 149-160. *In:* F. Brun and G. Buttoud. [eds.] Proceedings of the International Research Course on the Formulation of Integrated Management Plans for Mountain Forests, Bardonecchia, 30 June-5 July 2002. European Observatory of Mountain Forests and the University of Turin, Grugliaso, Italy.

Van Dijk, A. and M.H. Hussein. 1994. Environmental Profile. North West Frontier Province, Pakistan. DHV Consultants BV, Amersfoort, The Netherlands.

24

Farmer Participation in Research and Extension: The Key to Achieving Adoption of More Sustainable Cassava Production Practices on Sloping Land in Asia and Their Impact on Farmers' Income

Reinhardt H. Howeler[1], Watana Watananonta[2] and Tran Ngoc Ngoan[3, 4]

[1]CIAT, Cassava Office for Asia, Department of Agriculture, Chatuchak, Bangkok 10900, Thailand. E-mail: r.howeler@cgiar.org.

[2]Field Crops Research Institute, Dept. of Agriculture, Chatuchak, Bangkok 10900, Thailand.

[3]Thai Nguyen University of Agriculture and Forestry, Thai Nguyen, Vietnam.

[4]The many other persons who contributed significantly to this project are acknowledged at the end of the paper.

ABSTRACT

Cassava (*Manihot esculenta* Crantz) is the third most important food crop in Southeast Asia, where it is usually grown by smallholders in marginal areas of sloping or undulating land. Farmers grow cassava because the crop will tolerate long dry periods and poor soils and will produce reasonable yields with minimum inputs. Most farmers realize, however, that cassava production on slopes can cause severe erosion, while production without fertilizer or manure inputs will lead to a gradual decline in soil productivity. Current production practices may thus not be sustainable.

Research has shown that cassava yields can be maintained for many years with adequate application of fertilizers or manures, and that there are various ways to reduce erosion. Adoption of erosion control practices, however, has been minimal as farmers generally see little short-term benefits, while initial costs of establishing these practices may be substantial.

In order to enhance the adoption of soil conserving practices and improve the sustainability of cassava production under a wide range of socioeconomic and biophysical conditions, a farmer participatory research (FPR) approach was used not only to develop the most suitable soil conservation practices, but also to test new cassava varieties, fertilization practices and cropping systems that tend to produce greater short-term benefits. The FPR methodology was initially developed in two or three sites each in China, Indonesia, Thailand and Vietnam. The methodology includes the conducting of rapid rural appraisals in each site, farmer evaluation of a wide range of practices shown in demonstration plots, trials with farmer-selected treatments on their own fields, field days with discussions to select the best among the tested practices, scaling up of selected practices to larger fields, and farmer participatory dissemination to neighbours and neighbouring communities. Based on the results of these trials, farmers in the pilot sites have readily adopted better varieties, fertilization and intercropping practices, and many farmers have planted contour hedgerows to control erosion.

In the second phase of this project, supported by the Nippon Foundation, the farmer participatory approach for technology development and dissemination was further developed in about 34 pilot sites each in Thailand and Vietnam and in 31 sites in southern China. Farmers were generally very interested in participating in these trials. After becoming aware of the seriousness of erosion in their cassava fields, they have shown a willingness to adopt simple but effective practices to reduce erosion while at the same time obtaining short-term benefits from the adoption of new varieties and other improved practices. The testing by farmers on their own fields of new cassava varieties and fertilization practices in addition to soil conservation practices was found to be of crucial importance for the adoption of more sustainable production practices. The resulting increases in cassava yields in Asia over the past 10 years have increased the annual gross income of cassava farmers by an estimated 270 million US dollars.

Keywords: cassava, erosion control, farmer participatory research (FPR) and extension (FPE), Thailand, Vietnam, China, impact assessment

INTRODUCTION

Cassava (*Manihot esculenta* Crantz) is the third most important food crop (after rice and maize) grown in Southeast Asia and is used for human consumption, animal feed and industrial purposes. It is usually grown by smallholders in upland areas with poor soils and low or unpredictable rainfall. In some countries the crop is grown on steep slopes, but in others it is grown mainly on gentle slopes; in both cases, soil erosion can be serious. Moreover, cassava farmers seldom apply adequate amounts of fertilizers or manures to replace the nutrients removed in the harvested products. Thus, both erosion and nutrient extraction can result in a decline in soil fertility and a gradual degradation of the soil resource.

The fact that farmers do not apply sufficient fertilizers and do not use soil conservation practices when the crop is grown on slopes is a socioeconomic rather than a technical problem. Research has shown many ways to maintain or improve soil fertility and reduce erosion, but farmers usually consider that these practices are too costly or require too much labour. To overcome these obstacles to adoption it is necessary to develop simple practices that are suitable for the local situation and that provide short-term benefits to the farmer as well as long-term benefits in terms of resource conservation. Being highly site specific, these practices can best be developed by the farmers themselves, on their own fields, in collaboration with research and extension personnel.

Thus, a project was initiated, with financial support from the Nippon Foundation in Tokyo, Japan, to develop a farmer participatory methodology for the development and dissemination of more sustainable production practices in cassava-based cropping systems, which would benefit a large number of poor farmers in the uplands of Asia. Figures 1-4 show some aspects of farmers' participatory research and extension in China, Thailand and Vietnam.

MATERIALS AND METHODS

First Phase (1994-1998)

The first phase of the project was conducted in four countries: China, Indonesia, Thailand and Vietnam. The project was coordinated by CIAT and implemented in collaboration with research and extension organizations in each of the four countries. During an initial training course on farmer participatory research (FPR) methodologies, each country designed a work plan to implement the project. The steps in the process, from diagnosing the problem to adoption of suitable solutions, are shown in Fig. 5. The outstanding feature of this approach is that farmers participate in every step and make all important decisions.

Pilot Site Selection

Suitable pilot sites were selected in areas where cassava is an important crop, where it is grown on slopes, and where erosion is a serious problem. Detailed information obtained through Rapid Rural Appraisals (RRA) in each site have been reported by Nguyen The Dang et al. (1998), Utomo et

Fig. 1 FPR erosion control trial showing severe soil loss in the check plot, Saiyoke District, Kanchanaburi Province, Thailand.

Fig. 2 A vetiver grass hedgerow planted on a farmer's field in Kongba Village, Hainan, China, resulted in terrace formation after one year.

Fig. 3 The all-women farmers' group in Nhu Xuan, Thanh Hoa Province, Vietnam, conduct an FPR soil erosion control trial.

Fig. 4 During the field day a farmer from Khut Dook Village Thailand, explains the results of his FPR trial to other visiting farmers.

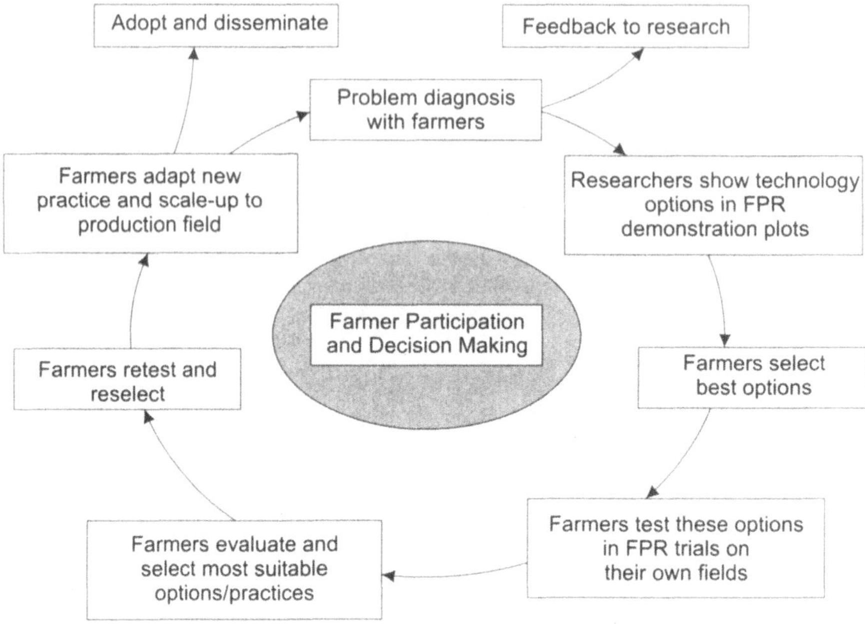

Fig. 5 Farmer participatory model used for the development of sustainable cassava-based cropping systems in Asia.

al. (1998), Vongkasem et al. (1998), and Zhang Weite et al. (1998). Table 1 is an example of information obtained from RRAs conducted in Vietnam, while Table 2 shows a summary of information obtained from RRAs conducted in several pilot sites in four countries. The detailed information from each site can serve as baseline data to monitor progress and evaluate the impact of newly adopted technologies. After the RRAs were conducted, the most suitable pilot sites (villages or sub-districts) were selected to work with farmers in the development and dissemination of new varieties and production practices.

Demonstration Plots

Each year demonstration plots were laid out on an experiment station or a farmer's field to show the effect of many alternative treatments on yield, income and soil erosion. Farmers from the selected pilot sites visiting the trial were asked to discuss and score the usefulness of each treatment. From this range of many options farmers usually selected three to four treatments that they considered most useful for their own conditions. Table 3 shows that farmers from different sites have different priorities and thus rank options quite differently. Some farmers then volunteered to test these treatments in FPR trials on their own fields.

Table 1 Cropping systems, varieties and agronomic practices, as determined from RRAs conducted in four FPR pilot sites in Vietnam in 1996/97.

Province	*Hoa Binh*	*Phu Tho*		
District	*Luong Son*	*Thanh Ba*	*Thai Nguyen*	
Village		*Phuong Linh*	*Pho Yen*	
Hamlet	*Dong Rang*	*Kieu Tung*	*Tien Phong*	*Dac Son*
Cropping system[a]				
-upland	tea C+T C monoculture peanut, maize	C monoculture C+P tea, peanut maize	C+P or C+B or 2 yr C rotated with 2 yr fallow sweet potato	C monoculture or C-P rotation or C-B, C-SP sweet potato
Varieties				
-rice	CR 203, hybrids from China	DT 10, DT 13, CR 203	DT 10, DT 13 CR 203	CR 203 DT 10, DT 13
-cassava	Vinh Phu, local	Vinh Phu, local	Vinh Phu	Vinh Phu
Cassava practices				
-planting time	early March	early March	Feb/March	Feb/March
-harvest time	Nov/Dec	Nov/Dec	Nov/Dec	Nov/Dec
-plant spacing (cm)	100x80	80x80; 80x60	100x50	100x50
-planting method	horiz./inclined	horizontal	horiz./inclined	horizontal
-land preparation	buffalo/cattle	by hand/cattle	buffalo	buffalo
-weeding	2 times	2 times	2 times	2 times
-fertilization	basal	basal + side[b]	basal + side[c]	basal + side[d]
-ridging	mounding	flat	flat	flat
-mulching	rice straw	peanut residues	peanut residues	peanut residues
-root chipping	hand chipper	knife	small grater	small grater
-drying	3-5 days	3-5 days	2-4 days	2-4 days

(Table 1 Contd.)

Fertilization				
-cassava				
-pig manure (t/ha)	5	5	3-5	8-11
-urea (kg/ha)	0	50-135	83	83-110
-SSP (18% P_2O_5) (kg/ha)	50-100	0	140	0-280
-KCl (kg/ha)	0	0	55	0-280
-rice				
-pig/buffalo manure (t/ha)	5	0	-	-
-urea (kg/ha)	120-150	80	-	-
Yield (t/ha)				
-cassava	11-12	8-15	8.5	8.7
-rice (per crop)	3.3-4.2	4.2	3.0-3.1	2.7-3.0
-taro	1.9-2.2	-	-	-
-sweet potato	-	-	8.0	3.3
-peanut	0.8-1.2	0.5-1.1	1.4	1.3
-pigs (kg live weight/yr)	100-120	-	-	-

[a]C = cassava, P = peanut, B = black bean, T = taro, M = maize, C+P = cassava and peanut intercropped, C-P = cassava and peanut in rotation.
[b]urea at 2 MAP.
[c]urea when 5-10 cm tall; NPK+FYM when 20 cm tall.
[d]NPK when 30 cm tall; hill up.

Table 2 Characteristics of eight pilot sites for the FPR trials in Asia in 1994/95.

	Thailand		*Vietnam*			*China*	*Indonesia*	
	Soeng Saang	*Wang Nam Yen*	*Pho Yen*	*Thanh Ba*	*Luong Son*	*Kongba*	*Malang*	*Blitar*
Mean temp. (°C)	26-28	26-28	16-29	25-28	16-29	17-27	25-27	25-27
Rainfall (mm)	950	1,400	2,000	~1,800	~1,700	~1,800	>2,000	~1,500
Rainy season	Apr-Oct	Apr-Nov	Apr-Oct	Apr-Nov	May-Oct	May-Oct	Oct-Aug	Oct-June
Slope (%)	5-10	10-20	3-10	30-40	10-40	10-30	20-30	10-30
Soil	± fertile loamy Paleustult	± fertile clayey Haplustult	infertile sandy loam Ultisol	very infertile clayey Ultisol	± fertile clayey Paleustult	± fertile sandy cl.l. Paleudult	infertile clay loam Mollisol	infertile clay loam Alfisol
Main crops	cassava rice fruit trees	maize soybean cassava	rice sw. potato maize	rice cassava tea	rice cassava taro	rubber cassava sugarcane	cassava maize rice	maize cassava rice
Cropping system[a]	C monocrop	C monocrop	C monocrop	C monocrop	C+T	C monocrop	C+M	C+M
Cassava yield (t/ha)	17	17	10	4-6	15-20	20-21	12	11
Farm size (ha)	4-24	3-22	0.7-1.1	0.2-1.5	0.5-1.5	2.7-3.3	0.2-0.5	0.3-0.6
Cassava (ha/hh)	2.4-3.2	1.6-9.6	0.07-0.1	0.15-0.2	0.3-0.5	2.0-2.7	0.1-0.2	0.1-0.2

[a]C = cassava, T = taro, M = maize.

Table 3 Ranking of conservation farming practices selected from demonstration plots as most useful by cassava farmers from several pilot sites in Asia in 1995/96.

	Thailand		*Vietnam*		*China*	*Indonesia*	
	Soeng Saang	*Wang Nam Yen*	*Pho Yen*	*Thanh Hoa*	*Baisha*	*Blitar*	*Dampit*
Farmyard manure				2			
Medium NPK	5						
High NPK					2		
Farmyard manure + NPK				1			
Cassava residues incorporated			5				
Reduced tillage	4						
Contour ridging		2					
Up-and-down ridging					5		
Maize intercropping	2					1	1
Peanut intercropping		5			4		2
Mungbean intercropping					3		
Black bean intercrop + *Tephrosia* hedgerows			1	4			
Tephrosia green manure			3	5			
Tephrosia hedgerows			4				
Gliricidia sepium hedgerows						2	4
Vetiver grass barriers	1	1	2	3			
Brachiaria ruziziensis barriers	3	4					
Elephant grass barriers						3	3
Lemon grass barriers		3					
Stylosanthes barriers					1		

In both the demonstration plots and FPR erosion control trials on farmers' fields, a simple methodology was used to measure soil loss due to erosion in each treatment. Plots were laid out carefully and exactly along the contour on a uniform slope; it is important that runoff water does not enter the plots either from above or from the sides. Along the lower side of each plot a ditch was dug and covered with plastic (Fig. 6); small holes in the plastic allowed runoff water to seep away, while eroded sediments remained on the plastic. These sediments were collected and weighed monthly or at least two or three times during the cropping cycle. After

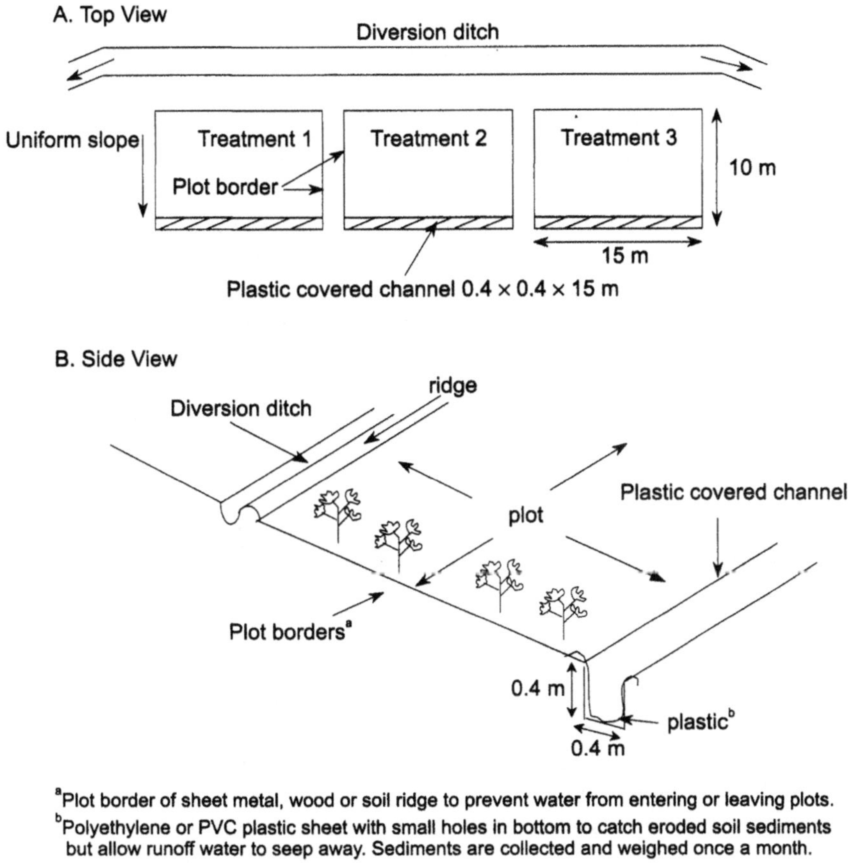

Fig. 6 Experimental layout of simple trials to determine the effect of soil/ crop management practices on soil erosion.

correcting for moisture content, the amount of dry soil loss per hectare was calculated for each treatment. This simple methodology gives a visual as well as a quantitative indication of the effectiveness of the various practices in controlling erosion (Howeler, 2001, 2002).

Farmer Participatory Research Trials

The FPR trials involved not only soil conservation practices but also new varieties, intercropping systems and fertilization, with the objective of developing a combination of practices that would increase farmers' income, reduce erosion and improve soil fertility. The trials usually had four to six treatments, with one treatment representing the farmer's traditional variety or practice. Plot size varied from 30 m^2 to 100 m^2. Treatments were not replicated, but wherever possible, farmers within one village conducting the same type of trial were encouraged to use the same treatments so that each trial could be considered a replication and results could be averaged over those replications. This increased the confidence in the reliability of the results.

During the first phase of the project, farmers in the four countries conducted a total of 177 erosion control trials, 157 variety trials, 98 fertilizer trials and 35 intercropping trials, all amounting to 467 trials. At time of harvest, field days were organized in each site for the participating farmers and their neighbours to harvest the various trials. The yields of cassava and intercrops, the dry soil loss due to erosion, as well as the gross income, production costs and net income were calculated for each treatment and presented to the farmers. Farmers and extension workers from the area discussed the results and then indicated their preferences for a particular treatment or production practice by raising their hands.

After one or more years of testing in small plots, farmers quickly identified the best varieties and production practices for their area and started using those on larger areas of their production fields (Howeler, 2002).

Second Phase (1999-2003)

The second phase of the project was conducted in collaboration with five institutions in Thailand, six in Vietnam, and three in China (Table 4). During the second phase the emphasis shifted from the development and use of FPR methodologies to farmer participatory extension (FPE) in order to reach more farmers and achieve more widespread adoption.

Once farmers had selected certain practices and wanted to adopt those on their fields, the project staff tried to help them, for instance, in setting out contour lines to plant hedgerows for erosion control or in obtaining seed or vegetative planting material of the selected hedgerow species, intercrops or new cassava varieties.

During both the first and second phase of the project some collaborative research continued on station in order to solve problems identified at the farm level, or to develop better technologies that farmers could later test on their own fields.

RESULTS AND DISCUSSION

First Phase (1994-1998): Farmer Participatory Research

Farmer Participatory Research Trials

Table 5 shows a typical example of an FPR erosion control trial conducted by six farmers having adjacent plots on about 40% slope. Contour

Table 4 Partner institutions collaborating in the second phase of the Nippon Foundation cassava project in Asia.

Research and extension organizations in Thailand
- Department of Agriculture
- Department of Agricultural Extension
- Land Development Department
- Kasetsart University
- The Thai Tapioca Development Institute
Research and extension organizations in Vietnam
- Thai Nguyen University of Agriculture and Forestry
- National Institute for Soils and Fertilisers
- Vietnam Agricultural Science Institute
- Hue University of Agriculture and Forestry
- Institute of Agricultural Sciences of South Vietnam
- Tu Duc University of Agriculture and Forestry
Research and extension organizations in China
- Chinese Academy for Tropical Agricultural Sciences
- Guangxi Subtropical Crops Research Institute
- Honghe Animal Husbandry Station of Yunnan

hedgerows of vetiver grass, *Tephrosia candida* or pineapple reduced erosion to about 30% of that in the check plot, while intercropping with peanut and planting vetiver hedgerows also markedly increased net income. Farmers clearly preferred those treatments that were most effective in both increasing net income and reducing soil erosion, such as hedgerows of vetiver grass or pineapple. Results of many other FPR trials have been reported by Nguyen The Dang et al. (2001), Huang Jie et al. (2001), Utomo et al. (2001) and Vongkasem et al. (2001).

Scaling Up and Adoption

After having selected the most promising varieties and production practices from FPR trials, farmers generally like to test some of these on small areas of their production fields, making adaptations if necessary. Some practices may look promising on small plots but are rejected as impractical when applied on larger areas; this may be due to lack of sufficient planting material (like vetiver grass) or lack of markets for selling the products (like pumpkin or lemon grass). Also, to be effective, hedgerows need to follow the contour rather precisely; otherwise they can cause serious gully erosion by channeling runoff water to the lowest spot. Contour hedgerows also force farmers to plough along the contour, which

Table 5 Effect of various crop management treatments on the yield of cassava and intercropped peanut as well as the gross and net income and soil loss due to erosion in an FPR erosion control trial conducted by six farmers in Kieu Tung village of Thanh Ba district, Phu Tho province, Vietnam in 1997 (third year).

Treatment[a]	*Slope (%)*	*Dry soil loss (t/ha)*	*Yield (t/ha)* cassava[b]	peanut[b]	*Gross income[c]*	*Prod. costs*	*Net income*	*Farmers ranking*
					(mil. dong/ha)			
1. C monoculture, with fertilizer, no hedgerows (TP)	40.5	106.1	19.17	-	9.58	3.72	5.86	6
2. C+P, no fertilizer, no hedgerows	45.0	103.9	13.08	0.70	10.04	5.13	4.91	5
3. C+P, with fertilizer, no hedgerows	42.7	64.8	19.23	0.97	14.47	5.95	8.52	-
4 C+P, with fertilizer, *Tephrosia* hedgerows	39.7	40.1	14.67	0.85	11.58	5.95	5.63	3
5. C+P, with fertilizer, pineapple hedgerows	32.2	32.2	19.39	0.97	14.55	5.95	8.60	2
6. C+P, with fertilizer, vetiver hedgerows	37.7	32.0	23.71	0.85	16.10	5.95	10.15	1
7. C monoculture, with fertilizer, *Tephrosia* hedgerows	40.0	32.5	23.33	-	11.66	4.54	7.12	4

[a]Fertilizers = 60 kg N + 40 P_2O_5, + 120 K_2O/ha; all plots received 10 t/ha pig manure. TP=farmer traditional practice.

[b]Cassava: fresh roots; peanut: dry pods.

[c]Prices: cassava (C) dong 500/kg fresh roots. peanut (P) dong 5,000/kg dry pods. US$1 = approx. 13,000 dong.

is more difficult and more costly. Moreover, it makes planting in neat straight lines, using tight strings as a guide, impossible. Thus, there are very practical reasons why farmers may be reluctant to adopt some of these soil conservation practices. Table 6 shows the particular technologies that farmers had adopted in the four countries at the end of the first phase of the project.

Table 6 Technological components selected and adopted by participating farmers from their FPR trials conducted from 1994 to 1998 in four countries in Asia.

Technology	*China*	*Indonesia*	*Thailand*	*Vietnam*
Varieties	SC8013*** SC8634* ZM9247* OMR35-70-7*	Faroka*** 15/10* OMM90-6-72*	Kasetsart 50*** Rayong 5*** Rayong 90**	KM60*** KM94* KM95-3*** SM1717-12*
Fertilizer practices	15-5-20 + Zn + chicken manure 300 kg/ha*	FYM 10 t/ha (TP) + 90 N + 36 P_2O_5 + 100 K_2O**	15-15-15 156 kg/ha***	FYM 10 t/ha (TP) + 80 N + 40 P_2O_5 + 80 K_2O**
Intercropping	monoculture (TP) C + peanut*	C + maize (TP)	monoculture (TP) C + pumpkin* C + mungbean*	monoculture (TP) C + taro (TP) C + peanut***
Soil conservation	sugarcane barrier* vetiver barrier*	*Gliricidia* barrier* *Leucaena* barrier* contour ridging**	vetiver barrier*** sugarcane barrier**	*Tephrosia* barrier*** vetiver barrier* pineapple barrier*

*some adoption
**considerable adoption
***widespread adoption
TP = traditional practice; FYM = farmyard manure.

Second Phase (1999-2003): Farmer Participatory Research and Extension

Since the objective of the second phase was to achieve widespread adoption of more sustainable production practices by as large a number of farmers as possible, it was necessary to markedly expand the number of pilot sites and to develop FPE methodologies to disseminate the selected practices and varieties to many more farmers.

Farmer Participatory Research

Implementing the project in collaboration with many different institutions in China, Thailand and Vietnam (Table 4), and with generous financial support from the Nippon Foundation, it was possible to expand the number of pilot sites each year. In 2001 the project was working in about 50 sites, and this further increased to 99 sites by the end of the project in 2003 (Fig. 7). Once the benefits of the new technologies became clear, the number of sites increased automatically, as neighbouring villages also wanted to participate in order to increase their yields and income.

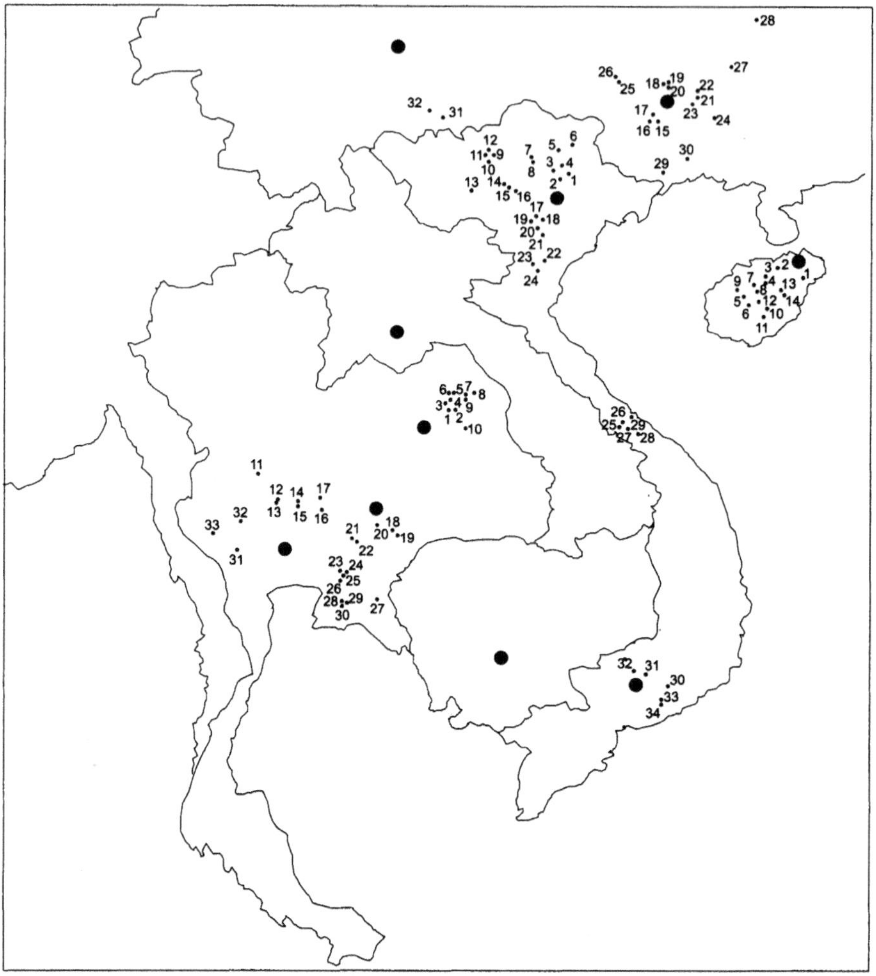

Fig. 7 Location of FPR pilot sites in China, Thailand and Vietnam in the Nippon Foundation cassava project in 2003.

Whenever the project extended to a new site, the process outlined above was re-initiated, i.e. an RRA was conducted, interested farmers visited demonstration plots and/or made a cross-visit to an already established site, and they conducted FPR trials, discussed results and eventually adopted those varieties or practices they had selected as most suitable for their own conditions. Table 7 shows the number and types of FPR trials conducted in China, Thailand and Vietnam during the second phase of the project. While initially farmers were mainly interested in testing new varieties, fertilization, intercropping and erosion control

Table 7 Number of FPR trials conducted in the second phase of the Nippon Foundation Project in China, Thailand and Vietnam.

Country	*Type of FPR trial*	*1999*	*2000*	*2001*	*2002*	*2003*	***Total***
China	Varieties	9	9	20	69	20	**127**
	Erosion control	3	5	8	17	-	**33**
	Fertilization	-	-	-	4	-	**4**
	Intercropping	-	-	-	9	-	**9**
	Pig feeding	-	-	-	59	-	**59**
		12	**14**	**28**	**158**	**20**	**232**
Thailand	Varieties	11	16	16	19	25	**87**
	Erosion control	14	10	6	-	11	**41**
	Chemical fertilizers	16	6	23	17	17	**79**
	Chem. + organic fertilizers	-	-	10	11	11	**32**
	Green manures	-	-	13	11	15	**39**
	Weed control	-	-	17	5	10	**32**
	Plant spacing	-	-	3	-	2	**5**
	Intercropping	-	-	16	7	-	**23**
		41	**32**	**104**	**70**	**91**	**338**
Vietnam	Varieties	12	31	36	47	35	**161**
	Erosion control	16	28	29	30	23	**126**
	Fertilization	1	23	36	24	24	**108**
	Intercropping	-	14	32	31	26	**103**
	Weed control	-	3	-	-	3	**6**
	Plant spacing	-	1	7	19	8	**35**
	Leaf production	-	-	2	2	1	**5**
	Pig feeding	-	-	11	16	13	**40**
		29	**100**	**153**	**169**	**133**	**584**
Total		**82**	**146**	**285**	**397**	**244**	**1,154**

practices, during the later part of the project they also wanted to test the use of organic or green manures, weed control, plant spacing and even leaf production and pig feeding. During the five years of the second phase of the project a total of 1,154 FPR trials were conducted by farmers on their own fields. Tables 8 to 12 are just a few examples of the various types of FPR trials conducted by farmers in different sites in Thailand and Vietnam.

Farmer Participatory Extension

The following FPE methods were found to be very effective in raising farmers' interest in soil conservation, disseminating information about

Table 8 Results of an FPR variety trial conducted by a farmer in Am Thang commune, Son Duong district, Tuyen Quang, Vietnam in 2002.

Treatments[a]	*Cassava yield (t/ha)*	*Gross income*	*Production costs ('000 dong/ha)*	*Net income*	*B/C*	*Farmers' preference*[b] *(%)*
1. Vinh Phu (local)	20.70	10,350	4,330	6,020	2.39	7.9
2. La Tre (SC205) (local)	21.40	10,700	4,330	6,370	2.47	10.5
3. KM60	29.20	14,600	4,330	10,270	3.37	21.0
4. KM94	37.50	18,750	4,330	14,420	4.33	94.7
5. KM95-3	32.80	16,400	4,330	12,070	3.79	26.3
6. KM98-7	25.40	12,700	4,330	8,370	2.93	10.5

[a]fertilized with 1,100 kg/ha of 7-4-7 fertilizers = 1.43 mil. dong/ha
[b]out of 38 farmers

Table 9 Average results of three FPR erosion control trials conducted by farmers in Suoi Rao and Son Binh villages, Chau Duc district, Baria-Vungtau, Vietnam in 2003/04.

Treatments	*Dry soil loss (t/ha)*	*Cassava yield (t/ha)*	*Maize + hedgerow yield (t/ha)*	*Gross income*[a]	*Prod. costs*[b] *('000 dong/ha)*	*Net income*	*Farmers' preference (%)*
1. Cassava monoculture, no hedgerows	77.12	26.34	-	10.536	6.079	4.457	20
2. C + pineapple hedgerows	11.65	27.02	-	10.808	6.279	4.529	0
3. C + *Paspalum atratum* hedgerows	12.18	30.13	11.40	12.052	6.279	5.773	65
4. C + vetiver grass hedgerows	9.94	28.33	8.84	11.332	6.279	5.053	15
5. C + maize intercrop	14.30	17.86	3.25	10.394	7.969	2.425	0

[a]Prices:	cassava		400 dong/kg fresh roots
	maize		1,000 dong/kg dry grain
[b]Costs:	labour		20,000 dong/manday
	cassava fertilizers		1,279,000 dong/ha
	maize fertilizers		550,000 dong/ha
	cassava stakes		500,000 dong/ha
	maize seed		440,000 dong/ha
	labour for cassava without HR (210 md/ha)	=	4.2 mil. dong/ha
	labour for maize (40 md/ha)	=	0.8 mil. dong/ha
	labour for fertilizer application (5 md/ha)	=	0.1 mil. dong/ha
	labour for hedgerow cutting/maintenance	=	0.2 mil. dong/ha

Table 10 Results of an FPR fertilizer and manure trial conducted in Khut Dook village, Baan Kaw, Daan Khun Thot, Nakhon Ratchasima, Thailand in 2002/03.

Treatments[a]	*Root yield (t/ha)*	*Starch content (%)*	*Gross income*[b]	*Fertilizer cost*[c]	*Production costs*[c]	*Net income*
			('000 B/ha)			
1. No fertilizers or manure	18.75	25.0	21.56	0	10.87	10.69
2. Chicken manure + rice hulls, 400 kg/rai	30.42	26.2	34.98	2.50	17.15	17.83
3. Pelleted chicken manure, 100 kg/rai	26.70	21.1	30.71	2.00	15.39	15.32
4. 15-7-18 fertilizer, 50 kg/rai	29.68	24.1	34.13	2.66	16.73	17.40
5. 13-13-21 fertilizer, 50 kg/rai	32.22	27.4	37.05	3.13	17.89	19.16
6. 16-20-0 fertilizer, 50 kg/rai	26.08	25.9	29.99	2.50	15.61	14.38
7. 15-15-15 fertilizer, 50 kg/rai	30.36	26.9	34.91	2.81	17.07	17.84

[a]1ha = 6.25 rai

[b]Prices:	cassava	baht 1.15/kg irrespective of starch content
[c]Costs:	chicken manure	1.0/kg
	pelleted chicken manure	3.20/kg
	15-7-18	8.50/kg
	13-13-21	10.0/kg
	16-20-0	8.0/kg
	15-15-15	9.0/kg
	harvest + transport costs	270/ton
	cassava production without fertilizer or harvest	12,757/ha

improved varieties and cultural practices, and enhancing adoption of soil conserving practices:

Cross-visits. Farmers from new sites were usually taken to visit older sites that had already conducted FPR trials and had adopted some soil conserving technologies. These cross-visits, in which farmers from the older site could explain their reasons for adopting new technologies, were a very effective means of farmer-to-farmer extension. After these cross-visits, farmers in some new sites decided to adopt some technologies immediately, while others decided to conduct FPR trials in their own fields first. In both cases, the FPR teams of the various collaborating institutions, together with provincial, district or sub-district extension staff, helped farmers to establish the trials, or they provided seed or planting materials required for the adoption of the new technologies.

Table 11 Average results of four FPR intercropping trials conducted by farmers in Tran Phu commune, Chuong My district, Ha Tay, Vietnam in 2003.

Treatments	*Cassava yield (t/ha)*	*Intercrop yield (t/ha)*	*Gross income*[a]	*Seed costs*[b]	*Production costs*[b]	*Net income*
			('000 d/ha)			
1. Cassava monoculture	24.54	-	9,816	0	5,460	4,356
2. C + 1 row peanut	21.93	1.187	14,707	480	8,115	6,592
3. C + 2 rows peanut	22.52	2.000	19,008	960	8,595	10,413
4. C + 2 rows mungbean	21.42	0	8,568	2,000	9,635	-1,067
5. C + 2 rows soybean	21.28	0.162	9,322	800	8,435	887

[a]Prices:	cassava:	dong	400/kg fresh roots	
	peanut:		5,000/kg dry pods	
	soybean		5,000/kg dry seed	
[b]Costs:	labour:	dong	15,000/manday	
	NPK fertilizers		= 0.86 mil. dong/ha	
	peanut seed (80 kg/ha)		12,000 /kg	= 0.96 mil. dong/ha for 2 rows
	mungbean seed (80 kg/ha)		25,000 /kg	= 2.00 mil. dong/ha for 2 rows
	soybean seed (80 kg/ha)		10,000 /kg	= 0.80 mil. dong/ha for 2 rows
	labour for cassava monoculture without fertilizers			= 4.5 mil. dong/ha (300 md/ha)
	labour for cassava intercropping without fertilizers			= 6.675 mil. dong/ha (445 md/ha)
	labour for cassava fertilizer application			= 0.10 mil. dong/ha

Table 12 Average results of five FPR pig feeding trials on adding ensiled cassava leaves to the diet, conducted by farmers in Huong Ha commune, A Luoi, Thua Thien-Hue, Vietnam in 2001/02.

Treatments	*No. of pigs*	*Live weight (kg)*		*LWG*[a] *(g/day)*	*FCR*[b] *(kg DM/kg gain)*	*Feed cost*[e] *(VND/kg gain)*
		initial	*3 months*			
Control diet[c]	6	24.30	52.50	313.3	4.83	10,745
Control +13% ECL[d]	6	26.92	57.75	342.5	4.36	7,862
F test						*

[a]LWG = live weight gain

[b]FCR = feed conversion ratio

[c]Control diet of rice bran, ensiled cassava roots (32% as DM), fish meal and sweet potato (SP) vines

[d]13% ensiled cassava leaves replaced part of fish meal, all SP vines; cassava leaves had been ensiled with 20% fresh grated cassava roots

[e]Prices:	rice bran	dong	2,000/kg
	fish meal		6,000/kg
	cassava roots		320/kg
	fresh SP vines		400/kg
	cassava leaves		3,000/20 kg

Field Days. At time of harvest, field days were organized at the site in order to harvest the trials and discuss the results. Farmers from neighbouring villages were usually invited to participate in these field days, to evaluate each treatment in the various trials and to discuss the pros and cons of the various practices or varieties tested.

In a few cases, large field days were also organized with participation of hundreds of neighbouring farmers, school children, local and high-level officials, as well as representatives of the press and television. The broadcasting and reporting of these events also helped to disseminate the information about suitable technologies. During the field days farmers explained the results of their own FPR trials to the other visiting farmers, while extension pamphlets and booklets about the farmer-selected technologies were distributed.

Training. Research and extension staff involved in the project had previously participated in Training-of-Trainers courses in FPR methodologies, including practical training sessions with farmers in some of the pilot sites. While some participants were initially skeptical, most course participants became very enthusiastic about this new approach once they started working more closely with farmers.

In addition, two or three key farmers from each site together with their local extension agent were invited to participate in FPR training courses. The objective was to learn about the various FPR methodologies, the basics of doing experiments as well as the implementation of commonly selected technologies, such as setting out contour lines or the planting, maintenance and multiplication of hedgerow species. By spending several days together in these courses, the farmers and extensionist got to know each other well, and they were encouraged to form a local FPR team to help other farmers in their community conduct FPR trials or adopt the new technologies.

Community-based Self-help Groups. Realizing that effective soil conservation practices, such as planting of contour hedgerows, can best be done as a group, farmers from some sites decided to form their own soil conservation group. These community-based self-help groups are similar to Landcare units that have been very effective in promoting soil conservation in Australia and the Philippines. Subsequently, the Department of Agricultural Extension in Thailand encouraged farmers to set up these groups as a way of organizing themselves, to conduct FPR trials, to implement the selected practices, and to manage a rotating credit fund, from which members of the group can borrow money for production inputs. Thus, by 2003, a total of 21 "Cassava Development Villages" had been set up in the pilot sites in Thailand. Each group needed to have at least 40 members, elect five officers to lead the group, and establish its own bylaws about membership requirements, election of officers, use of the rotating fund, etc. The formation of these groups helped to decide on collective action and to strengthen the community, while people gained confidence and the group became more self-reliant. When necessary, the group could request help from local or national extension services, obtain information about certain production problems, or get planting material of vetiver grass or other species for hedgerows or green manures. Some groups started their own vetiver grass nurseries to have planting material available when needed.

Effect of New Technologies on Cassava Yield and Soil Loss by Erosion

Farmers are interested in testing new technologies only if those technologies promise substantial economic benefits over their traditional practices. Thus, strategic and applied research needs to continue to produce and select still better varieties, better production practices and new utilization options. Some collaborative research in the area of agronomy and soil management therefore continued.

Long-term Fertility Maintenance

Long-term NPK trials were continued in four locations, one each in north and south Vietnam, one in Hainan island of China, and one in southern Sumatra of Indonesia. Figure 8 shows the effect of annual applications of various levels of N, P, and K on the yield and starch content of two varieties during the 14th year of continuous cropping in Hung Loc Agricultural Research Center in south Vietnam. It is clear that, similar to most other locations, the main yield response was to the application of K, while there were minor responses to the application of N and P and mainly in the higher yielding variety SM 937-26. The combined application of 160 kg N, 80 P_2O_5 and 160 K_2O/ha increased yields from about 10 t/ha to 30 t/ha.

Effect of Various Soil Conservation Practices on Cassava Yield and Soil Loss by Erosion

Table 13 shows the average effect of various soil conservation practices on relative cassava yields and dry soil loss by erosion from numerous trials conducted in Thailand from 1994 to 2003. Closer plant spacing, lemon grass hedgerows and contour ridging were the most effective in both increasing yields and decreasing erosion. Most other contour hedgerow species, including vetiver grass, decreased cassava yields - mostly by

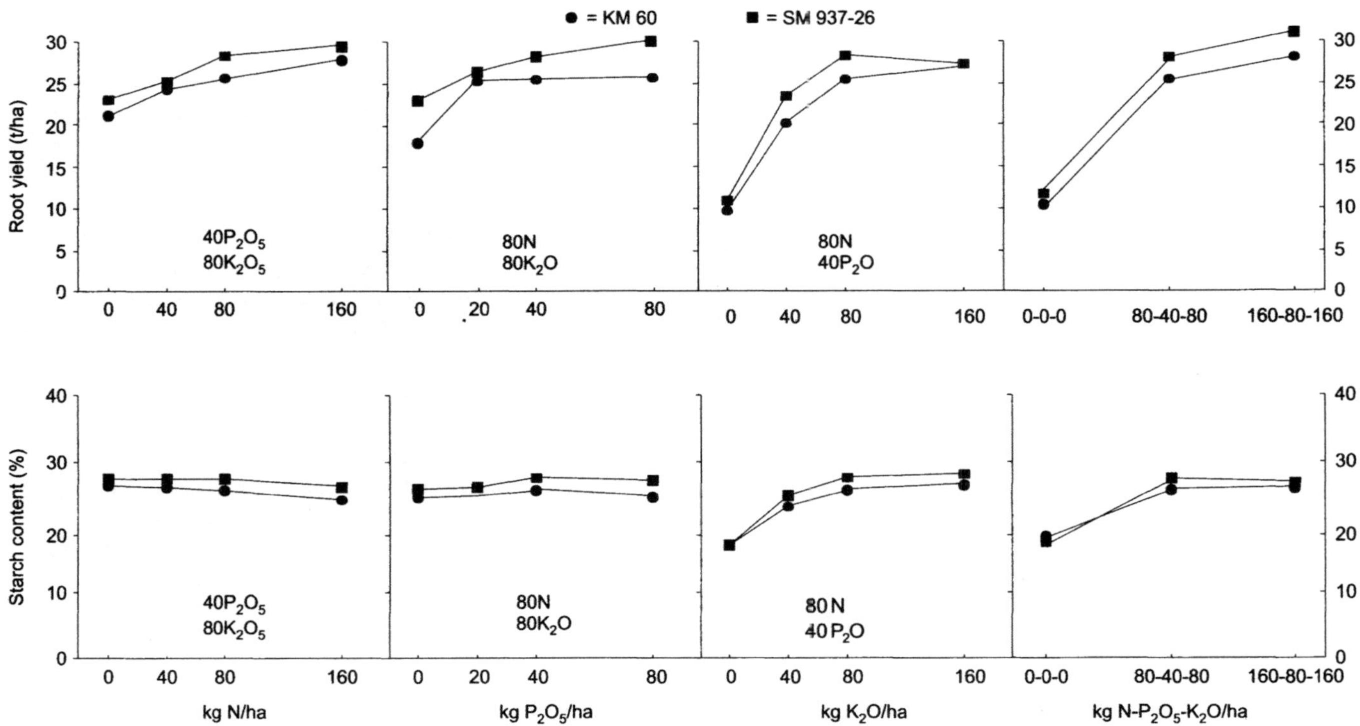

Fig. 8 Effect of annual applications of various levels of N, P and K on the root yield and starch content of two cassava varieties grown at Hung Loç Agric. Research Center in Thong Nhat, Dong Nai, Vietnam in 2003/04 (14th year).

Table 13 Effect of various soil conservation practices on the average[a] relative cassava yield and dry soil loss due to erosion as determined from soil erosion control experiments, FPR demonstration plots and FPR trials conducted in Thailand from 1994 to 2003.

Soil conservation practices[b]	*Relative cassava yield (%)*	*Relative dry soil loss (%)*
1. With fertilizers; no hedgerows, no ridging, no intercrop (check)	100	100
2. With fertilizers; vetiver grass hedgerows, no ridging, no intercrop**	90 (25)	58 (25)
3. With fertilizers; lemon grass hedgerows, no ridging, no intercrop**	110 (14)	67 (15)
4. With fertilizers; sugarcane for chewing hedgerows, no intercrop	99 (12)	111 (14)
5. With fertilizers; *Paspalum atratum* hedgerows, no intercrop**	88 (7)	53 (7)
6. With fertilizers; *Panicum maximum* hedgerows, no intercrop	73 (3)	107 (4)
7. With fertilizers; *Brachiaria brizantha* hedgerows, no intercrop*	68 (3)	78 (2)
8. With fertilizers; *Brachiaria ruziziensis* hedgerows, no intercrop*	80 (2)	56 (2)
9. With fertilizers; elephant grass hedgerows, no intercrop	36 (2)	81 (2)
10. With fertilizers; contour ridging, no hedgerows, no intercrop**	108 (17)	69 (17)
11. With fertilizers; up-and-down ridging, no hedgerows, no intercrop	104 (20)	124 (20)
12. With fertilizers; closer spacing, no hedgerows, no intercrop**	116 (10)	88 (11)
13. With fertilizers; C + peanut intercrop	72 (11)	102 (12)
14. With fertilizers; C + pumpkin or squash intercrop	90 (13)	109 (15)
15. With fertilizers; C + sweetcorn intercrop	97 (11)	110 (14)
16. With fertilizers; C + mungbean intercrop*	74 (4)	41 (4)
17. No fertilizers; no hedgerows, up/down or no ridging	96 (9)	240 (10)

[a]Number in parenthesis indicates the number of experiments/trials from which the average values were calculated.

[b]C = Cassava.

** = most promising soil conservation practices; * = promising soil conservation practices.

reducing the area available for cropping and by competition with nearby cassava - but were very effective in reducing soil loss by erosion. Most effective in reducing erosion were vetiver grass, *Paspalum atratum* and lemon grass, which reduced erosion by 33% to 47%. Intercropping was usually not effective in reducing erosion, while up-and-down ridging and especially the lack of fertilization markedly increased erosion. Similar results were obtained in Vietnam where hedgerows of vetiver grass, *Tephrosia candida* and *Paspalum atratum* all decreased erosion by about 50%, while also increasing cassava yields 10-13% (Howeler et al., 2005a,b).

The beneficial effects of contour hedgerows tend to increase markedly over time. Figure 9 shows the long-term effect of contour hedgerows of vetiver grass and *Tephrosia candida* on relative cassava yields and soil loss as compared to the check plot without hedgerows; data are average values from three FPR erosion control trials conducted by farmers for nine consecutive years in north Vietnam. Although the results are rather variable, there is a clear trend that the two types of hedgerows caused a 20-40% increase in cassava yields and reduced soil losses by erosion to 20-40% of those in the check plots without hedgerows. Vetiver grass tended to become more effective in reducing soil losses than *Tephrosia*, first because the grass is more effective in filtering out suspended soil sediments and second because *Tephrosia* hedgerows need to be replanted every three to four years, in contrast to vetiver grass, which is a more or less permanent barrier. While farmers claim that *Tephrosia* improves the

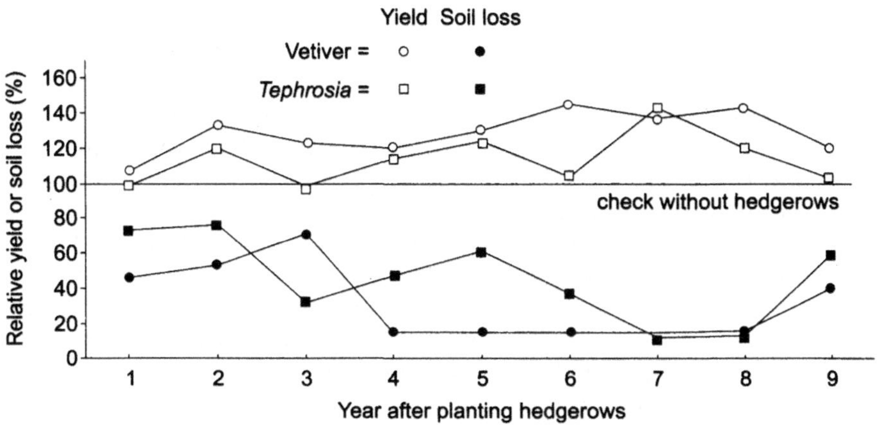

Fig. 9 Trend in relative yield and relative soil loss by erosion when cassava was planted with contour hedgerows of vetiver grass or Tephrosia candida during nine consecutive years of cassava cropping. Data are average values for one FPR erosion control trial in Kieu Tung and two trials in Dong Rang in north Vietnam from 1995 to 2003.

fertility of the soil more than vetiver grass, the data show that vetiver grass increased cassava yields more than *Tephrosia*, probably by reducing losses of top soil and fertilizers and improving water infiltration and soil moisture content.

ADAPTATION

After two to three years of testing various options in FPR trials, slowly narrowing down the number of best options, farmers started to adopt some of the tested varieties or practices on their bigger production fields. In some cases they adapted the practices to suit a larger scale. For instance, in Thailand farmers planted contour hedgerows of vetiver grass on their fields but left enough space between hedgerows (usually 30-40 m) to facilitate land preparation by tractor. In some cases, especially in Vietnam, farmers planted hedgerows on plot borders rather than along contour lines. This reduces the amount of land occupied by hedgerows, but also reduces their effectiveness in controlling erosion.

While contour hedgerows of vetiver grass are usually the most effective in reducing soil losses by erosion in experiments and FPR trials conducted in small plots on a uniform slope, the results are often disappointing when this practice is scaled up to a larger production field. In areas of rolling terrain large amounts of runoff water may accumulate and run down-slope in natural drainage ways. The force of the water is likely to wash out vetiver grass recently planted along the contour across the drainage way, and this may result in serious gully erosion. Attempts to repair these gullies by placing sand bags or other obstacles across them have usually failed as these obstacles are washed away, too. Over the past few years farmers and project staff have experimented informally with ways to reduce the speed of water in these gullies. They found that it is most effective to place a row of soil-filled plastic fertilizer bags across the gully in line but slightly below the washed out vetiver hedgerow. The bags need to be secured in place by pounding bamboo stakes into the soil behind them (Fig. 10). Once eroded soil is deposited in the gully above the soil bags, vetiver grass can be planted in this moist and fertile sediment. When the vetiver grass is well established across the gully and in line with the rest of the hedgerow, this will further slow the speed of runoff water, resulting in further deposition of sediments in the gully above the vetiver hedgerow. This allows weeds to re-establish in the gully bottom, protecting the gully from further erosion. With the next ploughing along the contours, parallel to the hedgerows, the gully will generally be filled up again with soil, while the hedgerow prevents further gully formation (Fig. 10). In some sites in Thailand, terraces up to 1 m height were formed

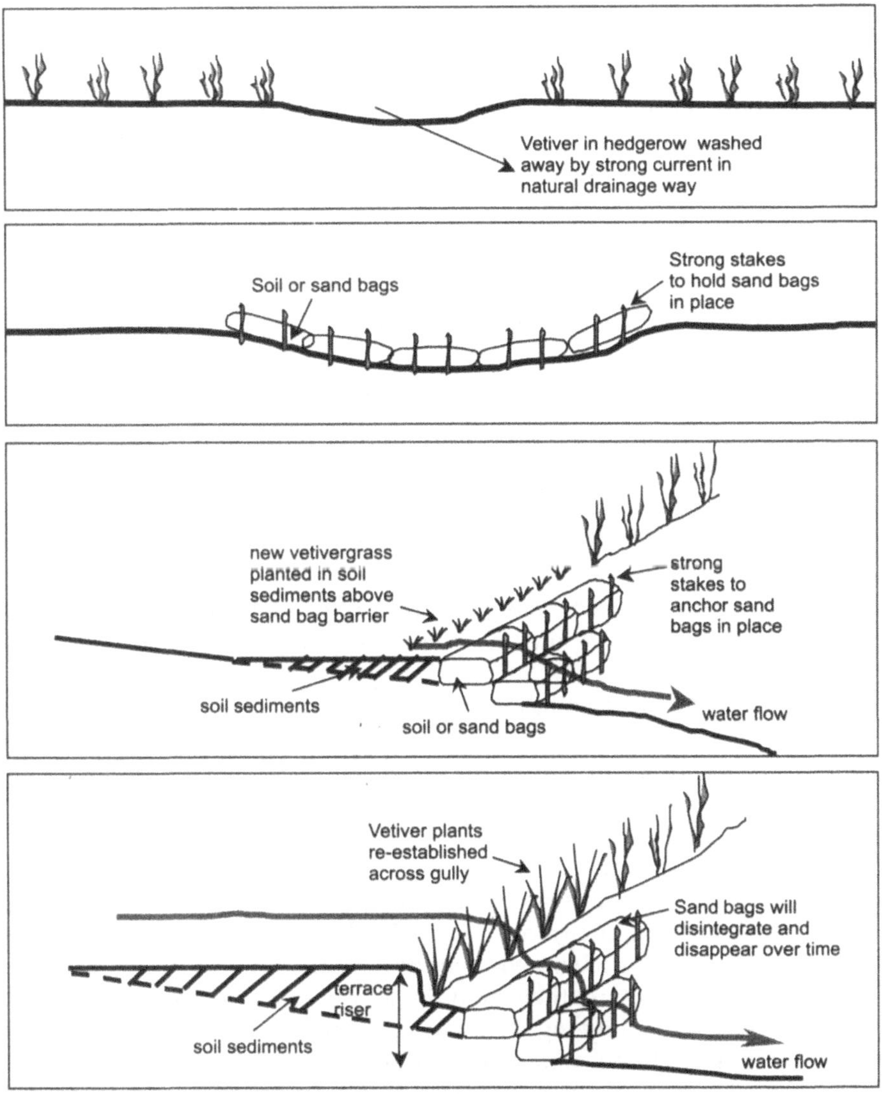

Fig. 10 Simple and effective way to repair gullies by placing soil bags across gully and planting vetiver grass in the soil sediments accumulating above the barrier.

within two years by the placing of soil bags and planting of vetiver hedgerows across the gully. This local adaptation of the traditional contour hedgerow system markedly increased its effectiveness under real field conditions.

ADOPTION

After conducting their own FPR trials, or after a cross-visit to another village where those trials were being conducted, farmers often decided to adopt one or more technologies on their production fields with the hope of increasing yields or income and protecting the soil from further degradation.

In Thailand, practically all of the cassava area is now planted with new varieties and about 75% of farmers apply some chemical fertilizers (TTDI, 2000), although usually not enough or not in the right proportion. As a result of the FPR fertilizer trials, farmers started to apply more K, while the official fertilizer recommendation for cassava was changed from an NPK ratio of 1:1:1 to 2:1:2. After trying various ways of controlling erosion, most farmers selected the planting of vetiver grass contour hedgerows as the most suitable. By the end of 2003, about 1,038 farmers had planted a total of 1.63 million vetiver plants, corresponding to about 145 km of hedgerows (Howeler et al., 2003, 2004, 2005a,b; Vongkasem et al., 2003).

In August 2002 a participatory monitoring and evaluation (M&E) was conducted in four pilot sites in Thailand where the project had been initiated at least four years earlier. Using focus group discussions and participatory evaluation methodologies, data were collected on the extent of adoption of the various technologies and the reasons for adoption or non-adoption. Table 14 shows that new varieties had been adopted in 100% of the cassava-growing areas in all four sites. Application of chemical fertilizers varied from 79% to 100%, vetiver hedgerows were planted in 22-55% of the cassava area, green manures were planted in 0-50%, and intercropping was not adopted at all, mainly due to lack of labour for managing intercrops. Table 15 shows in more detail how the various technologies changed over the years, mainly as a result of conducting FPR trials on their own fields. While in most sites some new varieties (Rayong 3, Rayong 60, Rayong 90) were already planted before the project started, the mix of new varieties changed over the years as higher-yielding varieties were released, tested and adopted. The data also indicate how the use of chemical fertilizers not only increased over time, but also changed from the standard 15-15-15 to various formulations high in N and K and low in P.

Table 16 shows how in Vietnam the number of households in the pilot sites adopting the various technology components increased over time, with most farmers adopting new varieties. This is partly due to the testing in FPR variety trials but is also due to the planting of new varieties by non-participating farmers in or near the pilot sites. For instance, during 2002

Table 14 Extent of adoption[a] of various cassava technology components in four pilot sites in Thailand in 2002 as a result of the Nippon Foundation project.

Technology component	*Baan Khlong Ruam Sra Kaew*		*Thaa Chiwit Mai Chachoengsao*		*Sapphongphoot Nakhon Ratchasima*		*Huay Suea Ten Kalasin*	
	(ha)	*(%)*	*(ha)*	*(%)*	*(ha)*	*(%)*	*(ha)*	*(%)*
Varieties	480	100	469	100	396	100	228	100
Chemical fertilizers	480	100	469	100	364	92	180	79
Vetiver grass hedgerows	139	29	94	20	218	55	89	39
Green manures	72	15	0	0	0	0	114	50
Intercropping	0	0	0	0	0	0	0	0

[a]Estimated by farmers in each site during participatory M&E in Aug 2002.

and 2003, farmers in Van Yen district of Yen Bai province in north Vietnam planted a total of 500 km of double hedgerows of *Tephrosia candida* or *Paspalum atratum* to control erosion, and they planted about 3,000 ha of new cassava varieties with improved fertilizer practices. This increased average yields from 10 t/ha to about 30 t/ha. Figure 11 shows how the number of farmers in the pilot sites adopting various soil conservation measures increased year after year, initially mostly in Thailand but subsequently also in Vietnam.

Data in Table 17 indicate that adoption of soil conservation practices in all sites in Vietnam increased yields, ranging from 13.5% in 2000 to 23.7% in 2002. As a result of the adoption of soil conservation practices, gross income, both per hectare and per household, also increased very markedly over time. Results from both FPR trials and on-station research also indicate that the beneficial effect of contour hedgerows in terms of increasing yields and decreasing erosion increased over time (Fig. 9) (Howeler et al., 2005a). This is mainly because the planting of contour hedgerows, almost independent of the species used, will result in natural terrace formation, which over time reduces the slope and enhances water infiltration, thus reducing runoff and erosion. Well-established hedgerows also become increasingly more effective in trapping eroded soil and fertilizers. Unfortunately, most FPR erosion control trials are conducted for only one to two years at the same site, so farmers do not quite appreciate the increases in beneficial effects that result over time. This, coupled with the fact that planting and maintaining hedgerows requires additional labour (and sometimes money for seed or planting material), while hedgerows take some land out of production and have initially little beneficial effect on yield, has hampered the more widespread acceptance and adoption of these soil conservation practices.

Table 15 Change in the use of new cassava production technologies[a] in four pilot sites[b] in Thailand from 1995 to 2002[c] as a result of the Nippon Foundation project.

Technology	*Baan Khlong Ruam*			*Thaa Chiwit Mai*			*Sapphongphoot*			*Huay Suea Ten*		
component	*1993*	*1995*	*2002*	*1995*	*1997*	*2002*	*1995*	*1997*	*2002*	*1995*	*1997*	*2002*
Varieties	R90 (60%) R3 (30% R60 (10%)	R90 (60%) R5 (20%) KU50 (20%)	R5 (67%) R90 (19%) KU50 (12%) R72 (2%)	R1 (94%) R60 (3%) R5 (3%)	KU50 (41%) R60 (32%) R5 (22%) R90 (5%)	KU50 (81%) R5 (18%) R72 (1%)	R1 R60 R90	KU50 R5 R90	KU50 (91%) R90 (5%) R72 (3%) R5 (1%)	R1 R90 KU50	KU50 R5 R90	KU50 (54%) R5 (20%) R90 (15%) R72 (11%)
Chemical fertilizers	not apply	15-15-15 13-13-21	15-15-15 (35%) 13-13-21 (17%) 21-4-21 (13%) 14-4-24 (10%) 16-20-0 (5%) other (20%)	not apply	15-15-15	15-15-15 (50%) 13-13-21 (38%) other (12%)	not apply or 15-15-15 (little)	15-15-15 46-0-0	15-15-15 (44%) 46-0-0 (27%) 13-13-21 (4%) other (25%)	not apply or 15-15-15 (little)	15-15-15 and 16-8-8 mixed at 2:1 ratio	15-15-15 (47%) 16-8-8 (33%) 21-0-0 (12%) 46-0-0 (7%) 13-13-21 (1%)
Vetiver grass	not plant	46%	29%	not plant	3%	20%	not plant	70%	55%	not plant	32%	39%
Green manures	not plant	not plant	*Canavalia* (little) cowpea (little)	not plant	not plant	*Canavalia* (little)	not plant	not plant	*Canavalia* (little) *Crotalaria* (little)	not plant	*Canavalia* (20%)	*Canavalia* (50%)

[a]Data collected from participatory M&E with farmers in August 2002; percentages are in terms of cassava area.
[b]Baan Khlong Ruam village, Wang Soombuun district, Sra Kaew province; Thaa Chiwit Mai village, Sanaam Chaikhet district, Chachoengsao province; Sapphongphoot village, Soeng Saang district, Nakhon Ratchasima; Huay Suea Ten village, Sahatsakhan district, Kalasin province.
[c]Nippon Foundation project started in these pilot sites around 1997, except in Baan Khlong Ruam where it started in 1995.

Table 16 Trend of adoption of new cassava technologies in the Nippon Foundation project sites in Vietnam from 2000 to 2003.

Technology component	*Number of households adopting* 2000	2001	2002	2003
New varieties	88	447	1,637	14,820
Improved fertilization	64	123	157	1,710
Soil conservation practices	62	200	222	831
Intercropping	127	360	689	4,250
Pig feeding with cassava root silage	-	759	967	1,172

[a]Number of project sites: 1999 = 9; 2000 = 15; 2001 = 22; 2002 = 25; 2003 = 34.
Source: Tran Ngoc Ngoan, 2003.

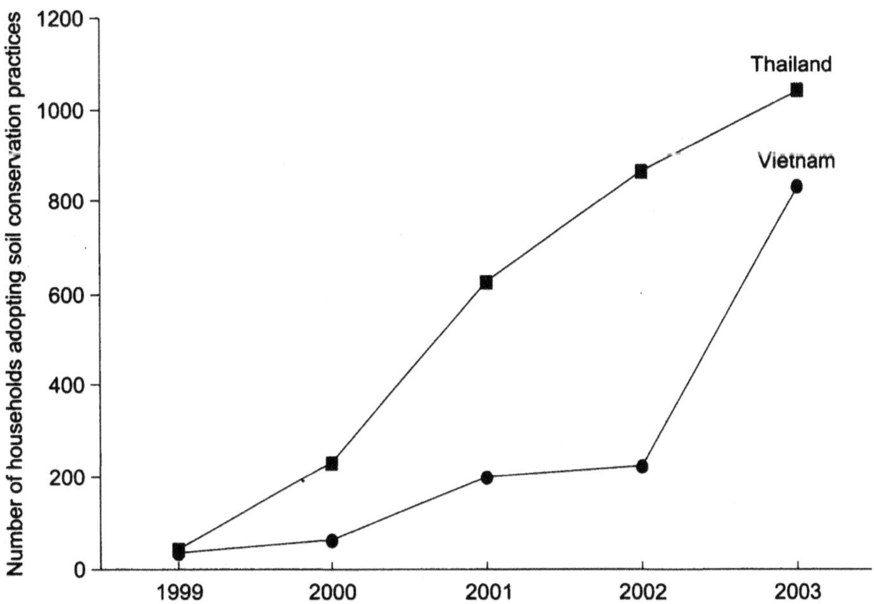

Fig. 11 Number of farmers adopting soil conservation measures in their cassava fields in FPR pilot sites in Thailand and Vietnam from 1999 to 2003.

Table 18 shows in more detail how the adoption of various technologies increased over time in one commune in Pho Yen district of Thai Nguyen province where the project first started working in 1994. Since 1995 farmers have conducted FPR trials on new varieties, more balanced fertilization, intercropping, and erosion control. After some years of testing, farmers initially adopted new varieties and intercropping in small areas of their land. This was followed by better fertilization and

Table 17 Extent of adoption of soil conservation practices and the estimated increase in yield and gross income of farmers in the FPR pilot sites in Vietnam from 2000 to 2003.

Year	Number of farmers	Area with soil conservation (ha)	Cassava yield (t/ha) Farmers' practice[a]	With soil conservation	Percent yield increase	Increase in gross income (mil VND)[b] per ha	total	per household
2000	62	21.12	12.11	13.75	13.5	0.574	12.123	0.196
2001	200	59.87	16.50	19.95	20.9	1.112	66.596	0.333
2002	222	88.85	20.60	25.48	23.7	1.952	173.728	0.782
2003	831	612.00	20.60[c]	25.48[c]		1.561	955.699	1.150
Total	**831**	**612.00**					**1,208.146**	

[a]Farmers' practice includes most new technologies except soil conservation.
[b]Fresh root price: in 2000, 350 VND/kg; in 2001, 350 VND/kg in north, 200 in central and 290 in south; in 2002, 400 VND/kg; in 2003, 320 VND/kg (estimated).
[c]Yields estimated from 2002.
Source: Tran Ngoc Ngoan, 2003.

erosion control; the latter was adopted by only a small number of farmers as most cassava fields in the commune are on gentle slopes or on terraced land. It is clear that the adoption of new technologies increased yields significantly, of both the local variety Vinh Phu and the new varieties, mainly KM 95-3 and KM 98-7. The gradual increases in yield, from 8.5 t/ha in 1994 (Table 1) to 36.8 t/ha in 2003, were accompanied by an increase in area planted using new technologies, resulting in about a 20-fold increase in net income and marked improvements in the livelihood of farmers in this commune.

Table 19 summarizes the extent of adoption of new cassava technologies in FPR pilot sites in 15 provinces of Vietnam in 2003 and the resulting increase in gross income due to higher yields obtained. Although balanced fertilization produced the greatest yield increase, it was not adopted over a very wide area. New varieties were most widely adopted resulting in the greatest increase in gross income. The total annual increase in gross income due to adoption of new technologies in the FPR sites was estimated at 1.67 million US dollars or $72.92 per household.

ASSESSMENT OF IMPACT

In order to determine more precisely the effect of this project on adoption of new technologies, an impact assessment was made by an outside consultant. He organized focus group discussions and collected data from

Table 18 Impact of the adoption of new cassava varieties and improved production practices on the livelihoods of farmers in Tien Phong commune, Pho Yen district of Thai Nguyen, Vietnam.

Year	Variety or practice[a]	No. of farmers	Cassava area (ha)	Cassava yield (t/ha)	Peanut yield (t/ha)	Gross income[b]	Production costs (mil. dong/ha)	Net income	Total net income (mil. dong)
1994[c]	Vinh Phu	115	50	8.5	-	3.40	2.93	0.47	23.50
	New varieties	0	-	-	-	-	-	-	-
			50						**23.50**
2000	Vinh Phu	NA[d]	NA	21.5	-	NA	NA	NA	NA
	New varieties	25	1.31	30.9	-	15.45	4.36	11.10	14.54
	Intercropping	37	2.59	29.3	0.81	18.70	6.16	12.54	32.48
	Erosion control	4	0.20	24.7	-	12.35	4.66	7.69	1.54
			>4.10						**>48.56**
2001	Vinh Phu	61	2.17	22.7	-	11.35	4.36	6.99	15.17
	New varieties	122	4.70	29.0	-	14.50	4.36	10.14	47.66
	Intercropping	40	3.38	26.2	0.77	16.94	6.16	10.78	36.44
	Erosion control	4	0.20	NA	-	NA	NA	NA	NA
			10.45						**>99.27**
2002	Vinh Phu	18	0.64	25.4	-	12.70	4.33	8.37	5.36
	New varieties	100	5.16	33.7	-	16.85	4.33	12.52	64.60
	Intercropping	118	3.69	32.3	1.73	24.80	6.13	18.67	68.89
	Balanced fert.	48	2.95	33.4	-	16.70	4.83	11.87	35.02
	Erosion control	5	0.18	25.4	-	12.70	4.63	8.07	1.45
			12.62						**175.32**

(Table 18 Contd.)

(Table 18 Contd.)

2003	Vinh Phu	NA	NA	NA	-	NA	NA	NA	NA
	New varieties	225	17.00	36.8	-	18.40	4.33	14.07	239.19
	Intercropping	120	11.00	36.0	0.67	21.35	6.13	15.22	167.42
	Balanced fert.	54	3.40	33.6	-	16.80	4.83	11.97	40.70
	Erosion control	5	0.60	27.0	-	13.50	4.63	8.87	5.32
			>32.00						**>452.63**

[a]In Tien Phong farmers traditionally grow mainly Vinh Phu variety but have now largely changed to KM 95-3 and KM 98-7; the new practices include intercropping with peanut, balanced fertilization of 10 t/ha of pig manure plus 80N-40P_2O_5-80 K_2O, and erosion control by contour hedgerows of *Tephrosia candida*.

[b]Price of cassava in 1994: 400 VND/kg fresh roots. Price of cassava in 2000-2003: 500 VND/kg fresh roots. Price of peanut in 2000-2003: 5,000 VND/kg dry pods.

[c]Data from RRA at the start of project.

[d]NA = data not available.

Table 19 Extent of adoption of new cassava production technologies in FPR pilot sites in 15 provinces of Vietnam in 2003/04, the effect on cassava yields, and the increase in gross income resulting from the yield increase in those sites.

Technology component	*No. of households*	*Area (ha)*	*Cassava yield (t/ha)*		*Increase in gross income ('000 US$)*[b]
			Farmers' practice[a]	*Improved technology*	
1. New varieties	14,820	7,849	19.93	28.95	1,462
2. Balanced fertilization	1,710	607	21.37	30.50	114
3. Soil conservation practices	831	612	20.60	25.48	62
4. Intercropping	4,250	160	29.95	28.94	15[d]
5. Root and leaf silage for pig feeding	1,172	-[c]	-	-	12
Total	**22,833**	**9,228**			**1,665**

[a]Farmers' practice usually includes most new technologies except the technology being tested.
[b]Based on a price of 320 VND/kg fresh roots in 2003/04; 1 US$ = 15,500 VND.
[c]3,370 pigs.
[d]Increase in gross income from the harvest of intercrops.
Source: Tran Ngoc Ngoan, 2003.

farmers in eight representative project sites - four sites in Thailand and four in Vietnam - as well as from farmers living within 10 km of those sites, who had not participated in the project. Table 20 shows the percentage of households (out of 767) that had adopted various technologies. New varieties were adopted[5] by nearly all cassava farmers in the eight sites in Thailand and by 70% of farmers in Vietnam; the use of chemical fertilizers had been adopted by 85-90% of households in the eight sites in each country; intercropping was adopted by nearly 60% of households in Vietnam, but by only 13% in Thailand. Contour ridging was adopted by about 30% of households in both Vietnam and Thailand, while contour hedgerows were adopted by 23% of households in Thailand and 25% in Vietnam; in Thailand these hedgerows were almost exclusively vetiver grass, while in Vietnam most farmers preferred the planting of *Tephrosia candida* or *Paspalum atratum*, as these are easier to plant (from seed) and can also serve as a green manure and animal feed, respectively. Thus, it is clear that adoption of specific practices varies from site to site, depending on local conditions and traditional practices. Table 20 also indicates that there were highly significant differences in the adoption of almost all the technologies between participating and non-participating farmers (with the exception of contour ridging and the use of chemical fertilizers in

[5]Planted in 50% or more of the farmer's total cassava area.

Table 20 Extent of adoption (percentage of households)[a] of new technologies by participating and non-participating farmers in the cassava project in Thailand and Vietnam in 2003 (n = 767).

	Thailand			*Vietnam*			*Full sample*		
	Partic.	*Non-partic.*	*Total*	*Partic.*	*Non-partic.*	*Total*	*Partic.*	*Non-partic.*	*Total*
Varieties									
- 100% improved varieties	100	88.0	91.1***	50.0	38.8	42.9***	73.2	67.3	69.1***[b]
- 75% improved varieties	0	11.7	8.6	5.6	6.7	6.3	3.0	9.6	7.6
- 50% improved varieties	0	0.3	0.2	26.2	18.3	21.1	14.0	7.9	9.8
- 25% improved varieties	0	0	0	4.0	5.4	4.9	2.1	2.3	2.2
- no improved varieties	0	0	0	14.3	30.8	24.9	7.7	13.0	11.3
Soil conservation practices									
- contour ridging	52	22	30***	35	31	33	43	26	31***
- hedgerows	60	10	23***	50	12	25***	54	11	24***
- vetiver grass	60	10	23***	10	3	5**	33	7	15***
- *Tephrosia candida*	0	0	0	38	6	18***	20	3	8***
- *Paspalum atratum*	1	0	0*	12	2	6***	7	1	3***
- pineapple	0	0	0	2	1	1	1	0	1
- sugarcane	2	1	1	0	0	0	1	0	1
- other hedgerows	3	0	1*	7	1	3***	5	1	2***
- no soil conservation	21	72	59***	23	58	45***	22	67	53***
Intercropping	28	8	13***	79	49	59***	55	25	34***
- with peanut	1	1	1	47	33	38***	26	14	18***
- with beans	0	0	0	27	29	29	14	12	13
- with maize	3	10	5***	2	3	3	6	3	4*
- with green manures	19	4	8***	0	0	0	9	2	4***
- with other species	3	2	2	39	15	24***	22	7	12***
Fertilization									
- chemical fertilizers	98	86	89***	85	86	86	91	86	87***
- farmyard or green manure	55	25	33***	74	60	65**	65	40	48***
- no fertilizer	0	13	9***	12	8	9	6	11	9*

[a] Percentages may total more than 100% as households can adopt more than one type of technology simultaneously.

Significant differences between participants and non-participants: * P<=0.10 ** P<=0.05 *** P<=0.01.

[b] Level of significance in this case refers to differences between participants and non-participants in terms of the categorical distribution, *not* the adoption levels.

Vietnam); participating farmers had a greater extent of adoption than non-participating farmers. In this case, "participants" were defined as farmers who had conducted at least one FPR trial and/or had participated in an FPR training course, while "non-participants" had done neither but may have attended a farmer field day organized by the project. It can be seen that new varieties and the use of chemical fertilizers were readily adopted by both participants and non-participants, while adoption of soil conservation practices and intercropping was both less widespread and largely limited to participating farmers. This clearly points to the difficulty of achieving spontaneous and widespread adoption of soil conservation practices.

But how does adoption of these new technologies translate into higher yields and income? Figure 12 shows the cassava yields that farmers reported before and after the project, corresponding more or less to the second phase of the project, or from 1999 to 2003. In Thailand the yields of participating farmers increased from 19.4 to 25.8 t/ha (33%), while yields of non-participating farmers increased from 15.5 to 20.3 t/ha (31%); in Vietnam project participants increased yield from 13.7 to 28.2 t/ha (106%), while non-participants increased their yields from 14.3 to 23.9 t/ha (67%) (Lilja et al., 2005). Thus, in both countries yields increased very markedly,

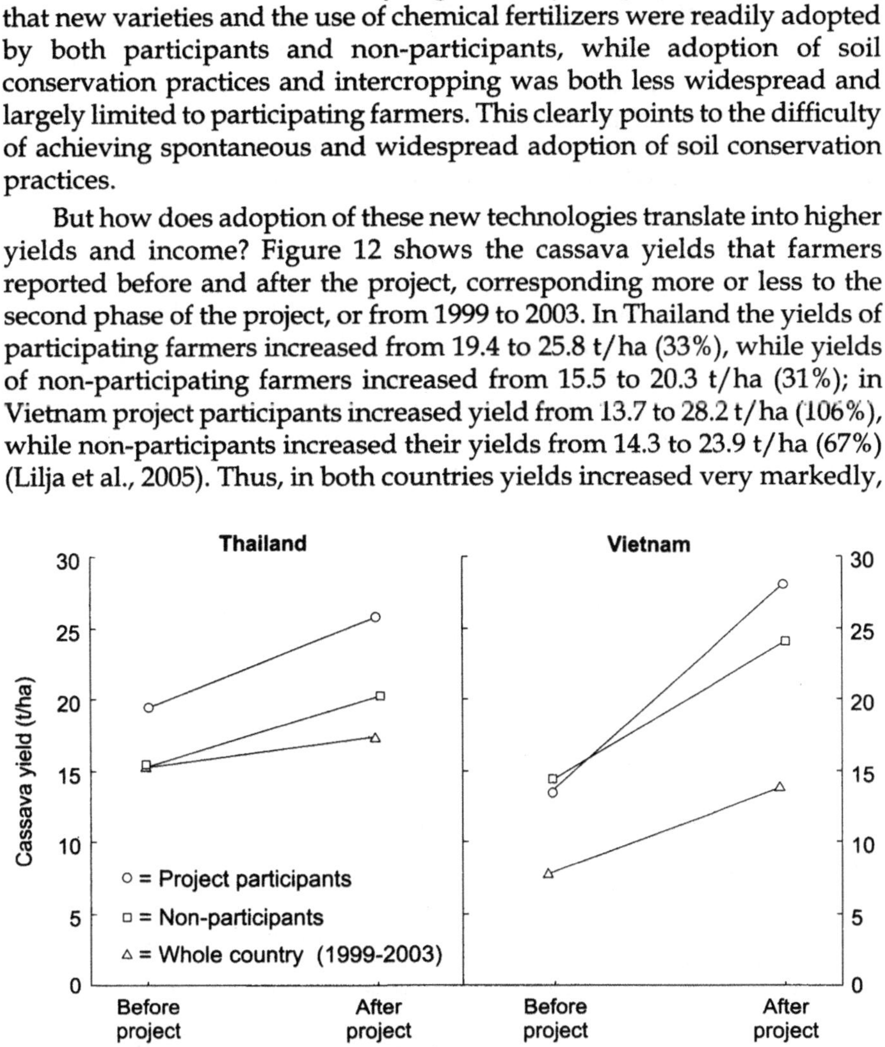

Fig. 12 Average cassava yields of farmers participating in the Nippon Foundation cassava project or of nearby but non-participating farmers, before the project started and at the end of the project. Data are from focus group census forms collected from 417 households in Thailand and 350 households in Vietnam. For comparison the national average cassava yields in 1999 (before) and 2003 (after) are also shown.

but these increases were greater for participants than for non-participants, especially in Vietnam. For comparison, Fig. 12 also shows the increase in yield for the whole country, as reported by FAO during approximately the same time period. Yields for the whole of Vietnam are considerably below those reported by the farmers in the focus groups; but the yield increases are similar to those reported by the non-participants. In Thailand the initial yields in the country were similar to those of non-participating farmers, but after-project yields were much higher for participants as well as nearby non-participants than for the country as a whole. This indicates that participating farmers benefited most from their experiences but that nearby farmers also benefited indirectly from the project.

Table 21 shows that during the past nine years the average cassava yields in all three countries increased; this increase ranged from 1.03 t/ha in China to 5.65 t/ha in Vietnam. The increased yields resulted in annual increases in gross income received by farmers of about 145 million US dollars in the three countries, and about 270 million US dollars in all of Asia. In addition, farmers in Thailand received higher prices due to the higher starch content of the new varieties. This was achieved not only by this project, but by the collaborative effort of many researchers, extensionists, factory owners and farmers, with strong support from national governments.

Table 21 Estimation of the annual increase in gross income due to higher cassava yields resulting from the adoption of new cassava varieties and improved practices, in China, Thailand and Vietnam, as well as in Asia as a whole.

Country	*Total cassava area (ha)*[a]	*Cassava yield (t/ha)*[a]		*Yield increase (t/ha)*	*Cassava price (US$/ton)*	*Increased gross income due to higher yields (mil. US$)*
		1994	*2003*			
China	240,110	15.22	16.25	1.03	27	6.7
Thailand	1,050,000	13.81	17.55	3.74	22	86.4[b]
Vietnam	371,900	8.41	14.06	5.65	25	52.5
Asia total	3,463,460	12.93	16.04	3.11	25	269.3

[a]Data from FAOSTAT, 2004; according to Chinese officials, the FAO data shown above grossly underestimate the actual cassava area and yield in 2003.

[b]Farmers also benefited from higher prices due to higher starch content.

RATE OF RETURN ON THE RESEARCH INVESTMENT

To calculate the internal rate of return (IRR) on investment of this project, we need to calculate the total costs and the total benefits that can be attributed directly to the project. The total costs of the project in Thailand and Vietnam were calculated as two-thirds of the Nippon Foundation annual budget over a 10-year period, plus contributions for salaries of national staff and other expenses provided by the two national governments. These costs totaled about US$3.5 million (Lilja et al., 2005).

Benefits were calculated by adding up the incremental yield increases obtained as a result of participation in the project (9.1 t/ha), by the adoption of contour hedgerows (2.7 t/ha) or of new varieties (up to 6.3 t/ha depending on the extent of adoption) multiplied by the average area in each village affected by either participation or the particular technology adopted. According to these calculations each village on average increased its cassava production by 1,895 tons as a result of the project. Since there were 67 project villages in Thailand and Vietnam and the price of fresh cassava roots was about US$25 per ton, this translates into a total annual benefit of US$3.2 million. If we assume a linear rate of adoption between 1998 and 2004 the project had an IRR of 33% over that period, or an IRR of 37% if we assume that adoption will continue at a similar rate until 2008 (Lilja et al., 2005).

CONCLUSIONS

Research on sustainable land use conducted in the past has mainly concentrated on finding solutions to the biophysical constraints, and many solutions have been proposed for improving the long-term sustainability of the system. Still, few of these solutions have actually been adopted by farmers, mainly because they ignored the human dimension of sustainability. For new technologies to be truly sustainable they should not only maintain the productivity of the land and water resources, but also be economically viable and acceptable to farmers and the community. To achieve those latter objectives farmers must be directly involved in the development, adaptation and dissemination of these technologies. A farmer participatory approach to technology development was found to be very effective in developing locally appropriate and economically viable technologies, which in turn enhances their acceptance and adoption by farmers.

The conducting of FPR trials is initially time consuming and costly, but once more and more people are trained and become enthusiastic about the use of this approach – including participating farmers – both the methodology and the selected improved varieties or cultural practices will spread rapidly. The selection and adoption of those farming practices that are most suitable for the local environment and in tune with local traditions will improve the long-term sustainability of the cropping system, to the benefit of both farmers and society as a whole.

ACKNOWLEDGEMENTS

The following persons in national programmes were also actively involved in the second phase of the project and contributed significantly to its success:

Thailand

Dr. Surapong Charoenrath, Mr. Danai Suparhan, Mr. Somphong Katong, Mr. Samnong Nual-on, Mr. Suttipun Brohmsubha, Mr. Kaival Klakhaeng, Mrs. Wilawan Vongkasem, Mr. Suwit Phomnum, Mr. Apichart Chamroenphat, Mr. Chanchai Wiboonkul, Ms. Sunan Muuming, Ms. Naruemol Pukpikul, Mrs. Chonnikarn Jantakul, Ms. Methinee Keerakiat, Mr. Somchit Chinno, Mrs. Nuttaporn Jaruenjit, Mr. Chaipipop Yotachai, Mr. Thinnakorn Withayakorn, Mr. Nava Takraiklaang, Mrs. Anurat Srisura, Mr. Choosak Aksonvongsin, Mr. Wirawudhi Kaewpreechaa, Mr. Chatphon Wongkaow, Mr. Sanga Saengsuk, Mr. Parinya Phaithuun, Mr. Sonsing Srisuwan, Mr. Numchai Phonchua, Mr. Prayoon Kaewplod, Mr. Sanit Taptanee, Mr. Sanit Phuumphithayanon, Mr. Banyat Vankaew, Mr. Preecha Petpraphai, Mrs. Supha Randaway, Mrs. Kittiporn Srisawadee, Mr. Decha Yuphakdee and Dr. Somjate Jantawat.

Vietnam

Dr. Nguyen The Dang, Mr. Nguyen Viet Hung, Mr. Nguyen The Nhuan, Dr. Thai Phien, Mr. Tran Minh Tien, Mr. Nguyen Hue, Mrs. Trinh Thi Phuong Loan, Mr. Hoang Van Tat, Mrs. Nguyen Thi Cach, Mrs. Nguyen Thi Hoa Ly, Dr. Hoang Kim, Mr. Nguyen Huu Hy, Mr. Vo Van Tuan, Mr. Tong Quoc An, Mr. Tran Cong Khanh, Dr. Tran Thi Dung and Mrs. Nguyen Thi Sam.

China

Mr. Li Kaimian, Mr. Huang Jie, Mr. Ye Jianqiu, Mr. Tian Yinong, Mr. Li Jun, Mr. Ma Chongxi, Mrs. Chen Xian Xiang, Mrs. Pan Huan and Mr. Liu Jian Ping.

We also want to express our gratitude to the Nippon Foundation in Tokyo, Japan, which generously supported this project over a 10-year period, from 1994 to 2003.

References

FAOSTAT, 2004. http://apps.fao.org

Howeler, R.H. 2001. The use of farmer participatory research (FPR) in the Nippon Foundation Project: Improving the sustainability of cassava-based cropping systems in Asia. pp. 461-489. *In*: R.H. Howeler and S.L. Tan. [eds.] *Cassava's*

Potential in Asia in the 21st Century: Present Situation and Future Research and Development Needs. Proc. 6th Regional Workshop, held in Ho Chi Minh City, Vietnam. 21-25 February 2000.

Howeler, R.H. 2002. The use of a participatory approach in the development and dissemination of more sustainable cassava production practices. pp. 42-51. *In*: M. Nakatani and K. Komaki. [eds.] *Potential of Root Crops for Food and Industrial Resources.* Proc. 12th Symp. Intern. Soc. Trop. Root Crops, held in Tsukuba, Japan, 11-16 September 2000.

Howeler, R.H. 2004. A participatory and inter-institutional project to enhance the sustainability of cassava production in Thailand, Vietnam and China: Its impact on soil erosion and farmers' income. Paper presented at Intern. Conf. on Interdisciplinary Curriculum and Research Management in Sustainable Land Use and Natural Resource Management, held in Bangkok, Thailand. 17-19 August 2004. Paper distributed on CD.

Howeler, R.H., W. Watananonta, W. Vongkasem, K. Klakhaeng, S. Jantawat, S Randaway and B. Vankaew. 2003. Working with farmers: The key to adoption of vetiver grass hedgerows to control erosion in cassava fields in Thailand. pp. 12-22. *In:* P. Truong and Xia Hanping. [eds.] *Vetiver and Water.* Proc. 3rd Intern. Conf. on Vetiver and Exhibition, held in Guangzhou, P.R. China, 6-9 October 2003.

Howeler, R.H., W. Watananonta and Tran Ngoc Ngoan. 2004. Farmers decide: A participatory approach to the development and dissemination of improved cassava technologies that increase yields and prevent soil degradation. *In:* Proc. 13th Symp. Intern. Soc. Tropical Root Crops, held in Arusha, Tanzania, 10-14 November 2003 (in press).

Howeler, R.H., W. Watananonta, W. Vongkasem and K. Klakhaeng. 2005a. The challenge of achieving adoption of more sustainable cassava production practices on sloping land in Asia. pp. 217-238. *In*: I. Kheoruenromne, J.A. Riddell and K. Soitong. [Eds.] Proc. of SSWM 2004 International Conference on Innovative Practices for Sustainable Sloping Land and Watershed Management, held in Chiangmai, Thailand, 5-9 September 2004.

Howeler, R.H., W. Watananonta and Tran Ngoc Ngoan. 2005b. Working with farmers: The key to achieving adoption of more sustainable cassava production practices on sloping land in Asia. Paper presented at UPWARD Network Meeting, held in Hanoi, Vietnam. 19-21 January 2005. Paper distributed on CD.

Huang Jie, Li Kaimian, Zhang Weite, Lin Xiong and R.H. Howeler. 2001. Practices and progress in farmer participatory research in China. pp. 413-423. *In*: R.H. Howeler and S.L. Tan [eds.] *Cassava's Potential in Asia in the 21st Century: Present Situation and Future Research and Development Needs.* Proc. 6th Regional Workshop, held in Ho Chi Minh City, Vietnam. 21-25 February 2000.

Lilja, N., N. Johnson, T. Dalton, R.H. Howeler and P. Calkins. 2005. Impact of participatory natural resource management research in cassava-based cropping systems in Vietnam and Thailand. Paper submitted for publication to the Systemwide Participatory Impact Assessment (SPIA). 31 pp.

Nguyen The Dang, Tran Ngoc Ngoan, Le Sy Loi, Dinh Ngoc Lan and Thai Phien. 1998. Farmer participatory research in cassava management and varietal dissemination in Vietnam pp. 454-470. *In*: R.H. Howeler [ed.] *Cassava Breeding, Agronomy and Farmer Participatory Research in Asia.* Proc. 5th Regional Workshop, held in Danzhou, Hainan, China. 3-8 November 1996.

Nguyen The Dang, Tran Ngoc Ngoan, Dinh Ngoc Lan, Le Sy Loi and Thai Phien. 2001. Farmer participatory research in cassava soil management and varietal dissemination in Vietnam - Results of Phase 1 and plans for Phase 2 of the Nippon Foundation project pp. 383-401. *In*: R.H. Howeler and S.L. Tan [eds.] *Cassava's Potential in Asia in the 21st Century: Present Situation and Future Research and Development Needs.* Proc. 6th Regional Workshop, held in Ho Chi Minh City, Vietnam. 21-25 February 2000.

Thai Tapioca Development Institute (TTDI). 2000. Cassava production situation in 1999/2000, according to a survey of farmer groups' leaders. 27 pp. (mimeograph, in Thai)

Tran Ngoc Ngoan. 2003. Evolution of FPR methodologies used and results obtained in Vietnam. Paper presented at the End-of-Project Workshop, held in Thai Nguyen, Vietnam. 27-30 October 2003.

Utomo, W.H., Suyamto, H. Santoso and A. Sinaga. 1998. Farmer participatory research in soil management in Indonesia. pp. 471-481. *In*: R.H. Howeler (ed.), *Cassava Breeding, Agronomy and Farmer Participatory Research in Asia.* Proc. 5th Regional Workshop, held in Danzhou, Hainan, China. 3-8 November 1996.

Utomo, W.H., Suyamto and A. Sinaga. 2001. Implementation of farmer participatory research (FPR) in the transfer of cassava technologies in Indonesia. pp. 424-435. *In*: R.H. Howeler and S.L. Tan [eds.] *Cassava's Potential in Asia in the 21st Century: Present Situation and Future Research and Development Needs.* Proc. 6th Regional Workshop, held in Ho Chi Minh City, Vietnam. 21-25 February 2000.

Vongkasem, V., K. Klakhaeng, S. Hemvijit, A. Tongglum, S. Katong and D. Suprahan. 1998. Farmer participatory research in soil management and varietal selection in Thailand. pp. 412-437. *In*: R.H. Howeler [ed.] *Cassava Breeding, Agronomy and Farmer Participatory Research in Asia.* Proc. 5th Regional Workshop, held in Danzhou, Hainan, China. 3-8 November 1996.

Vongkasem, V., K. Klakhaeng, S. Hemvijit, A. Tongglum, A. Katong, D. Suparhan and R.H. Howeler. 2001. Reducing soil erosion in cassava production systems in Thailand - A farmer participatory approach. pp. 402-412. *In*: R.H. Howeler and S.L. Tan [eds.] *Cassava's Potential in Asia in the 21st Century: Present Situation and Future Research and Development Needs.* Proc. 6th Regional Workshop, held in Ho Chi Minh City, Vietnam. 21-25 February 2000.

Vongkasem, V., K. Klakhaeng, W. Watananonta and R.H. Howeler. 2003. The use of vetiver for soil erosion prevention in cassava fields in Thailand. Paper presented 3rd Intern. Conf. on Vetiver and Exhibition, held in Guangzhou, P.R. China, 6-7 October 2003.

Zhang Weite, Lin Xiong, Li Kaimian and Huang Jie. 1998. Farmer participatory research in cassava soil management and varietal dissemination in China. pp. 389-411. *In*: R.H. Howeler and S.L. Tan [eds.] *Cassava's Potential in Asia in the 21st Century: Present Situation and Future Research and Development Needs.* Proc. 6th Regional Workshop, held in Ho Chi Minh City, Vietnam. 21-25 February 2000.

25

WOCAT: A Framework for Monitoring and Evaluation of Soil and Water Conservation Initiatives

William Critchley[1] and Hanspeter Liniger[2]

[1]Centre for International Cooperation, Vrije Universiteit Amsterdam, de Boelelaan 1105-2G, Amsterdam, The Netherlands. E-mail: WRS.Critchley@dienst.vu.nl

[2]Centre for Development and Environment, University of Berne, Berne, Switzerland. E-mail: hanspeter.liniger@cde.unibe.ch

ABSTRACT

The original rationale for the World Overview of Conservation Approaches and Technologies (WOCAT) was that there is a wealth of know-how that is simply not documented. This is one of the reasons why soil degradation persists, despite decades of effort throughout the world. WOCAT was formally established in 1992 as a global network of specialists in soil and water conservation (SWC). This chapter demonstrates how WOCAT also plays an important role in supporting and enhancing monitoring and evaluation (M&E) of SWC. That includes the process of spot check self-evaluation, where individuals use questionnaires to compile and compare information and knowledge. In addition, WOCAT contributes to M&E through taking specific questions and using these to regularly monitor specific parameters. These two functions are described here. One other objective in this chapter is to show and analyse what WOCAT has in its database regarding the M&E processes of SWC initiatives worldwide. The advantages of M&E through WOCAT are basically three. First, there is a comprehensive array of questions covering most important aspects of SWC initiatives. Second, WOCAT provides a consistent methodology. Third, this consistency means that SWC programmes can both put their achievements in the "marketplace" that the databases provide and make comparative assessments at the same time.

INTRODUCTION

The rationale for the World Overview of Conservation Approaches and Technologies (WOCAT) is that both land users and soil and water conservation (SWC) specialists have developed (and continue to do so) a wealth of know-how related to land management, improvement of soil fertility, and protection of soil resources. *Most of this valuable knowledge, however, is simply not documented* or, if it is, then it remains poorly accessible, and comparison of different types of experience is difficult. Much SWC knowledge therefore remains a local, individual resource, unavailable to others working in similar situations, seeking to accomplish similar tasks. This is one of the reasons why soil degradation persists, despite decades of effort throughout the world and high investments in SWC. In this context, WOCAT was conceived during the 6th International Soil Conservation Organisation (ISCO) conference, held in Australia in 1992, and was formally established as a global network of SWC specialists. WOCAT is organized as an international consortium, coordinated by an international management group and supported by a secretariat located at the Centre for Development and Environment in Berne, Switzerland. Apart from compiling knowledge and making it available, it has developed methods and tools in collaboration with many national and international institutions.

Since that conference, WOCAT has grown from the efforts of just a few SWC specialists to become a truly global network of collaborating national and international institutions and individual experts. WOCAT is one of the oldest international SWC programmes in existence; in fact, after 14 years it cannot be considered a programme anymore, but an institution. The original vision was to produce by 1995 a world map showing SWC measures that are undertaken all over the world that prevent further land degradation or rehabilitate degraded land; the whole process is taking much longer. To date, WOCAT tools have been used to document over 300 SWC technologies and 200 approaches in almost 40 countries in Africa, Asia, the Middle East, Europe and South America. Over 30 national and international WOCAT workshops have been held to develop and improve the methodology, train users, and enhance the network. While WOCAT provides a common platform and standardized methodology for the exchange of information and experience, it aims to allow flexible use of its outputs so that they can be adapted to different end users and different biophysical, socioeconomic and institutional environments. The methods and tools developed and the experience garnered by WOCAT to date are available in books, reports and papers (Liniger and Schwilch, 2002; Liniger et al., 2002; van Lynden et al., 2002) and are accessible in three languages (English, French and Spanish) on the Internet and on CD-ROMs (WOCAT, 2001, 2004). Some WOCAT tools have been translated and are now available in six additional languages. An overview that presents case studies of SWC technologies and approaches, along with an analysis of how and where they work and their strengths and weaknesses is now available (WOCAT, 2007).

USING AND ADAPTING THE WOCAT TOOLS

This chapter demonstrates how WOCAT can support or enhance monitoring and evaluation (M&E) of SWC. This includes both SWC *technologies* and the *approaches* (Box 1: Definitions). WOCAT is sometimes mistakenly thought of as simply a means of documenting SWC. It is actually much more than this. WOCAT is *a framework for evaluation of SWC*. That includes, first, the process of spot check self-evaluation, whereby different persons involved in SWC (stakeholders) use the WOCAT questionnaires to compile and compare information and knowledge. The compiling of activity enables them to assess impact and effectiveness of an SWC-related initiative. The comparing activity allows the comparison of a technology or an approach with similar ones, through the WOCAT database. WOCAT also can contribute to M&E on an individual basis through taking specific questions and using these to regularly monitor specific parameters. These two functions are described here. One more, and rather different, objective in this chapter is to show and analyse what WOCAT has in its database regarding the M&E processes of SWC initiatives worldwide. The vision and mission of WOCAT are stated in Box 2.

WOCAT AS A TOOL FOR MONITORING

Looking at the two questionnaires covering technologies (QT) and approaches (QA) respectively, it becomes clear that many of the questions in the first part of each (General Information) and several in the other sections are essentially snapshots of background data. These need only to be filled in once. Names, altitude, soil types do not change. But the answers to many other questions *would* change from year to year, season to season, such as area coverage or economic analysis (costs/benefits).

Box 1. Definitions

Soil and water conservation in the context of WOCAT is defined as activities at the local level that maintain or enhance the productive capacity of the land in areas affected by or prone to degradation. It includes prevention or reduction of soil erosion, compaction and salinity, conservation or drainage of soil water, and maintenance or improvement of soil fertility.

SWC technologies are agronomic, vegetative, structural and management measures that control land degradation and enhance productivity in the field.

SWC approaches are ways and means of support that help to introduce, implement, adapt and apply SWC technologies on the ground.

Box 2. WOCAT's vision and mission

The *vision* of WOCAT is that knowledge on local sustainable land management is shared and used globally to improve livelihoods and the environment.

WOCAT's *mission* is to support decision making and innovation in sustainable land management by connecting stakeholders, enhancing capacity, and developing and applying standardized tools for documenting, monitoring, evaluating, sharing and using knowledge in SWC. WOCAT mainly addresses SWC specialists, but also land users and extensionists at the field level and decision makers at the planning level.

Questions about training are an example: perhaps initially workshops are held for field staff, then after a few years when results from the field are tangible, there are seminars for decision-makers. Questions about downstream impacts of SWC are another example: the answer in early years (perhaps little impact) will differ from answers several years later (possibly significant impact). Such questions cover parameters that require regular *monitoring*. Thus, Table 1 gives an overview of the main questions, within the WOCAT questionnaires, that relate to regular monitoring. While there are many other systems of monitoring projects and programmes in the arena of development, the advantage of this WOCAT framework is that it highlights information that will be useful to an SWC initiative both in the short term and long term and will guide internal improvements - as well as contributing to a consistent national and international database. One other advantage is that the WOCAT framework includes questions on various elements that are often overlooked in M&E systems; these include *Benefits, advantages and disadvantages* (a group of questions under QT 3.1) and *Acceptance or adoption* (under QT 3.4) as well as *Impact analysis* (under QA 3.2).

The questionnaires are formulated for a one-off data input exercise, and thus for regular monitoring it is necessary - having established which questions are relevant - to develop a monitoring format. That precise

Table 1 Selected monitoring questions within WOCAT questionnaires.

QUESTIONNAIRE SWC TECHNOLOGY	*QUESTIONNAIRE SWC APPROACH*
1 GENERAL INFORMATION 1.3 Area information *1.3.1 Define area where technology (T) has been applied*	1 GENERAL INFORMATION 1.3 Area information *1.3.1 Define area where approach (A) is implemented*
2 SPECIFICATION OF TECHNOLOGY (T) 2.2 Purpose and classification *2.2.2.5 How does the T combat land degradation?* *2.3 Status* *2.3.6* Has appearance of T changed over time?* 2.4 Design, technical & management specifications *(series of questions related to agronomic, vegetative, structural and management measures)* 2.5 Natural environment *2.5.8 Soil fertility* *2.5.10 Topsoil organic matter* *2.5.11 Soil drainage* 2.7 Costs *2.7.1 Establishment and recurrent costs*	2 SPECIFICATION OF APPROACH (A) 2.1 Description, objectives, operation *2.1.3.3 List constraints hindering implementation* 2.2 Participation *2.2.3 Involvement of land users* 2.3 Financing *(over monitoring period)* *2.3.1 Source* *2.3.2 Budget* 2.4 Indirect subsidies *2.4.1.2 Was training provided to land users?* *2.4.1.3 What form of training was provided?* *2.4.1.4* What training activities to which groups?* *2.4.2.4* Which were target groups for extension and which activities?* *2.4.3.2 Was applied research part of the approach?* 2.5 Direct subsidies *2.5.1.1 Was labour rewarded; if so how?* *2.5.1.2 What was financed and under what conditions?*
3 ANALYSIS OF TECHNOLOGY *3.1 Benefits, advantages and disadvantages* *3.1.1 Measurements production, soil loss and runoff* *3.1.2.1 Production/socioeconomic benefits* *3.1.2.2 Socio-cultural benefits*	3 ANALYSIS OF APPROACH 3.1 Methods used for monitoring and evaluation *3.1.3.1 Were there changes in the approach as a result of monitoring and evaluation?* *3.2.1.1 Did the approach help improve SWC?*

(Table 1 Contd.)

(Table 1 Contd.)

3.1.2.3	*Ecological benefits*	3.2.1.2 – 3.2.1.5	*Series of questions asking if the approach led to changes in (crop/grazing/ forest etc.) land management*
3.1.3	*Off-site benefits*	3.2	Impact analysis
3.1.5.1	*Production/socioeconomic disadvantages*	3.2.2*	*Implementation progress*
3.1.5.2	*Socio-cultural disadvantages*	3.2.3.3	*Did other projects/land users adopt the A?*
3.1.5.3	*Ecological disadvantages*	3.2.4.1	*How effective was training for different groups?*
3.1.6	*Off-site disadvantages*	3.2.4.2	*How effective was extension for different groups?*
3.2	*Economic analysis*	3.2.4.3	*How effective was research?*
3.2.3*	*Production increase/decrease without SWC in 10 years*	2.3.6.2*	*If incentives were used, were they changed over time?*
3.2.4–3.2.8	*A series of questions relating to costs and economic benefits, short term/ long term*	3.3	Concluding statements
3.3	Adaptation	3.3.1	*List major strengths/ advantages of the A*
3.3.1*	*What changes have been made to T?*	3.3.2	*List major weaknesses/ disadvantages of the A*
3.3.2*	*Who made which modifications?*		
3.4	Acceptance or adoption		
3.4.1.1	*How many implemented T with incentives?*		
3.4.1.2	*Which groups implemented T with incentives?*		
3.4.1.3*	*Which groups didn't adopt T with incentives?*		
3.4.2.1	*How many implemented T without incentives?*		
3.4.2.2	*Which groups adopted spontaneously?*		
3.4.2.3*	*Which groups did not adopt spontaneously?*		
3.4.2.4	*What is the adoption trend?*		
3.5	Concluding statements		
3.5.1	*List major strengths/ advantages of the T*		
3.5.2	*List major weaknesses/ disadvantages of the T*		

Note: Many original questions are abridged to fit framework above.
*These questions are exclusive to the WOCAT Professional questionnaires; all others are in the Basic as well as Professional questionnaires.

format, or the most appropriate method of data collection, or the people who should be involved depends on specific circumstances. However, it is clear that the data required to answer some questions are very specific and quantitative; others are more qualitative in nature. The first group can be readily measured (as in "conventional M&E"); the latter lend themselves to participatory processes of assessment ("participatory M&E"). Thus, for example, under technologies (QT), the questions *What changes have been made to T* (QT 3.3.1) which can be described in detail under *Design, technical & management specifications* (QT 2.4) and *Measurements production, soil loss and runoff* (QT 3.1.1) are quantitative questions requiring field measurements. Similarly, under approaches (QA), the responses to *Define area where approach is implemented* (QA 1.3.1) and *Was/ what form of/ training provided to land users?* (QA 2.4.1.2/ 3) are measurable. Other questions can be monitored better through participatory methods. These include *Socio-cultural benefits* (QT 3.1.2.2), *Which groups implemented/ didn't adopt/ T with incentives* (QT 3.4.1.2/ 3), *List constraints hindering implementation* (QA 2.1.3.3) and *Did the approach lead to changes in (crop/ grazing/ forest etc.) land management?* (QA 3.2.1.3 - 3.2.1.5). Generally, monitoring of approaches is more difficult, as changes in the approach often imply a completely new approach (which would mean the filling of a new QA). The questions listed under Table 1 are a selection, but depending on the technology/ approach and the focus of the monitoring, other or all questions could be selected.

EVALUATION THROUGH WOCAT

WOCAT is a tool for evaluation and indeed hopes to *stimulate* evaluation within SWC initiatives where, all too often, there is not only insufficient monitoring, but also a lack of critical analysis. It is helpful to separate this evaluation function into two different aspects. The first is *self-evaluation* (as has been introduced already) and the second is *learning from comparing experiences.*

Self-evaluation emanates simply from the process of collecting and analysing the data required to complete a WOCAT questionnaire. This is not just theoretical or wishful thinking; WOCAT users have remarked on the value of this self-evaluation time and again. The process actually enhances the capacity of individuals and institutions in both SWC technologies and approaches. What happens is that users find themselves challenged by various questions in QT and QA. Do they have the information required? How do they get it? Who has which piece of information and is there coherence in the information provided by different partners or are there contradictions and different perceptions? Why is information not matching up? Have they actually thought of the importance of the questions? Why does it matter? What often starts out as a frustrating and time-consuming exercise develops into a learning process. This is supported by the "assessment indicators" tool in the database, which helps the user to look at a set of questions grouped into various indicators and to evaluate their interrelation (e.g., *Which level of technical knowledge is required?* (QT 2.6.11) vs. *Is training provided and how effective was it?* (QA 2.4.1.2/ 3.2.4.1). The self-evaluation process involves different resource persons, such as SWC specialists and researchers with different backgrounds, as well as land users, and thereby builds capacity among all of them. Thus, self-evaluation can be thought of as WOCAT helping people evaluate not just "their own" SWC technologies or approaches, but *also* their own knowledge about that technology or approach.

The second element of evaluation, namely *learning from comparing experiences*, is a more straightforward concept. Assuming that the user has a particular area of interest, or to put it another way, is concerned in an aspect of a technology or an approach, then this - through WOCAT's large and growing global database (over 300 case studies) - can be compared with other experiences, using WOCAT's database facility called "search by criteria" (available on CD ROM or on-line at http://www.wocat.net). While the actual evaluative judgment is mostly left to the user, there will usually be a large enough sample for comparative purposes, and additionally the "assessment indicators" tool in the database helps compare the effectiveness of selected technologies or approaches in the line of a set of indicators (see above). So, under *technologies*, taking terracing, for example, there may be an interest in the *production/ socioeconomic benefits* (QT 3.1.2.1) or *establishment and recurrent costs* (QT 2.7.1). These questions can simply be looked up in the database and compared against other terracing case studies or indeed alternative types of conservation, to compare terraces with, for example, contour vegetative barriers. And under *approaches* the area of concern may be the levels of subsidies or incentives used. Here the question *What was financed and under what conditions?* (QA 2.5.1.2) is of interest for comparative purposes. More topically, what about M&E systems? How do these compare? Here the relevant questions are *Methods used for monitoring and evaluation* (QA 3.1) and *Were there changes in the approach as a result of monitoring and evaluation?* (QA 3.1.3.1).

RESULTS FROM THE WOCAT DATABASE

In this section we analyse information derived from the WOCAT database. Specifically we are interested here in examining how often SWC projects carry out evaluations and reviews and whether, as a result of M&E, changes were made to the projects in terms of approaches or technologies. Seventy-one case studies documented through WOCAT, from all over the world, were analysed with respect to these questions.

The information was obtained from the Approaches database (WOCAT 2004). We have selected only those cases that involved project initiatives and where these questions regarding M&E were therefore relevant.

Taking the question concerning the *external evaluation/review* (QA 3.1.2.1): in 24% of the cases there was no external evaluation or review carried out. In 31% of the cases there was an external evaluation every 1-2 years and in another 31% every 3-5 years. In 6% an external evaluation took place in every 6-10 years. In the remaining 8% of the cases no answer was given. With regard to the question concerning the *internal evaluation/ review* (QA 3.1.2.2), there was no such process in 10% of the case studies. However, 66% of all case studies had an internal evaluation or review every 1-2 years, 11% every 3-5 years, and 3% every 6-10 years (Fig. 1). The remaining 10% provided no information to this question.

This clearly indicates that most SWC projects take M&E quite seriously. Naturally this is a sample of relatively successful projects (since WOCAT focuses on sustainable land management options) but even so it is encouraging to note that two-thirds have a regular internal evaluation (every 1-2 years) supported by external evaluation (most of them every 1-5 years). In 28% of the case studies there was an external as well as an internal evaluation every 1-2 years, which indicates a very strong M&E system. In only 8% of the cases was there neither an internal nor an external evaluation.

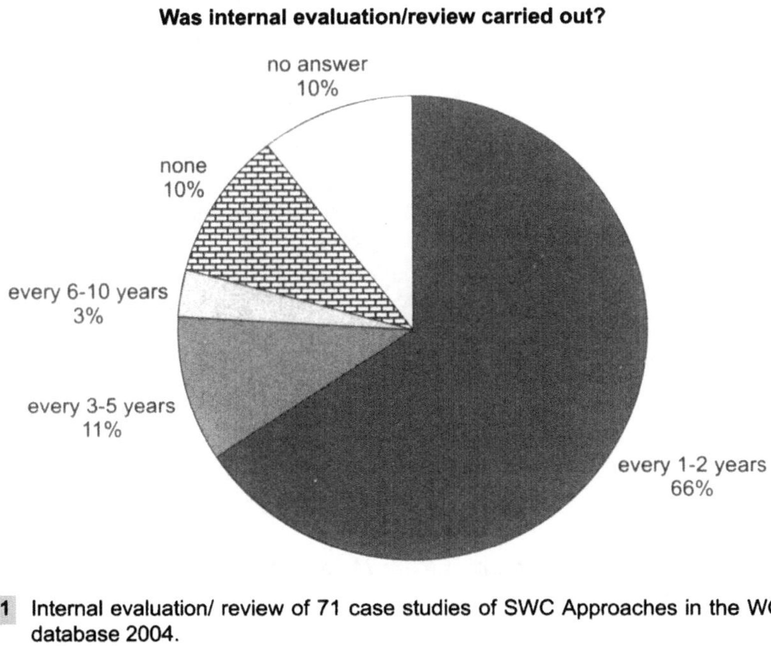

Fig. 1 Internal evaluation/ review of 71 case studies of SWC Approaches in the WOCAT database 2004.

The question regarding whether changes were made due to M&E (QA 3.1.3.1) revealed that 6% indicated many changes, 24% several, 33% few and 23% no changes. The remaining 14% gave no answer (Fig. 2). This shows that, in about two-thirds of the cases, there had been changes in response to M&E. These included modifications to the SWC approach, especially:

- increasing the involvement of land users;
- devolving responsibilities to land users;
- changing from mono-sector to multi-sector approaches (e.g. from a narrow focus of soil conservation to natural resource management and sustainable land management including a strong focus on water and vegetation and crop production); and
- changing the implementation mechanism, e.g. from a project- and government-driven to a land user-driven approach.

Changes in the SWC technology as a result of M&E included:

- putting more stress on productivity and economy of SWC, e.g. changing emphasis from structural to vegetative or cheaper agronomic measures, to include more economic crops such as fruit trees;

Were there changes in the approach as a result of monitoring and evaluation?

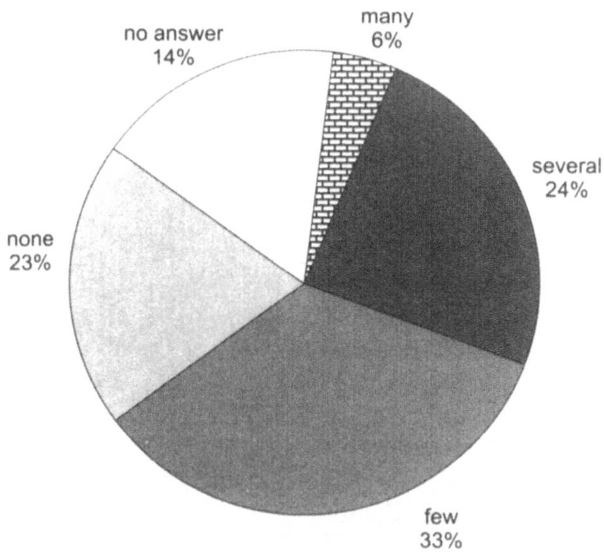

Fig. 2 Changes made as a result of M&E: 71 case studies of SWC Approaches in the WOCAT database 2004.

- improving the technology through modifications; and
- adapting the technology to farmers' needs and wishes.

This analysis of the WOCAT database indicates that M&E plays an important role in the successful implementation of SWC. The key seems to be a good interaction with the different stakeholders and to adapt SWC to the land users' needs and to their situations.

THE WAY FORWARD

WOCAT emphasizes the documentation, monitoring, evaluation and dissemination of SWC experiences that have potential value in other situations. Thus, the WOCAT database is mainly derived from examples of successful sustainable land management. The analysis in this chapter suggests that M&E has been an important component of most of these success stories, and that major beneficial changes and adaptations have been made as a result of M&E. Land users must continuously adapt to a changing environment, especially growing populations and pressures on resources, evolution in land use systems, and dynamic economic and market conditions, at the local as well as at the global level (this now includes the impacts of climate change). This is a challenge not only directly for them, but also to extension workers, project specialists and researchers providing services to them. Thus, all stakeholders have a vested interest in an effective M&E system. Monitoring and evaluation, together with associated reporting, as well as documentation and dissemination of knowledge, is demanding and needs special investment. But these are commonly afforded a low priority. Because time and funds are limited, M&E suffers. This is a major constraint to knowledge management, and more directly to improvements in land management and adaptation to specific local situations. The value of M&E needs to be reassessed and certainly requires more investment and commitment.

This chapter has sought to highlight the importance of M&E in effective SWC initiatives through information drawn from the WOCAT database. But the second, and equally important, message is that *WOCAT supplies a format for M&E* that SWC initiatives can draw on. The advantages of M&E through WOCAT are basically three. First, there is a comprehensive array of questions covering most important aspects of SWC initiatives. Second, WOCAT provides a consistent methodology, which has been developed to fit the widest range of situations, across the globe. Third, this consistency means that SWC programmes have a platform not only to put their achievements in the marketplace that the databases provide, but also to make comparative assessments at the same time.

Sustainable land management is complex, involving all natural resources - water, soil, vegetation, animals and people. Not all factors can be monitored and evaluated; only a selection of indicators can be addressed. WOCAT offers three levels for the evaluation of SWC, from Professional (the most comprehensive) to Basic to Light (the simplest) (Box 3). In order to ease the exchange between different users, it is essential to maintain a standard set of core information for M&E - and this core is common to all three levels. At the same time, *the system must remain flexible* for individual additions according to local, national and regional needs. WOCAT experience confirms that there is a great need for training on M&E and a hunger for it as well - and indeed such training has been given already in over 40 countries worldwide since 1995 (Fig. 3). The WOCAT tools as well as the training can be combined with other monitoring methods described in this book (e.g. Herweg, Chapter 4 of this volume).

Fig. 3 Land users and SWC specialists discussing and learning from each other on a traditional hillside terrace in Nepal. The questions help them to think about issues for M&E that they might not have considered otherwise. Apart from proper documentation this stimulates discussions about problems and possible improvements. (Extract from WOCAT poster "WOCAT questionnaires: Solution-oriented learning". Photos: H.-P. Liniger.)

Box 3. The WOCAT M&E tools

Three questionnaires and examples of questions related to M&E. The Basic versions of the Technologies and Approaches questionnaires are depicted here. There are also Professional (more comprehensive) and Light (shorter than Basic) versions.

3.1.2 Evaluation/review procedures

3.1.2.1 Was external evaluation/review carried out?

no ☐ yes, every 1-2 years ☐ yes, every 3-5 years ☐ yes, every 6-10 years ☐

3.1.2.2 Was internal evaluation/review carried out?

no ☐ yes, every 1-2 years ☐ yes, every 3-5 years ☐ yes, every 6-10 years ☐

3.1.3 Changes due to monitoring and evaluation

3.1.3.1 Were there changes in the approach as a result of monitoring and evaluation?

no ☐ few ☐ several ☐ many ☐

Explain and give examples where possible: ______________________

As we have seen, WOCAT has evolved from that original project-based concept in 1992 of mapping SWC worldwide to being an essential institution within the overall framework of global land management. As

M&E has evolved according to changing needs, WOCAT is developing also. It remains responsive to the needs and ideas of individuals and institutions concerned with the use of M&E in soil and water conservation and sustainable land management.

ACKNOWLEDGEMENTS

The authors would like to thank all who contributed to the WOCAT programme so far, the major donors and collaborators SDC, DANIDA, Syngenta Foundation, FAO, and ISRIC, the national and regional collaborating institutions, as well as the numerous SWC specialists and land users, who have contributed with their vast knowledge.

References

Herweg, K., and H.-P. Liniger. 2003. Soil Erosion Control - An integral part of sustainable land management. pp. 23-32. *In:* M. Zlatic, S. Kostadinov and N. Dragovic. [eds.] Natural and socio-economic effects of erosion control in mountainous regions. Proceedings International Year of Mountain Conference, 10-13 December 2002, Faculty of Forestry, University of Belgrade, Belgrade.

Hurni, H., H.-P. Liniger and U. Wiesmann. 2005: Research partnerships for mitigating syndromes in mountain areas. In: U.M. Huber, M.A. Reasoner and H. Bugmann. [eds.], Global Change and Mountain Regions: A State of Knowledge Overview. Kluwer Academic Publishers, Dordrecht.

Liniger, H.-P. and G. Schwilch. 2002: Better decision making based on local knowledge - WOCAT method for stainable soil and water management. Mt. Res. Dev. J. 22(1), February 2002.

Liniger, H.-P. and G. Schwilch. 2004: The WOCAT methodology as a practical tool for capacity development. Proceedings of the 5th international DANIDA workshop on watershed development, 10-19 November 2003, Indore, Madhya Pradesh, India.

Liniger, H.-P., D. Cahill, D.B. Thomas, G.W.J. van Lynden and G. Schwilch. 2002b. Categorization of SWC Technologies and Approaches - a global need? Proceedings of ISCO Conference 2002, Vol. III, pp. 6-12, Beijing.

Liniger, H.-P., M. Douglas and G. Schwilch. 2004. Towards sustainable land management - "Common sense" and some of the other key missing elements (the WOCAT experience). Proceedings of ISCO Conference 2004, Brisbane.

Liniger, H.-P. and G. van Lynden. 2005. Building up and sharing knowledge for better decision making on soil and water conservation in a changing mountain environment - the WOCAT experience. pp. 103-114. In: M. Stocking, H. Hellemann and R. White. [eds.] Renewable Natural Resources Management for Mountain Regions. Hill Side Press (P) Ltd, Kathmandu.

Liniger, H.-P., G. van Lynden and G. Schwilch. 2002. Documenting field knowledge for better land management decisions - Experiences with WOCAT tools in local, national and global programs. Proceedings of ISCO Conference 2002, Vol. I, pp. 259-267, Beijing.

Schwilch, G., H.-P. Liniger and G. van Lynden. 2004. Towards a global map of soil and water conservation achievements: A WOCAT initiative. Proceedings of ISCO Conference 2004, Brisbane.

Van Lynden, G.W.J., H.-P. Liniger and G. Schwilch. 2002. The WOCAT map methodology, a standardized tool for mapping degradation and conservation. Proceedings of ISCO Conference 2002, Vol. IV, pp. 11-16, Beijing.

WOCAT. 2001. Knowledge makes a difference. CD-ROM Video, FAO Land and Water Digital Media Series 16, Rome.

WOCAT. 2003a. Questionnaire on SWC Technologies. A Framework for the Evaluation of Soil and Water Conservation (revised). Centre for Development and Environment, Institute of Geography, University of Berne, Berne, 63 pp. (Officially available in English, French and Spanish but also translated into Chinese, Arabic, Russian, German. Contact www.wocat.net or WOCAT secretariat, wocat@giub.unibe.ch)

WOCAT. 2003b. Questionnaire on SWC Approaches. A Framework for the Evaluation of Soil and Water Conservation (revised). Centre for Development and Environment, Institute of Geography, University of Berne, Berne, 40 pp. (Officially available in English, French and Spanish but also translated into Chinese, Arabic, Russian, German. Contact www.wocat.net or WOCAT secretariat, wocat@giub.unibe.ch)

WOCAT. 2003c. Questionnaire on the SWC Map. A Framework for the Evaluation of Soil and Water Conservation. Centre for Development and Environment, Institute of Geography, University of Berne, Berne, 14 pp. (Officially available in English, French and Spanish but also translated into Chinese, Arabic, Russian, German. Contact www.wocat.net or WOCAT secretariat, wocat@giub.unibe.ch)

WOCAT. 2004. CD-ROM V3: World Overview of Conservation Approaches and Technologies: Introduction - Network - Questionnaires - Databases - Tools - Reports. FAO Land and Water Digital Media Series 9 (revised). FAO, Rome.

WOCAT. 2005a. BASIC Questionnaire on SWC Technologies. A Framework for the Evaluation of Soil and Water Conservation (revised). Centre for Development and Environment, Institute of Geography, University of Berne, Berne, 42 pp. (Currently available in English, French and Chinese)

WOCAT. 2005b. BASIC Questionnaire on SWC Approaches. A Framework for the Evaluation of Soil and Water Conservation (revised). Centre for Development and Environment, Institute of Geography, University of Berne, Berne, 23 pp. (Currently available in English, French and Chinese)

WOCAT. 2007. Where the Land is Greener: case studies and analysis of soil and water conservation initiatives worldwide. Editors: H.-P. Liniger and W.R.S. Critchley. CTA, FAO, UNEP and CDE. 364 pp.

Epilogue

Reading through all chapters of the book, and focusing in particular on the conclusions drawn by the respective authors, the editors have identified five main challenges in the M&E of SCWD activities. These challenges are discussed hereunder, whereby as much as possible use is made of authors' own words.

THE CHALLENGE OF DETERMINING THE SHAPE AND DIRECTION OF INFORMATION FLOWS

The first challenge concerns the eventual target groups to be informed about the results of M&E. While in small projects information flows will be primarily aimed at the project management and the local stakeholders, in particular farmers; in larger projects many more stakeholders are likely to be interested in the data generated by M&E of SCWD, e.g. project donors, research institutes and policymakers. Several authors feel that M&E data should therefore be much more widely disseminated.

The increasing importance of appropriate M&E data about SCWD projects, both nationally and internationally, is clearly expressed by Sheng: "As resources become more and more limited and competition for investment becomes more red-hot nationally and internationally, proper monitoring and evaluation of such projects to show their real benefit and impact will be a necessary task".

The international dissemination and comparison of M&E data is also a major objective of WOCAT (Critchley and Liniger). It emphasizes the documentation, monitoring, evaluation and dissemination of soil and water conservation (SWC) experiences that have potential value in other situations, and it supplies a format for M&E that SWC initiatives can draw on.

Jayakody et al. stress the importance of data collection at national level: "A national level compilation of critical data and progress information through a central M&E body ... provides a sense of responsibility to the people and institutions ... at all levels. ... This type of central pooling of information would be more beneficial for a broad spectrum of users than having those in isolation at project offices satisfying only the donor agencies."

It is not only a matter of making sure that data are made more amply available to the various stakeholders at different levels, but this may also require more regular patterns of information flows. While M&E data have often been supplied on an ad-hoc basis, there is certainly a need to institutionalize it. "The overriding advantage of institutionally embedded M&E is to have a permanent, stable and uncontested M&E system in place, which serves all stakeholders, has more scope for coordination with other organizations and can be improved over time (Nibbering).

Both elements, the need for a wider dissemination and the institutionalization, can be found in the key findings in the evaluation of the Landcare programme in Australia. In his conclusions, Curtis states that "such a nationwide programme requires an evaluation framework, whereby the roles of respective stakeholders should be clearly defined, and that should set out to learn and to sustain, and that should strive towards informing both managers and policymakers."

THE CHALLENGE OF IMPROVING PARTICIPATION FOR BETTER USE OF INFORMATION

Many authors have stressed the importance of involving the beneficiaries or participants as much as possible in M&E of SCWD, since the information is in most cases meant to be used to increase adoption of appropriate measures and either directly or indirectly to improve their situation and well-being.

Walle and Workman express this as follows: "With active monitoring and evaluation components, community participation in SWC can be increased and quality of conservation impacts improved. ... Monitoring the direct effects of SWC practices and incorporating them into larger watershed actions helped assure that these important project activities continue and will improve the livelihoods of residents"

In a similar way Sang-Arun et al. state: "participatory M&E is recommended for enhancing people's perceptions of the project needs that may strongly affect adoption and non-adoption of project innovation."

In a more general sense Howeler et al. conclude from their research in several Asian countries, that "A farmer participatory approach to technology development was found to be very effective in developing locally appropriate and economically viable technologies, which in turn enhances their acceptance and adoption by farmers".

While the importance of participatory approaches is generally accepted, it is realized by several authors that it will only be possible to properly involve local stakeholders when these people are ready for it. The involvement of communities is a must and just, but it can only happen with the maturity of a particular community towards Watershed Development" (Kotru). And the key to success of soil and water conservation strategies is "Laying first a solid foundation for sustainable rural development", aiming at a progress-driven attitude at village level (Kessler and Ago).

However there may be situations in which public and private interests in soil conservation and watershed development can not be properly reconciled, even after successive trials and lengthy discussions to convince farmers of the need for certain SCWD activities. In such cases other approaches may be required. Studies in the Darby Creek watershed in Ohio (USA, by Napier) revealed that the voluntary approach towards adoption of SCWD had failed. The author therefore suggests that "One of the approaches that should be considered in the future is the use of public conservation policies combined with command and control mechanisms for enforcement".

THE CHALLENGE OF COMBINING "SCIENTIFIC" AND "TESTIMONY" INFORMATION

Several of the chapters emphasise combining information gained through physical measurements with information gathered from direct participants/ beneficiaries in the form of testimonies - either through household structured or semi-structured questionnaires or records of group meetings. For instance:

"The assessment of various outcomes has largely been based on beneficiary response and participatory appraisals besides those determined by scientific measurements." (Das et al.)

"In addition to the beneficiary responses and collective assignment of a qualitative description or giving an agreed quantified value to a specific impact, determination of the degree of impact would call for application of some scientific principles as well as tools and techniques." (Gupta and Das).

As we have already noted in the previous section, the gathering of testimony information does not only serve M&E, but also encourages ownership of the activities by the people giving the testimonies. It may even lead to the adaptation of the activities themselves to ensure support by people being directly affected.

Most authors of the chapters in Parts 1, 2 and 4 explicitly or implicitly suggest that projects that neglected the voices of people directly affected by soil conservation and watershed development activities in the past were weakened by that neglect. They also suggest that engaging people in the information system through listening to their testimonies as part of giving them voice has M&E value.

But the ambiguity of combining participation and information also raises problems for M&E. Testimonies are bound to be influenced by the values and interests of those making them. In addition, even if values and interests are minimal, the data is only as accurate as people's memories allow. Generally the chapters in this book that discuss testimony information use it to test robust hypotheses. The information is presented in the form of comparative rankings, large percentage differences and tendencies in direction of change, rather than quantitative point estimates.

Despite these concerns, testimony evidence will continue to feature in M&E of soil conservation and watershed development activities. But there are clear limits to how far it can displace physical measurements. Partially this is due to the limitations of testimony information, but more significantly it is due to the advances in physical observation techniques described in Part 3 of the book. For instance:

- Evans describes a field-based approach, in which the focus is on estimating the size of rills and gullies. This is a rapid and realistic way to assess erosion, yielding results that can be compared across a wide range of environments, making it a useful tool in M&E of SCWD projects.
- Bergsma and Farshad show how by recording the presence of seven types of microtopographic features during a certain rainy period, erosion intensity can be compared for sites representing the situation "with and without" conservation practices.
- Nichols describes how long-term rainfall, runoff and erosion research programmes play a critical role in interpreting data of short-term monitoring and evaluation efforts, particularly in semi-arid lands providing baseline conditions, temporal data with its variability and a framework for developing and testing new measurement methods.
- Abdelaziz et al. discuss the striking advances that have been made in the last two decades in the development and implementation of geomatics technologies, such as GIS, RS, GPS and telemetry, and claim the use of these technologies in M&E of SCWD may soon become routine.

Therefore the challenge is not to replace physical observation with testimony information, or, on the other hand, dismiss testimony information entirely as unscientific, but to combine them in imaginative ways that utilises the mix of methods that are appropriate to the goals and scales of the activities being undertaken, and the significance of people's engagement with those activities as potential cost-payers and beneficiaries outside elite policy, managerial and technical expert circles.

THE CHALLENGE OF CHOOSING TIME AND SPACE DIMENSIONS FOR M&E

The chapters in this book discuss M&E in relation to soil conservation and watershed development activities that range in spatial scales from village to regional eco-system to national to global. Cultural, ecological and political boundaries criss-cross and no scale has any unambiguous claim to superiority as more appropriate for M&E focus. Choice of scale does have implications for choice of observation technique - though this does not distinguish physical measurement from testimony information as both can be conducted at all potential scales. Physical measurement tends to become more hi-tech with increasing scale and testimony will tend to be gathered from more powerful individual key informants with increasing scale. Both share the challenge that the larger the scale being considered, the greater the need for smaller scale ground-truthing - and this applies to statements by political leaders as well as satellite images. M&E needs techniques for all scales, but it does not need the same technique for all scales!

The issue of time is even more complex. Soil and watershed change processes are almost invariably long duration in time scale, both ecologically and socially. But M&E in practice tends to be driven by demands for rapid policy responses to a claim of crisis. The following statement suggests the time scale in which M&E might expect to identify changes:

"It is observed ... that after 6 years the M&E system shows important impacts and benefits in the following areas: habitat, organization and roadways. It is good to point out that the deforestation process has stopped. Nevertheless, it is a worrisome performance in the crop yields ... and difficulties in applying the fertiliser programme." (Hernandez)

Patience in these circumstances is therefore not only a virtue but a necessity.

Long duration is one temporal challenge. Another is uncertainty. Woodhill sums this up in general terms as follows:

"learning processes ... enable different individuals and groups to continually improve their performance, importantly recognising that they are working in highly dynamic and uncertain contexts. The challenge then is to design effective learning systems that can underpin management behaviours and strategies aimed at optimising impact, rather than simply delivering predetermined outputs." (Woodhill)

This in some ways solves both the duration and uncertainty questions as continuous learning imposes no time limit and can adjust quickly to changing circumstances. But it will not satisfy resource providers who expect blueprints to be implemented and an exit date for project completion. We need ethical solidarity among M&E practitioners in not claiming to offer quick and certain information on long and uncertain processes in order to get M&E consultancy contracts!

Other temporal challenges can be associated with internal processes inside M&E itself. Some observation techniques are still developing. For instance, Zapata concludes that the application of Fallout Radio Nuclide (FRNs) technologies, such as Cs-137 in soil conservation projects entails many challenges, among which the need for an integrated approach to assess both on-site and off-site effects of SCWD activities at the watershed level. Similarly routine application of the geomatics technologies advocated by Abdelaziz et al. is still in the future. Also testimony evidence techniques are being debated in the development studies literature.

How we observe is still dynamic flux and all of us engaged in M&E in this area need to be keen readers of state of the art literature on all observation techniques.

THE CHALLENGE OF ATTRIBUTION

The task of M&E is not only to record activities and link them to planned outputs and intended outcomes. M&E has a responsibility to identify how much of change is actually attributable to the activities as outcomes. The contribution of activities' outcomes may be less or more than intended. In the words of Herweg:

"impact monitoring should make it difficult to paint rosy scenarios and omit mention of negative effects ... as part of a learning process." (Herweg)

And attribution will not be self-evident in the outcome data alone as pointed out by Bodnar et al.:

"Monitoring of indicators at the goal level does not automatically allow the attribution of change to project interventions .... This became possible only after the M&E unit had adjusted their monitoring format and considered SWC [project management] output ... and purpose." (Bodnar et al.)

Attribution requires complex systems to be linked up in causal chains, Cameron suggests Social Cost-Benefit Analysis can help make these linkages and compare the strengths of influence by using a common numeraire and include other factors through sensitivity tests. Hessel et al. suggest that physical science can contribute through measurement and modelling of erosion providing information that can be used with local people's testimony to provide a wider picture. Similarly Kotru suggests combining a logical positivist epistemology with well-informed judgements by local people:

"progressive IM [Impact Monitoring] ... relies on one major activity or sectoral intervention and gradually interlinks to other sectors ... [therefore] quasi-experimental method of data collection can be a relevant option .... The involvement of communities is a must and just, but it can only happen with the maturity of a particular community towards WSD [Water Shed Development]." (Kotru)

The key challenge for modelling for watershed and soil conservation activities can be characterised as (in rough order of degree of abstraction from most to least):

- Irreducible complexity in numbers of variables and causal connections
- Multiple scales in space and time
- Multi-disciplinarity, respecting both more closed, physical science and more deliberative, discursive methodologies
- The incorporation of multiple human agencies
- Accessibility in terms of technical language and language of reporting
- Linking to Millennium Development Goals as mainstream developmental concerns
- Financial resourcing of Research and Development of models.

A key word is 'complexity' (see as example in Fig. 1). The main route to knowledge gains in the past three hundred years has been simplifying problems and distinguishing exogenous and endogenous variables so that controlled experiments or quasi-experiments can be designed and tests made. But the present conjuncture of environmental and developmental challenges on all scales of space and time crossing conventional disciplinary boundaries demands efforts at re-integration. Not surprisingly therefore, there is growing thinking on the epistemological principles and practices that could guide such re-integration. Some emerging principles for modelling derived from critical realism and complexity theory might be:

- Differentiation: do not rush to type and classify complex situations, rather reflect on differences seeking meaning.

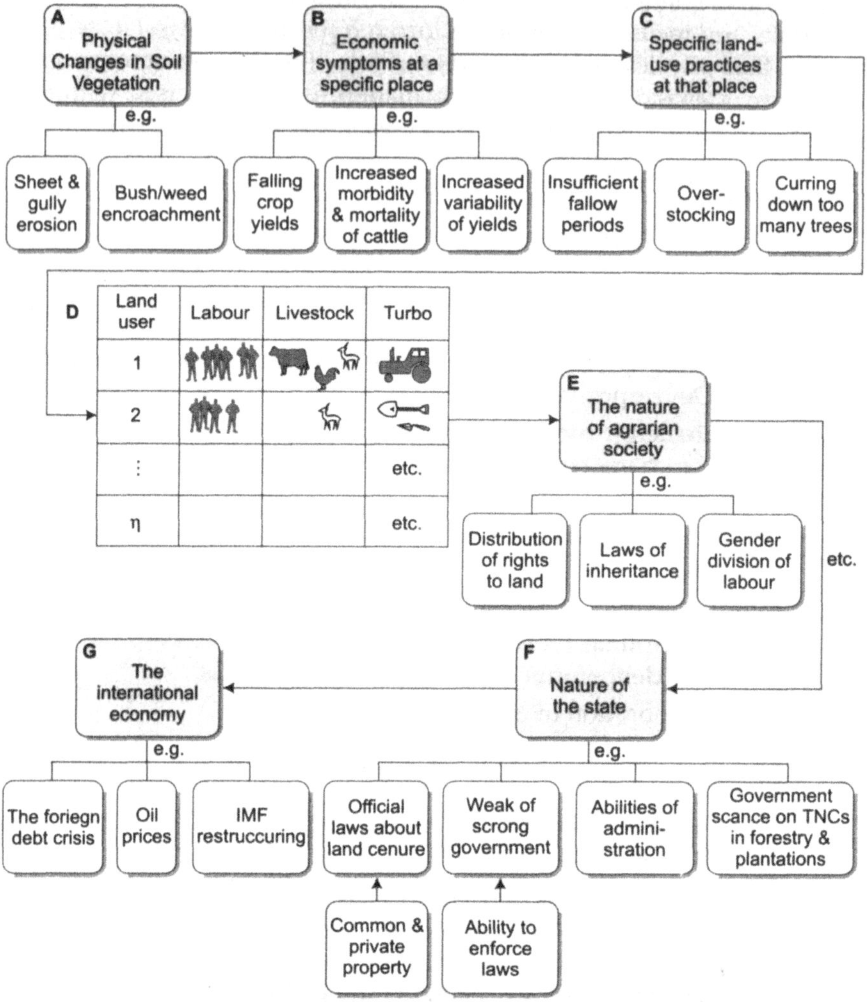

Source: Prof. Michael Stocking, University of East Anglia, Norwich, UK gave us permission to use this diagram which was developed by Prof. Piers Blaikie.

Fig. 1 A schematic image of a complex human/nature interface.

- Determinations: do not rush to exogenise variables and set boundaries around investigation, rather seek to internalise causalities through mutual causalities and feedbacks.
- Dialectics: do not impose equilibrium; rather look for the tensions that cause movement and instability.

- Accepting that both scientific and local people's indigenous knowledge make classifications that have claims to represent reality.
- Accepting that the search for knowledge cannot be totally holistic and boundaries have to be identified using relatively simple algorithms.
- Accepting that all complex systems tend to have periods of steady evolutionary change (called attractor periods in the literature), though punctuated by less stable periods of greater instability (called strange attractor periods).

So we should be ambitious in attempts to model complex systems combining ecological and social processes while modest in our claims to thereby know the future in our planning, the present in our monitoring, or indeed the past in our evaluating. Attribution will always involve elements of epistemological judgement.

But to conclude on a more optimistic note, here is an example of a global initiative that seeks to improve our understanding of complex, water stressed systems to which we can look for inspiration in all aspects on M&E at the people:nature interface, including the challenge of attribution. The LADA (Land Degradation Assessment in Drylands) programme is planned to be implemented over a 6-year project (2004-2010), is funded by the GEF, implemented by UNEP and executed by FAO that plans to contribute to combating land degradation by:

- developing **tools and methods** to assess the magnitude and rate of land degradation and its impact on ecosystems
- carrying out a **global assessment** of land degradation and more detailed assessments in six pilot countries
- building **assessment and monitoring capacities** at national, regional and global scales to enable the design of appropriate mitigation interventions and sustainable land management practices

LADA has started implementation in mid-2006 in six pilot countries: Argentina, China, Cuba, Senegal, South Africa and Tunisia. The participating countries have committed in-kind and some in cash. UNEP, FAO, ISRIC, WOCAT, GLCN and the UN University have committed cash and in-kind contributions to the project. The LADA project is summarised in Table 1 which brings out its potential networking role for M&E of soil conservation and watershed development activities looking towards 2010.

Table 1 The global LADA project http://LADA.virtualcentre.org/.

	LADA ACTIVITIES	*LADA OUTPUTS*
1.	Identification of Land Degradation Problems and Users Needs Assessment	Perception of Impacts and Economic & Ecological Costs/ Users Surveys
2.	Establishment of a LADA Task Force	Institution Building
3.	Stocktaking & Preliminary Analysis	Knowledge Base Gap Analysis Cost/Benefit Analysis
4.	Stratification and Sampling Strategy	Criteria and Indicators
5.	Field Surveys & Participatory Assessment	Field Data Bases
6.	Information Integration	Decision Support Tools
7.	Monitoring Strategies & Tools	LADA database and Decision Support Tools for Land Rehabilitation

Source: Prof. Michael Stocking, University of East Anglia, Norwich, UK.

Appendix – I

Training Courses on Monitoring and Evaluation of Development Projects

Annemarieke de Bruin
Wageningen University and Research Centre, Wageningen, The Netherlands
(The information has been updated by the organisations concerned.)

This appendix shows some of the courses, given around the world, about evaluation and monitoring. The courses have been listed in the order of length, beginning with the 3-day courses. For every course the place where it is situated, the language in which it is given, the profile of the participants, the training methods and the objectives are described. Website and e-mail address are added for more information.

- **PCM Group, Process Consultants Moderators. In Brussels, Belgium.**

  3 days: Course on Monitoring and Evaluation.

  This is a regular open course given in English and French. The course builds strongly on the concepts and techniques introduced in the Introductory module preceding the M&E course. Desk officers and consultants from various organisations could apply. A maximum of 12 participants can join. The training is highly participative, practical, based on case exercises in sub-groups with discussions on presentations. A certificate of attendance will be provided.

  More information can be found at www.pcm-group.com/pcm_cour_4_desc_en.jsp. Contact Erik Kijne as trainer: info@pcm-group.com.

  The objectives: In the Monitoring & Evaluation course the concepts and components of an M&E plan are identified and clarified and structured into a Logical Framework matrix. Each component will be discussed and special emphasis will be placed on WHAT to monitor or evaluate and HOW to organise this process. Elaborate practice with the identification of parameters for designing indicators of Results. Clarity on the information flow between the various actors will assist in the choice of information to receive and to send out.

- **International NGO Training and Research Centre (INTRAC). In North London, United Kingdom. 3 days: Participatory Methodologies for Development.**

  This is a regular course given in English. Staff from NGOs, donor organisations, southern and eastern civil society umbrella bodies and support agencies should apply. The training methods used are participatory, such as case studies, group work, role-play, and presentations on relevant theoretical frameworks. More information can be found at www.intrac.org and the contact address is training@intrac.org

  The objectives:

  - Develop conceptual understanding of the roots of participatory development and why it is still valid;
  - Reflect on participatory methodologies in the modern context;
  - Consider participation in your organisation: what it means to run a participatory project; implications for management and organisational capacity building;
  - Identify the basic principles that need to be respected in use of participatory techniques.

- **International NGO Training and Research Centre (INTRAC). In North London, United Kingdom. 5 days: Participatory Monitoring and Evaluation.**

  This is a regular course given in English. Staff from NGOs, donor organisations, southern and eastern civil society umbrella bodies and support agencies should apply. The training methods used are participatory, such as case studies, group work, role-play, and presentations on relevant theoretical frameworks. More information can be found at www.intrac.org and the contact address is training@intrac.org

  The objectives:

  - Examine the purposes of monitoring and evaluation;
  - Explore the uses and limitations of the hierarchy of objectives in both planning and monitoring of social development programmes and projects;
  - Discuss the evolution of approaches to monitoring and evaluation and the reasons for the emergence of participatory methods;

   - Look at the different information requirements of different stakeholders and how to involve the stakeholders;
   - Identify which participatory tools to use in different circumstances;
   - Look at issues around impact assessment and the monitoring and evaluation of intangibles;
   - Examine issues around managing the process;
   - Explore ways of ensuring institutional learning

- **MDF (the Management for Development Foundation) Training & Consultancy Pvt. Ltd, Ede, The Netherlands. 1 week: PCM_M&E-monitoring and evaluation of project portfolios**. In Ede, the Netherlands; Arusha, Tanzania; Hanoi, Vietnam; Colombo, Sri Lanka; Goma, Congo and Brussels, Belgium.

  This is a regular course given in English. Desk officers, programme officers, officers for monitoring and evaluation, resident representation officers, managers of financing and implementing agencies should apply if they are familiar with the Logical Framework. The training methods used are theoretical input, sharing experiences and group work. More information can be found at www.mdf.nl and the contact address is MDF@MDF.nl

  The objectives: To design a monitoring system to steer the portfolio and organise and steer evaluations so that the portfolio manager can make use of the results for the benefit of individual projects and for the portfolio as a whole.

- **Training for Development and Management, IMA International. 2 weeks: Monitoring and Evaluation in Development.** In Dubai, UAE; Pretoria, South Africa; Islamabad, Pakistan; Nairobi, Kenya; Brighton, United Kingdom; Cape Town, South Africa and Bangkok, Thailand.

  This is a regular course given in English. Managers, professionals from government, NGOs and the private sector, all sectors of development and from all around the world should apply. The training methods used are with active participation of the participants, such as case studies, short lectures, visits and group discussions. More information can be found at www.imainternational.com/courses dates.php or e-mail: post@imainternational.com

  The objectives: To extend skills and methods in project monitoring and evaluation that are appropriate to actual situations and to gain knowledge of participatory tools and basic theoretical understanding of different approaches.

- **International Institute of Rural Reconstruction (IIRR). 2 weeks: Participatory Planning, Monitoring and Evaluation.** In Nairobi, Kenya and Kampala, Uganda.

  These are regularly advertised courses given in English. Managers and leaders of development organisations, indigenous and international NGOs, community-based organisations, bilateral and multilateral programmes should apply. Until now mainly people from Africa have attended. The training methods used are participatory classroom exercises, simulations and daylong field practicum, which involves work in communities where IIRR has its learning community centres and agency visits to other development organisations. More information can be found at www.iirr.org and the contact address is training@iirr-africa.org

  The objectives: To share ideas, experiences and acquire skills in the use of various tools for participatory planning, monitoring and evaluation with a particular focus on the active involvement of beneficiaries in setting up and applying monitoring and evaluation systems.

  Customised courses are also organised for organisations that need to training more than 10 staff at one go. Such courses take 6-7 days depending on the need.

- **International Institute of Rural Reconstruction (IIRR). 3 weeks (e.g. August 13-31, 2007): Participatory Monitoring and Evaluation,** Y.C. James Yen Center, Silang, Cavite, Philippines.

  This course is intended for development leaders and executives who design and manage community-based development programmes or projects, and extension officers and field personnel that implement them. The course provides a facilitative learning environment conducive to a cooperative yet challenging self-assessment and shared discourse about conventional and participatory monitoring and evaluation (PM&E) concepts and practices. Simulated exercises and guided field practicum demonstrate practical methods and tools in designing and facilitating PM&E processes.

  The course also explores critical elements that support the establishment and maintenance of PM&E within development organisations. Participants are meaningfully engaged to reflect on the practical applications of PM&E within the context of their own organisation, programme, or project; and to identify ways to improve their current practices in monitoring and evaluation.

  For more information, please write to Education.Training@iirr.org.

- **Wageningen International. 3 weeks: International course on 'Participatory planning, monitoring and evaluation (PPM&E). Managing and learning for impact'. In Wageningen, The Netherlands.**

  This is a regular course given in English. Staff with distinct planning, monitoring and evaluation or management responsibilities and consultants and facilitators providing planning, monitoring and evaluation services should apply. At least 5 years of experience is required for the 24 participants who can join. The training methods used are interactive, with participant's experiences, theoretical inputs, practical sessions, workshop character, group assignments and field trips. More information can be found at www.wi.wur.nl/UK/newsagenda/agenda/Managing learning for impact PPME.htm and the contact address is training.wi@wur.nl

  The objectives: To gain new insights about the principles of participatory and learning oriented planning, monitoring and evaluation; current trends in the theory and practice and the requirements of funding bodies; creating a learning culture in teams and organisations. Furthermore strengthened competence to: Facilitate participatory planning and M&E processes; Design and manage M&E systems that support adaptive and flexible project and programme management in complex, multi-stakeholder environments; Provide support in assessing and enhancing impact of the project / programme / organisation / network. Lastly clear ideas for: Improving planning and M&E in your organisation; Further strengthening your own competence to facilitate participatory planning and M&E processes.

- **Asian Institute of Technology, AIT Extension. 4 weeks: Project Monitoring and Evaluation. In Bangkok, Thailand.**

  This is a regular course given in English. Project personnel, managers from public and private organisations, development officers and monitoring and evaluation officials should apply. The training methods used are classroom discussions and study visits. More information can be found at www.extension.ait.ac.th and the contact address is extension@ait.ac.th

  The objectives: To analyse the impact assessment of development projects key variables and indicators to conduct monitoring and evaluation of projects and to learn various methods and techniques of project implementation, monitoring and evaluation. To be able to design an effective monitoring and evaluation system. Logframe, CPM, appraisal techniques and PPAR are discussed.

- **Overseas Development Group (ODG), University of East Anglia, Norwich. 4 weeks: Monitoring and Evaluating for Development Activities. E.g. 18 July-10 August 2007**

  This is a regular course given in English. In total 20 participants may apply which have a degree or an equivalent in the discipline of the course and which are people with a practical professional interest in building information systems for more effective decision-making. The training methods used are with inputs of participants, such as case-studies, small-group work and preparation of a work-related project. More information can be found at www.odg.uea.ac.uk and click on Professional Development. The contact e-mail is odg.train@uea.ac.uk

  The objectives: To provide skills, tools and concepts in specifying and implementing monitoring and evaluation systems to support decision-making across a wide variety of organisational and sectoral settings.

- **Overseas Development Group (ODG), University of East Anglia, Norwich. 2 weeks: Management Information Systems for Monitoring and Evaluating. E.g. 2-13 July 2007**

  This is a regular course given in English. In total 10 participants may apply. In Week One, the underlying structure for each participant MIS is designed. Week Two takes the design further and using available software packages (database, spreadsheet and World Wide Web software), develops the MIS software. Participants will take home a workable system, which can be field tested with live data. Participants should possess basic competency in commonly available software packages. More information can be found at www.odg.uea.ac.uk and click on Professional Development. The contact e-mail is odg.train@uea.ac.uk

  The objectives: This module provides professional managers with the opportunity to develop an IT Based Management Information System (MIS) in a two-week period.

Appendix – II
Publications Dealing with Monitoring and Evaluation

Annemarieke de Bruin
Wageningen University and Research Centre, Wageningen, The Netherlands

This appendix gives an overview of some of the manuals and guidelines about monitoring and evaluation. In the abstracts something is mentioned about the target group for whom the guidelines are written and the topics that are dealt with in the publication. At the end some additional literature are presented.

- **Team Consult, INWENT, 2003.** ***Manual for training course: Monitoring and Evaluation in Development Projects/ Programmes.*** **Team Consult, Hamburg.**

  This manual is part of a course with the overall goal to improve effective planning and implementation of participant's projects/ programmes, improving assessment of achievements and impacts, and increasing the attention towards social and gender aspects in project implementation and evaluation.

  Within this manual the following topics are dealt with. The development process; target groups and target population; indicators; methods of data collection; means of verification; planning, monitoring and steering of implementation; external and impact evaluation; impact and environmental monitoring; reporting the results of monitoring; project cycle management, projects information system and gender as a global development agenda.

- **UNDP, Evaluation Office, 2002.** ***Handbook on Monitoring and Evaluating for Results.*** **New York.**

  The growing demand for development effectiveness is largely based on the realization that producing good "deliverables" is not enough. Efficient or well-managed projects and outputs will lose their relevance if they yield no discernable improvement in development conditions and ultimately in peoples' lives. The United Nations Development Programme is therefore increasing its focus on results and on how it can better contribute to them.

www.undp.org/eo/, or www.intra.undp.org/eo/

- **World Bank, 2002.** ***Monitoring and Evaluation: Some tools, methods and approaches.*** **Washington D.C.**

  Because many developing countries would like to systematically monitor their progress in reducing rural poverty the rural score card could serve as a barometer for rural well-being. This publication describes the process of selecting a set of rural development indicators that was used to develop a rural Score Card for measuring rural well-being (poverty reduction).

  In order to ensure the reliance of the rural economy on the natural resource base, there is a need for increasing the efficiency of natural resources uses and conserving the base of natural resources available. In section five of the appendix, the indicators for management of the base of natural resources are presented.

  Also indicators for rural well-being, performance of agricultural and non-agricultural activities, development of markets, improvement of accessibility and communications and policy, decentralisation and governance are presented.

- **IFAD, 2002.** ***A guide for project monitoring & evaluation. Managing for impact in rural development.*** **Office of Evaluation and Studies, IFAD.** (www.ifad.org/evaluation/guide)

  This guide has been written to help project managers and M&E staff improve the quality of M&E in IFAD-supported projects. The guide focuses on how M&E can support project management and engage project stakeholders in understanding project progress, learning from achievements and problems, and agreeing on how to improve both strategy and operations. The main functions of M&E are ensuring improvement-oriented critical reflection, learning to maximise the impact of rural development projects, and showing this impact to be accountable. The guide provides comprehensive advice on how to set up and implement an M&E system, plus background ideas that underpin the suggestions.

  The guide is built up out of the following modules: Using M&E to manage impact; Linking project design, annual planning and M&E; Setting up the M&E system; Deciding what to evaluate; Gathering, managing and communicating information; Putting in place necessary capacities and conditions and Reflecting critical to improve action.

- **International Federation of Red Cross and Red Crescent Societies (IFRC), 2002. *Handbook for Monitoring and Evaluation.* Geneva, Switzerland.**

  While drafted for use by all stakeholders, it is particular mindful of the role of M&E from a National Society perspective. The handbook contains useful M&E tools and is supported by some theoretical background.

- **DANIDA, 2001. *Evaluation Guidelines.* 2nd Edition. Ministry of Foreign Affairs of Denmark.** (www.evaluation.dk)

  These guidelines shall be applied to evaluation of development activities financed by Danida. They should guide specific evaluations and serve as a source of general information for the staff of Danida, for concerned parties in partner countries, for external consultants, and for Danida's technical assistance personnel in the field.

  The guidelines are divided into eight chapters. The first two chapters discuss the basic principles of evaluation and different types of evaluation. Chapters 3-6 describe the evaluation process, criteria, methods, and quality assurance. The final two chapters outline reporting requirements and utilisation of evaluations.

- **European Commission, 2001. *Manual Project Cycle Management.* EuropAid Co-operation Office, General Affairs, Evaluation.**

  (http://europa.eu.int/comm/europeaid/evaluation/methods/PCM_Manual_EN-march2001.pdf)

  The European Commission adopted 'Project Cycle Management' (PCM), a set of project design and management tools based on the Logical Framework method of analysis. The central objective of the EC is poverty reduction with the following strategic areas: sustainable development, integration into the world economy, fight against poverty and democracy, human rights, and rule of law. This manual, which updates the original 1993 version, presents the main features of PCM.

  It explains the project cycle and its phases, the Logical Framework with its four parts of analysis (stakeholder and problem analysis and that of the objectives and strategies) and how to build the Logframe Matrix. Quality factors are discussed to ensure 'sustainability' throughout the project. When complete the Logical Framework Objectively Verifiable Indicators and Sources of Verification are identified. To develop activity and resources using the Logical Framework and to plan complex interventions by interlocking frameworks is being outlined.

- **World Bank, 2000.** ***Monitoring Rural Well-being: a Rural Score Card.*** **Washington D.C.**

  To strengthen awareness and interest in M&E this overview has been written. Of the following tools, methods and approaches the purpose, use, advantages and disadvantages, costs, skills and time requirements and key references are outlined. Performance indicators, the logical framework approach, theory-based evaluation, formal surveys, rapid appraisal methods, participatory methods, public expenditure tracking surveys, impact evaluation and cost-benefit and cost-effectiveness analysis.

- **Herweg, K., K. Steiger and J. Slaats, 1999.** ***Sustainable Land Management Guidelines for Impact Monitoring.*** **CDE, Bern.** (http://www.gtz.de/lamin/download/future/workbook.pdf)

  These guidelines assist programme and project co-ordinators and managers in initiating a monitoring procedure, selecting indicators and methods, assessing the results, and organising user-oriented outputs, presentation, dissemination and storage of the information gathered in the process of SLM-IM. The guidelines provide project specialists with tools to carry out monitoring. Two guideline documents have been published, the Workbook and the Toolkit.

  The Workbook contains a brief summary and three modules: The pathfinder, the Sustainable Land Management module with basic information on the importance of the SLM concept, and the SLM Impact Monitoring module.

  In the last mentioned module the seven steps of the monitoring procedure are described: 1. Identification of stakeholders, 2. Identification of core issues, 3. Formulation of impact hypotheses, 4. Identification and selection of indicator sets, 5. Selection and development of SLM-IM methods, 6. Data analysis and assessment of SLM, 7. Information management.

  The Toolkit contains methodological options corresponding to the steps 2, 3, 4, 5 and 6 of the monitoring procedure.

- **IFPRI, 1999.** ***Designing Methods for the Monitoring and Evaluation of Household Food Security Rural Development Projects.*** **Technical Guide No. 10. Washington, D.C. (www.ifpri.org)**

  This guide emphasises the design of quantitative impact evaluation exercises for Household Food Security and Nutrition, and provides development practitioners with the basic principles on why, when and how to choose and implement a particular evaluation system. According to this guide two of the key features of a good impact evaluation study are the availability of accurate base-line information and a properly thought-out control group, respectively allowing for before-after and with-without comparisons.

In the first section the reader is provided with the conceptual underpinnings for the choice of a particular design suited to the type and the level of accuracy of the information required by the different intended end-users. In the second part of the document, two of the evaluation methodologies used in the field in the course of projects focused on HFS and nutritional aspects of poverty alleviation projects are reported.

- **Sanders, J. 1998. *W.K. Kellogg Foundation Evaluation Handbook.* Collateral Management Company, Battle Creek.** (www.wkkf.org/Pubs/Tools/Evaluation/Pub770.pdf)

  This handbook provides a framework for thinking about evaluation as a relevant and useful programme tool. It is written for project directors who have direct responsibility for the ongoing evaluation of W.K. Kellogg Foundation-funded projects, for other project staff and external evaluators.

  The W.K. Kellogg Foundation Framework for Evaluation is built up out of the Project-level, Cluster and Policymaking and Program evaluation. The Project-level evaluation has three components; the context, implementation and outcome evaluation. To plan and implement a project-level evaluation the following steps should be followed: 1. Identifying stakeholders and establishing an evaluation team, 2. Developing evaluation questions, 3. Budgeting for an evaluation, 4. Selecting an evaluator, 5. Determining data-collection methods, 6. Collecting data, 7. Analysing and interpreting data, 8. Communicating findings and insights, 9. Utilising the process and results of evaluation.

- **Guijt, I. 1998. *Participatory Monitoring and Impact Assessment of Sustainable Agriculture Initiatives: an Introduction to the Key Elements*. SARL Discussion Paper 1. IIED, London, U.K.**

  This document is a practical methodological introduction to setting up a participatory monitoring process for sustainable agriculture initiatives, based on experience at the first stage of a research project on monitoring and impact assessment with small-scale producers, rural workers union, and NGOs engaged in sustainable agriculture in Brazil.

  The document introduces several central concepts and identifies key steps in developing a monitoring system. This is followed by a discussion on the complexity of indicator selection and choosing methods, showing a range of possible methods with examples from the agricultural sector. The paper ends by reflection on common pitfalls and specific difficulties faced in Brazil starting up a participatory monitoring system for sustainable agriculture. Annex 1 provides a description and visual examples of 20 participatory methods that can and/or have been used for monitoring change.

(Abstract taken from http://www.worldbank.org/participation/pme/webfiles/pme-latam.pdf)

- **IDRC, 1997. *Planning, Monitoring and Evaluation of Programme Performance, a Resource Book.* IDRC, Ottawa, Canada.**

  This is a resource book that is written to build a performance management system and to design an evaluation component for a project to systematically monitor or collect performance or other data. The performance evaluation plan consists of the following parts: developing performance targets, developing a performance framework, identifying evaluation priorities, preparing an evaluation plan and evaluation design. In this book the four elements of a performance target are explained: the impact, reach, activities/outputs and timeframe and within the evaluation the components: indicators, information required, the sources of information, the research methods, timing and special concerns.

- **Douglas, M. 1996. Monitoring and Evaluation of better land husbandry. Annex 4 and 5. In: *Formulation of a Soil Erosion Control Programme - Jamaica.* Technical report TCP/JAM/4451 (A). FAO, Rome.**

  Guidelines for the monitoring and evaluation of better land husbandry are presented in this annex 4. The monitoring and evaluation is built up out of three parts: the conservation effectiveness, the status, type and severity of the erosion and the land husbandry rating.

  In annex 5 an example is given of a status/monitoring report with the above mentioned parts of the monitoring. Each geographic area in which monitoring is to take place should be subdivided into separate Land Management Units defined by farmers' traditional criteria.

- **Granger, A., J. Grierson, R.T. Quirino and L. Romano. 1995. *Evaluation in Agricultural Research and Management.* Cali, Colombia: ISNAR. Module 4 of the series "Training in Planning, Monitoring and Evaluation for Agricultural Research Management.**

  This training module consists of a package of materials designed to facilitate the learning and technique of planning, monitoring and evaluation. It is part of a series of four modules. The module includes information about the target group and instruments, to assess the participants' expectations and their knowledge of PM&E. They also contain practical exercises and instructions as well as feedback sessions for each exercise.

The terminal objective of this module is that all participants after following this course can analyse the essential requirements for designing and implementing evaluation processes in an institution, using different criteria. The evaluation framework, the methodology and evaluation related to other institutional processes are discussed.

- **Rubin, F. 1995. *A Basic Guide to Evaluation for Development Workers*. Oxfam, U.K.**

  This guide is intended to help people understand the underlying principles of evaluation, to be clear about its uses and limitations. It does so by first defining evaluation, and then its place in projects. A brief history of 'traditional' evaluation and its characteristics is presented. The purpose and use of evaluation and the main questions that have to be addressed when evaluation is incorporated at an early stage in project planning are being discussed. Practical considerations are mentioned when planning a specific evaluation exercise, for example the Terms of Reference, and questions of feedback are given.

- **Marsden, D., P. Oakley and B. Pratt. 1994. *Measuring the Process: Guidelines for Evaluating Social Development, NGO Management and Policy*. INTRAC, Oxford, UK.**

  This book is an attempt to set out guidelines for evaluating social development processes, based on the results of an international workshop. It is intended primarily as a practical guide for undertaking the evaluation of social development projects and combines a theoretical overview of the concepts involved, with insights into planning and implementation of evaluation. Three substantial case studies of evaluations are provided from Colombia, India and Zimbabwe, and an extensive literature review is also included.

Other Literature

- Casley, D.J. and K. Kumar. 1987. *Project Monitoring and Evaluation in Agriculture*. John Hopkins Press, Baltimore.
- Casley, D.J. and K. Kumar. 1988. *The Collection, Analysis, and Use of Monitoring and Evaluation Data*. A World Bank Publication, John Hopkins Press, Baltimore.

- FAO. 1995. *Monitoring and Evaluation of Watershed Management Project Achievements*. FAO Conservation Guide 24. FAO, Rome.
- Guijt, I., 1999. *Participatory Monitoring and Evaluation for Natural Resource Management and Research*. Natural Resource Institute, Chatman, U.K.
- International Institute of Rural Reconstruction (IIRR). *Participatory Monitoring and Evaluation: Experiences and Lessons*. A Workshop Proceedings.
- Murphy, J. and L.H. Sprey. 1982. *Monitoring and Evaluation of Agricultural Change*. ILRI Publication 32. ILRI, Wageningen.
- Patton, M.Q. 1990. *Qualitative Evaluation and Research Methods*. 2nd Edition. Sage Publications, Newbury Park, U.S.A.
- Patton, M.Q. 2001. *Utilization-focused Evaluation: The New Century Text*. 3rd Edition. Sage Publications, Beverly Hills, U.S.A.
- Posavac, E.J. and R.G. Carey. 1989. *Program Evaluation Methods and Case Studies*. 3rd Edition. Prentice-Hall, New Jersey.
- Rossi, P.H. and H.E. Freeman. 1989. *Evaluation: a Systematic Approach*. 4th Edition. Sage Publications, Newbury Park, U.S.A.
- UN. 1984. Guiding principles for the design and use of monitoring and evaluation in rural development projects and programmes. ACC Task Force on Rural Development. FAO, Rome.

Index

About the Editors

Dr Jan de Graaff is an agricultural economist, specialized in the evaluation of soil conservation and watershed development activities. He has worked many years abroad, first for FAO, in several countries in Africa, in Rome and in the Caribbean, and later for the Royal Tropical Institute, among others, in Indonesia. In 1990 he returned to the Netherlands and since then he has been employed with the Erosion and Soil and Water Conservation group of Wageningen University. Besides lecturing, study coordination and supervising MSc and PhD students, he has been involved in agro-economic aspects of soil and water conservation research projects in Burkina Faso, northern Europe, northern Africa, southern Europe and eastern Africa, including Rwanda.

Dr John Cameron has worked as a researcher and consultant in the Monitoring and Evaluation field in eleven countries in Africa, Asia and the South Pacific. His employers have included FAO, ILO and the World Bank as well as bilateral aid agencies, national governments and NGOs. He has also trained more than 500 people from all over the world in Monitoring and Evaluation skills. In 2007 he is working for WHO on a project designing a global Social Cost Benefit framework for rural drinking water and sanitation interventions.

Dr Samran Sombatpanit had worked as a land development officer of the Land Development Department, Thailand, during the period 1964-1999, the last 18 years having spent for soil and water conservation. He created the Soil and Water Conservation Forum of Thailand in 1980 and served as a Vice President of WASWC for Asia in 1995, Deputy President for 1997-2001, President for 2002-2004, Acting President for January 2005 to mid-2006 and Past President for mid-2006 to December 2007. He has edited the book *Soil Conservation Extension* in 1997 and co-edited *Incentives in Soil Conservation* in 1999, *Response to Land Degradation* in 2001 and *Ground and Water Bioengineering for Erosion Control and Slope Stabilization* in 2004.

Dr Christian Pieri, a soil scientist trained in France (Institut National Agronomique, Paris) with PhD on carbon balance of tropical savannas, for more than 20 years he has worked on soil fertility issues, tillage and soil erosion control in tropical and sub-tropical countries in CIRAD (French Agricultural Research Centre for International Development). He joined the World Bank in 1992 and has worked extensively on natural resources management projects with special focus on sustainable land management, land degradation / desertification and integrated ecosystem management in the framework of GEF Focal Areas.

Dr Jim Woodhill is currently Director of the Programme for Capacity Development and Institutional Change (CD&IC) at Wageningen University and Research Centre in the Netherlands. He has over 20 years experience applying participatory and learning orientated approaches to rural development and natural resources management. He has developed monitoring systems and undertaken evaluation work for a wide range of organizations and co-authored *'Managing for Impact in Rural Development – a guide for project M&E'* produced by the IFAD. Previously, he worked in Africa as a planning monitoring and evaluation facilitator for the World Conservation Union and for 10 years was actively engaged with Australia's Landcare and integrated catchment management initiatives.

For Product Safety Concerns and Information please contact our EU representative GPSR@taylorandfrancis.com Taylor & Francis Verlag GmbH, Kaufingerstraße 24, 80331 München, Germany

Batch number: 10726380

Printed by Printforce, the Netherlands